Deep Learning for Pathology Detection and Diagnosis in Medical Imaging

Deep Learning for Pathology Detection and Diagnosis in Medical Imaging

Editors

Sergiu Nedevschi
Delia-Alexandrina Mitrea

Basel • Beijing • Wuhan • Barcelona • Belgrade • Novi Sad • Cluj • Manchester

Editors
Sergiu Nedevschi
Technical University of
Cluj-Napoca
Cluj-Napoca
Romania

Delia-Alexandrina Mitrea
Technical University of
Cluj-Napoca
Cluj-Napoca
Romania

Editorial Office
MDPI AG
Grosspeteranlage 5
4052 Basel, Switzerland

This is a reprint of articles from the Special Issue published online in the open access journal *Sensors* (ISSN 1424-8220) (available at: https://www.mdpi.com/journal/sensors/special_issues/dlvb).

For citation purposes, cite each article independently as indicated on the article page online and as indicated below:

Lastname, A.A.; Lastname, B.B. Article Title. *Journal Name* **Year**, *Volume Number*, Page Range.

ISBN 978-3-7258-2041-2 (Hbk)
ISBN 978-3-7258-2042-9 (PDF)
doi.org/10.3390/books978-3-7258-2042-9

© 2024 by the authors. Articles in this book are Open Access and distributed under the Creative Commons Attribution (CC BY) license. The book as a whole is distributed by MDPI under the terms and conditions of the Creative Commons Attribution-NonCommercial-NoDerivs (CC BY-NC-ND) license.

Contents

About the Editors vii

Preface ix

Ying Cui, Shangwei Ji, Yejun Zha, Xinhua Zhou, Yichuan Zhang and Tianfeng Zhou
An Automatic Method for Elbow Joint Recognition, Segmentation and Reconstruction
Reprinted from: *Sensors* **2024**, 24, 4330, doi:10.3390/s24134330 1

Jongwook Whangbo, Juhui Lee, Young Jae Kim, Seon Tae Kim and Kwang Gi Kim
Deep Learning-Based Multi-Class Segmentation of the Paranasal Sinuses of Sinusitis Patients
Based on Computed Tomographic Images
Reprinted from: *Sensors* **2024**, 24, 1933, doi:10.3390/s24061933 18

Domenico Amato, Salvatore Calderaro, Giosué Lo Bosco, Riccardo Rizzo and Filippo Vella
Metric Learning in Histopathological Image Classification: Opening the Black Box
Reprinted from: *Sensors* **2023**, 23, 6003, doi:10.3390/s23136003 30

Aurel Baloi, Carmen Costea, Robert Gutt, Ovidiu Balacescu, Flaviu Turcu and Bogdan Belean
Hexagonal-Grid-Layout Image Segmentation Using Shock Filters: Computational Complexity
Case Study for Microarray Image Analysis Related to Machine Learning Approaches
Reprinted from: *Sensors* **2023**, 23, 2582, doi:10.3390/s23052582 50

Delia-Alexandrina Mitrea, Raluca Brehar, Sergiu Nedevschi, Monica Lupsor-Platon, Mihai Socaciu and Radu Badea
Hepatocellular Carcinoma Recognition from Ultrasound Images Using Combinations of
Conventional and Deep Learning Techniques
Reprinted from: *Sensors* **2023**, 23, 2520, doi:10.3390/s23052520 67

Yiqing Liu, Qiming He, Hufei Duan, Huijuan Shi, Anjia Han and Yonghong He
Using Sparse Patch Annotation for Tumor Segmentation in Histopathological Images
Reprinted from: *Sensors* **2022**, 22, 6053, doi:10.3390/s22166053 96

Md. Robiul Islam, Md. Nahiduzzaman, Md. Omaer Faruq Goni, Abu Sayeed, Md. Shamim Anower, Mominul Ahsan and Julfikar Haider
Explainable Transformer-Based Deep Learning Model for the Detection of Malaria Parasites
from Blood Cell Images
Reprinted from: *Sensors* **2022**, 22, 4358, doi:10.3390/s22124358 113

Suliman Mohamed Fati, Ebrahim Mohammed Senan and Ahmad Taher Azar
Hybrid and Deep Learning Approach for Early Diagnosis of Lower Gastrointestinal Diseases
Reprinted from: *Sensors* **2022**, 22, 4079, doi:10.3390/s22114079 133

Rui Yan, Fei Ren, Jintao Li, Xiaosong Rao, Zhilong Lv, Chunhou Zheng and Fa Zhang
Nuclei-Guided Network for Breast Cancer Grading in HE-Stained Pathological Images
Reprinted from: *Sensors* **2022**, 22, 4061, doi:10.3390/s22114061 164

Radu Chifor, Tiberiu Marita, Tudor Arsenescu, Andrei Santoma, Alexandru Florin Badea, Horatiu Alexandru Colosi, et al.
Accuracy Report on a Handheld 3D Ultrasound Scanner Prototype Based on a Standard
Ultrasound Machine and a Spatial Pose Reading Sensor
Reprinted from: *Sensors* **2022**, 22, 3358, doi:10.3390/s22093358 179

Temitope Emmanuel Komolafe, Cheng Zhang, Oluwatosin Atinuke Olagbaju, Gang Yuan, Qiang Du, Ming Li, et al.
Comparison of Diagnostic Test Accuracy of Cone-Beam Breast Computed Tomography and Digital Breast Tomosynthesis for Breast Cancer: A Systematic Review and Meta-Analysis Approach
Reprinted from: *Sensors* **2022**, 22, 3594, doi:10.3390/s22093594 . . . . **193**

About the Editors

Sergiu Nedevschi

Sergiu Nedevschi is a full professor of image processing and pattern recognition at the Faculty of Automation and Computer Science, Computer Science Department, Technical University of Cluj-Napoca, Romania.

Beginning his career in 1976, he served as a researcher at the Research Institute for Computer Technologies in Cluj-Napoca until 1983, after which he joined TUCN. Rising through the ranks, he was appointed Full Professor of computer science in 1998, concurrently establishing and directing the Image Processing and Pattern Recognition Research Center. His administrative roles at TUCN include Head of the Computer Science Department (2000–2004), Dean of the Faculty of Automation and Computer Science (2004–2012), and Vice-Rector for Scientific Research (2012–2020). Since 2016, he has been a member of the Romanian Academy.

Throughout his career, Professor Nedevschi has been deeply engaged in research, spearheading over 80 projects and serving as coordinator for 65 of them. His collaborations with prominent companies like Volkswagen AG, Robert Bosch GmbH, and SICK AG, as well as research institutions such as VTT, INRIA, and universities like ETH Zurich, TU Braunschweig, Czech Technical University in Prague, and the University of Medicine and Pharmacy "Iuliu Hațieganu" in Cluj-Napoca, have been facilitated through funded research initiatives.

His scholarly and research pursuits span various domains, including image processing, pattern recognition, computer vision, machine learning, deep learning, and the development of advanced driving solutions, autonomous vehicles, and medical imaging for multiple modalities. With a prolific output, he has authored more than 400 scientific papers and edited over 20 volumes, including books and conference proceedings.

Delia-Alexandrina Mitrea

Delia-Alexandrina Mitrea received a bachelor's degree in computer science from the Faculty of Automation and Computer Science, Technical University of Cluj-Napoca, in 2003, and the Ph.D. degree in computer science (image processing and pattern recognition) in 2011. She attended post-doctoral studies with the Technical University of Cluj-Napoca, her research involving artificial intelligence techniques applied in medical imaging, being involved in many research projects in this field during her career. Currently an Associate Professor at the Computer Science Department, Faculty of Automation and Computer Science, Technical University of Cluj-Napoca, she has a wide experience in machine learning, data mining, image analysis and recognition, complex databases, her main research interests being therefore towards image and data intelligence. She is a member of the IEEE Computer Society and of the Institute for Systems and Technologies of Information, Control and Communication (INSTICC).

Preface

Severe pathologies, such as the diffuse liver diseases or tumors, can lead to the significant degradation of the human health and sometimes to lethal stages. The most reliable methods for the diagnosis of these affections, such as the classical biopsy or surgery, are invasive and dangerous. Advanced computerized methods are urgently needed to reduce invasiveness and enhance the information derived from medical images as much as possible by unveiling their subtle aspects, conducting virtual biopsy. Computer Vision and Machine Learning can be successfully employed to achieve this target. Thus, advanced image analysis combined with conventional machine learning, as well as the deep learning techniques, can lead to a highly accurate automatic diagnosis process. The corresponding features, together with the classification, segmentation, fusion of multiple image modalities, and 3D reconstruction techniques, can be involved in the achievement of appropriate 2D and 3D models for the considered affections, which are helpful in computer-aided diagnosis and surgery. The purpose of the special issue "Deep Learning for Pathology Detection and Diagnosis in Medical Imaging" is that of offering researchers the opportunity to disseminate valuable and original results achieved in the corresponding field, with focus on the latest deep-learning techniques, eventually compared and combined with conventional methods. The Convolutional Neural Networks (CNNs), as well as the transformers, were successfully involved, during the last decade, in Artificial Intelligence, positively impacting the field of pathology automatic detection and diagnosis based on medical images, as well. Performing, at the same time, image analysis, dimensionality reduction and classification, CNNs lead to a subtle disease characterization, considerably improving the accuracy of the diagnosis process, while the transformers, endowed with attention mechanisms, enhance even more the corresponding performance. Together with the Explainable Artificial Intelligence (XAI) methods, one can achieve a deeper understanding upon the whole process, but also upon the correlations between the imaging features and the morphological, physical, respectively chemical properties of the pathologic tissues and structures. The dep-learning based recognition, segmentation, as well as the 3D reconstruction methods have a real potential to enhance current medical technologies, as well as human health, being therefore emphasized in the current special issue.

Sergiu Nedevschi and Delia-Alexandrina Mitrea
Editors

Article

An Automatic Method for Elbow Joint Recognition, Segmentation and Reconstruction

Ying Cui [1,2], Shangwei Ji [3], Yejun Zha [3], Xinhua Zhou [4], Yichuan Zhang [1] and Tianfeng Zhou [1,2,*]

1. School of Mechanical Engineering, Beijing Institute of Technology, Beijing 100081, China; bitcuiying@163.com (Y.C.); zyc@bit.edu.cn (Y.Z.)
2. School of Medical Technology, Beijing Institute of Technology, Beijing 100081, China
3. Department of Orthopedic Trauma, Beijing Jishuitan Hospital, Beijing 100035, China; ddv950315@126.com (S.J.); yijune23@126.com (Y.Z.)
4. Department of Orthopedics, Beijing Jishuitan Hospital, Beijing 100035, China; bjzhouxinhua@sina.com
* Correspondence: zhoutf@bit.edu.cn; Tel.: +86-18810631729

Abstract: Elbow computerized tomography (CT) scans have been widely applied for describing elbow morphology. To enhance the objectivity and efficiency of clinical diagnosis, an automatic method to recognize, segment, and reconstruct elbow joint bones is proposed in this study. The method involves three steps: initially, the humerus, ulna, and radius are automatically recognized based on the anatomical features of the elbow joint, and the prompt boxes are generated. Subsequently, elbow MedSAM is obtained through transfer learning, which accurately segments the CT images by integrating the prompt boxes. After that, hole-filling and object reclassification steps are executed to refine the mask. Finally, three-dimensional (3D) reconstruction is conducted seamlessly using the marching cube algorithm. To validate the reliability and accuracy of the method, the images were compared to the masks labeled by senior surgeons. Quantitative evaluation of segmentation results revealed median intersection over union (IoU) values of 0.963, 0.959, and 0.950 for the humerus, ulna, and radius, respectively. Additionally, the reconstructed surface errors were measured at 1.127, 1.523, and 2.062 mm, respectively. Consequently, the automatic elbow reconstruction method demonstrates promising capabilities in clinical diagnosis, preoperative planning, and intraoperative navigation for elbow joint diseases.

Keywords: elbow computerized tomography (CT) image; elbow MedSAM; bone recognition; medical image segmentation; three-dimensional (3D) reconstruction

1. Introduction

Computerized tomography (CT) scans are widely applied in the medical imaging field, providing rapid and accessible imaging for the musculoskeletal system [1]. Moreover, CT scans have significant advantages in three-dimensional (3D) representation, allowing for the detailed description and quantification of complex anatomical regions. They are commonly employed in elbow examinations for various elbow pathologies [2–4], including acute elbow trauma [5,6], fracture-dislocation [7], degenerative changes [8], and elbow osteoarthritis [9,10]. Furthermore, elbow CT examinations assist in personalized implant design, precise alignment [11], elbow joint kinematics, dynamic analysis [12], and personalized treatment [13–15], thereby improving surgical accuracy and patient safety. However, it is essential for senior surgeons to perform bone reconstruction before diagnosing based on CT images.

In clinical practice, the reconstruction of bone surface models is often cumbersome. Initially, bone reconstruction was performed using engineering modeling software [16], including AutoCAD and SolidWorks. However, with recent advances in computer-aided medicine, medical analysis software has made bone reconstruction more scientific and reasonable. Willing et al. [12] generated 3D elbow joint models using threshold segmentation and performed wrapped and smoothed operations based on Mimics. Bizzotto et al. [17] created articular models by uploading images into OsiriX Dicom Viewer for initial processing and then creating 3D models using surface rendering tools. Antonia et al. [18]. used Mimics for elbow reconstruction as well as Geomagic Studio 9 (3D Systems, Morrisville, NC, USA) for further refining and mesh repairing to ensure the reasonability of elbow bone structure. Savic et al. [19] reconstructed 15 femurs using Mimics 18 (Materialize Inc., Leuven, Belgium) to establish a parameterized femur model, spending five days and achieving 97% accuracy compared to real bones. Grunert et al. [20] utilized D2P^{TM} (3D Systems Inc., Rock Hill, SC, USA) for bone segmentation, under the precision control of senior joint surgeons, to acquire the STL file for 3D printing of the elbow joint.

Overall, the recognition, segmentation, and reconstruction of the elbow joint have primarily been performed by engineers and senior surgeons. The accuracy of reconstruction was primarily influenced by the precision of segmentation, which commonly involves threshold segmentation, region-growing, pixel trimming, and supplementation. Due to the variations in bone mineral density (BMD) and the ambiguity of bone edges, ensuring segmentation precision is often challenging. Additionally, this process is time-consuming, labor-intensive, and subjective, requiring surgeons with extensive experience and specialized knowledge [21,22]. However, accurate image segmentation is paramount for precise elbow joint manipulation. To improve segmentation efficiency and accuracy, advanced image segmentation algorithms have been explored and continuously developed, especially with the emergence of convolutional neural networks (CNNs) and deep learning [23].

Image segmentation based on deep learning allows for the acquisition of complex image features and delivers accurate segmentation results in specific tasks [24]. Ronneberger et al. [25] proposed U-Net, which consists of an encoder and decoder. It has been widely applied in the field of biomedical image segmentation due to its effectiveness with small datasets. Based on U-Net, researchers improved the network structure by optimizing various components, such as the backbone network, skip connections, bottleneck structure, and residual design. These enhancements have led to the proposal of U-Net++ [26], U-Net3+ [27], AReN-UNet [28], and HAD-ResUNet [29]. Moreover, transformer architecture has been gradually applied in the field of medical image segmentation. It utilizes a mechanism known as self-attention, allowing it to effectively capture relationships between different words or elements in a sequence. Several researchers employed deep learning methods for the segmentation of elbow joints. Wei et al. [30] used Faster R-CNN to detect and locate elbow bones, and designed a global–local fusion segmentation network to solve the overlapping area identification problem. Xu et al. [31] applied PointNet++ to perform segmentation on the single-frame point cloud of the human body, enabling the generation of six parts of the human body for estimating human joints in various poses, and realizing the capture and simulation of human joint movement based on computer vision.

However, obtaining large-scale datasets is relatively difficult in some tasks, which poses a significant obstacle to training neural networks. To address this issue, foundation models have emerged and achieved remarkable progress in artificial intelligence (AI). Pre-trained on extensive datasets, foundation models often exhibit impressive generalization capabilities, showing promising performance across various downstream tasks [32]. In the field of computer vision (CV), the segment anything model (SAM) has emerged as a pivotal presence. Trained on over 11 million images, SAM serves as a foundational segmentation model, which is promptable, capable of zero-shot and few-shot generations, and demonstrates considerable potential in downstream image segmentation tasks [23,33]. However, due to significant disparities between natural and medical image fields, SAM encounters challenges in medical image segmentation [34,35]. In response, MedSAM has emerged as a foundational medical segmentation model [36]. Considering data privacy and security in medical images, MedSAM demonstrates considerable promise and importance in medical image segmentation.

The MedSAM model utilizes prompt boxes to assist in segmentation, contributing to the improvement of segmentation accuracy and efficiency. The prompt boxes provide the region of interest (ROI), guiding models to locate and segment targets more accurately. For elbow joint CT scans, multiple bones may be present in the same CT image, it is crucial to minimize interference from other bones during segmentation. Traditionally, the prompt boxes are marked manually by researchers. Undoubtedly, generating the prompt boxes automatically based on the results of bone recognition can improve automation and accuracy in bone segmentation and reconstruction.

Therefore, a method for recognizing, segmenting, and reconstructing elbow joint bones automatically is proposed. This method can accurately and objectively identify adjacent elbow bones and generate segmentation prompt boxes, thereby achieving automation in elbow joint bone segmentation and facilitating seamless 3D reconstruction. Furthermore, this method is expected to be applied even with limited training sample sizes in clinical applications. It will provide significant assistance in clinical diagnosis, preoperative planning, and intraoperative navigation.

2. Method

The method can be divided into three steps, as shown in Figure 1. The elbow CT images obtained through CT scanning are used as input [33]. Firstly, a spatial elbow joint recognition is introduced, involving threshold segmentation, region-growing techniques, and elbow anatomical knowledge, and the prompt boxes are generated. Subsequently, the original CT images undergo clipping and encoding by an image encoder, obtained via transfer learning from the MedSAM. The elbow bone masks are predicted by a segmentation model, with mask correction and reclassification designed based on bone recognition results. Finally, the three main bones at the elbow joint, namely the humerus, ulna, and radius, are reconstructed and smoothed using the marching cube algorithm and Laplace smooth operation.

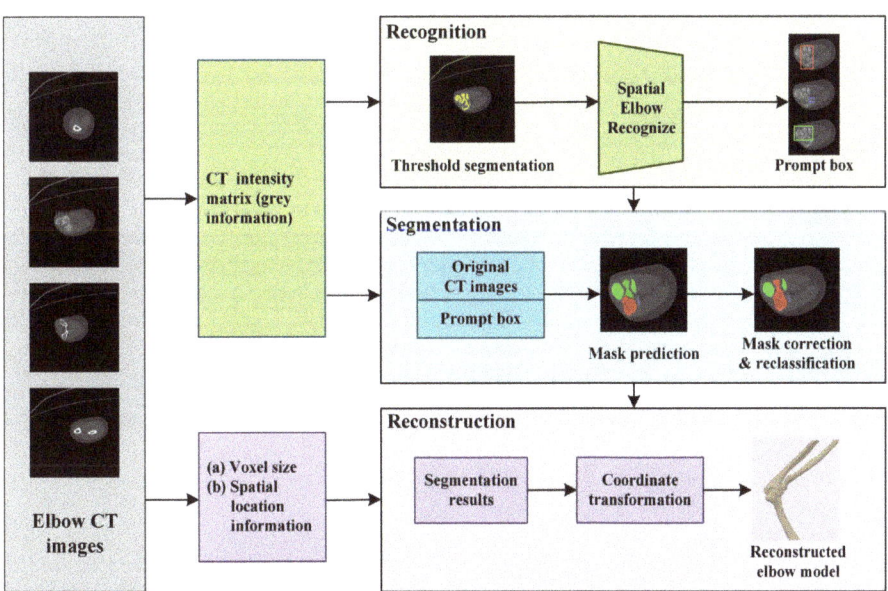

Figure 1. The elbow recognition, segmentation, and reconstruction method. In figure, the humerus, ulna and radius are shown in red, green, and blue, respectively.

2.1. Original Elbow CT Image Segmentation

MedSAM is a typical segmentation foundational model, generated by transfer learning based on SAM [33] deploying over 11 million medical images. Functioning as a prompt segmentation method, MedSAM employs prompt boxes to specify segmentation objects, thereby reducing segmentation ambiguity. Similar to SAM, MedSAM comprises an image encoder, a prompt encoder, and a mask decoder, as illustrated in Figure 2. The image encoder maps the original CT images to a higher dimensional space, and the prompt encoder converts the prompt bounding boxes into feature representation through positional encoding. The mask decoder utilizes image embeddings, and prompt embeddings to generate masks. MedSAM demonstrates comparable or even superior segmentation accuracy to professional models, as evidenced by its performance on over 70 internal tasks and 40 external validation tasks [36]. As a foundational segmentation model, it possesses robust feature extraction capabilities due to its extensive pre-training on a large number of medical images, implying that even in the absence of specific samples or with a limited dataset, the model can leverage its internal feature representations to perform effective inference. Consequently, the MedSAM model and the MedSAM model after transfer learning exhibit powerful generalization capabilities with small sample sizes.

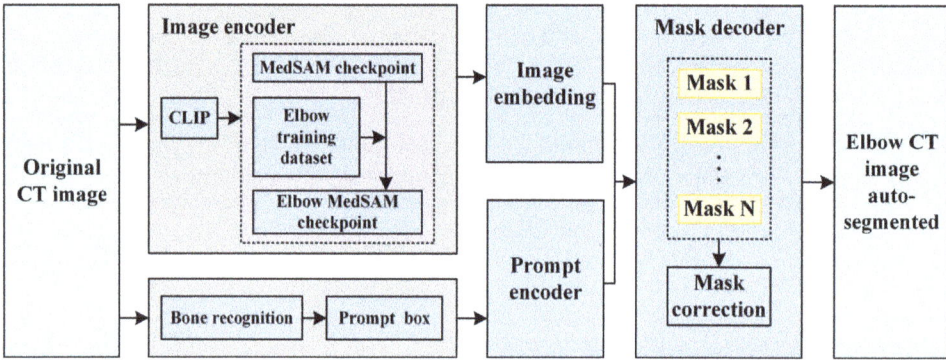

Figure 2. Automatic elbow CT image segmentation.

To improve the adaptability of MedSAM for elbow CT images, several series of CT images are utilized for transfer learning. For clarity, the fine-tuned MedSAM is referred to as elbow MedSAM. During dataset construction, several series of elbow CT images are randomly selected to generate the training dataset, and a CT image is considered as a sample. Specifically, each sample includes three parts: (1) original image, (2) image labeled by senior surgeons (binary matrix with the same shape as the original image), and (3) prompt box. The prompt box surrounding the target is automatically generated, by slightly expanding the narrow box with a random amount ranging from 0 to 30 pixels, which prevents the network from excessively relying on the boundary range of the rectangular box, which is shown in Figure 3.

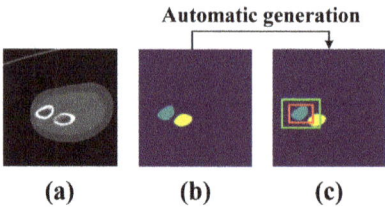

Figure 3. Composition of the training dataset. (**a**) Original image; (**b**) labeled image; (**c**) labeled image with automatically generated prompt box. The red and green rectangular boxes are narrow box and expanded box, respectively.

Manual drawing of prompt boxes is required for MedSAM, significantly impeding the efficiency of elbow joint segmentation and reconstruction. It is crucial to automatically recognize elbow bones from original CT images to improve efficiency. An initial threshold segmentation is performed, followed by the utilization of the region-growing technique to determine the connected component labels of each voxel within the segmented foreground by applying the breadth-first search algorithm (BFS). The connected components are then sorted based on the number of voxels they encompass. Specifically, the initial threshold setting should prevent bone adhesion from occurring at the elbow joint. Additionally, the initial threshold should be set as small as possible to encompass more bone tissue during the initial segmentation process.

In standard unilateral elbow joint CT scans, the four largest connected components obtained through the initial threshold calculation correspond to the CT bed, humerus, ulna, and radius, as shown in Figure 4. Considering that the joint space is smaller than the distance between the bone and the CT bed, the minimum distances between every pair of connected components are calculated using the KD tree algorithm, and the distance matrix is recorded as follows:

$$D_{min}^{4\times 4} = \{d_{ij}\}^{4\times 4}, \quad (1)$$

where d_{ij} is the minimum distance between the ith and jth connected components, and the index of CT bed can be determined as follows:

$$id_{CT\ bed} = \underset{i}{\mathrm{argmax}}\left(\sum_{i=1}^{4} d_{ij}\right), \quad (2)$$

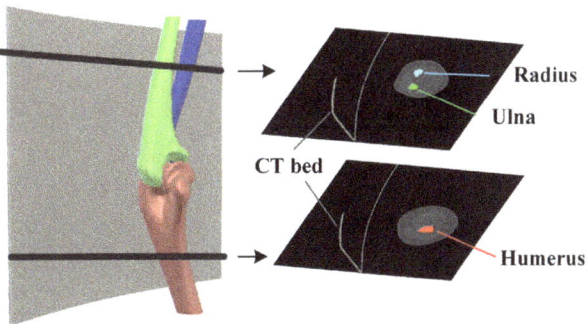

Figure 4. Main connected components after threshold segmentation and region-growing.

For the remaining three connected components, the relative distances are calculated. Several points are uniformly chosen for each connected component, and the distances from these points to other connected components are calculated. The average distance is considered as the relative distance between bones. Since the relative distance between the ulna and radius is the smallest among the three groups, the relative distance matrix $D_{mean}^{3\times 3}$ can be utilized to select the humerus from the three bones, calculated similarly to Equation (2).

The forearm shaft orientation ($\vec{n}_f$) is estimated based on a vector defined by point A (the distal point of the humerus) and point B (the farthest point from point A in the other two connected components), as shown in Figure 5. The proximal projection location of the ulna and radius in this direction can be calculated according to the following:

$$p_p = \max\left(\mathbf{C} \times \vec{n}_f\right) \quad (3)$$

where $\mathbf{C} \in \mathbf{R}^{N\times 3}$ represents the connected component of the ulna or radius. The connected component with the largest p_p is recorded as the ulna.

The prompt box is generated based on the results of bone recognition. The voxels of connected components are projected onto each layer. A narrow prompt box can be obtained:

$$Box_{narrow} = [x_{min} \quad x_{max} \quad y_{min} \quad y_{max}] \quad (4)$$

Bone recognition results cannot encompass the entire bone, it is necessary to expand the narrow prompt box with an interval d (Figure 6). Therefore, the expanded prompt box can be expressed as follows:

$$Box_{expand} = [x_{min} - d \quad x_{max} + d \quad y_{min} - d \quad y_{max} + d] \quad (5)$$

The expanded prompt boxes are input into the segmentation model and encoded by the prompt encoder. Finally, segmentation masks are predicted layer by layer.

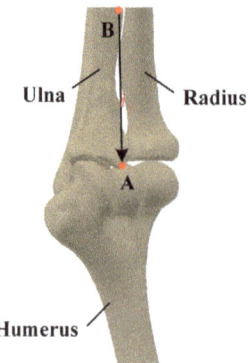

Figure 5. Evaluation of forearm shaft orientation.

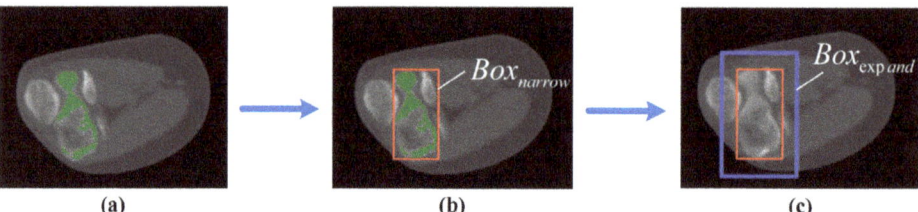

Figure 6. Automatic prompt box generation and expansion. (**a**) Bone recognition; (**b**) narrow prompt box; (**c**) expanded prompt box.

2.2. Mask Correction and Reclassification

Several details need to be supplemented to reduce segmentation errors. Tiny holes within segmentation bones are observed, which are caused by the inhomogeneity of BMD within the bone marrow cavity. These holes should be filled using a hole-filling algorithm.

Near the elbow joint, multiple bones may appear in the same image with a narrow spacing. The mask predicted from the prompt box may not only include the target bone but also other bones. To address this issue, reclassification and validation operations are performed for the predicted mask based on the bone recognition result. Considering that a single bone might comprise multiple separated regions on a single layer, the bone reclassification result can be expressed as follows:

$$C = \bigcup_{i=1}^{N_C} C_i \cdot sign(\|C_i \cap C_{recognize}\|) \quad (6)$$

where C_i denotes the connected component calculated based on the segmentation mask, $C_{recognize}$ denotes the point set corresponding to the bone recognition result, and N_c indicates the number of connected components in the prompt box. Finally, the target bone can be accurately predicted, as shown in Figure 7.

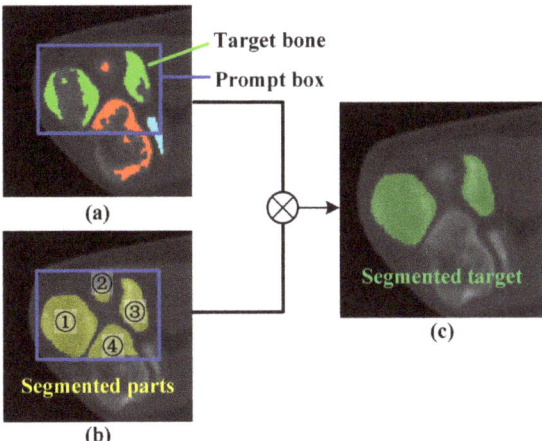

Figure 7. Schematic of mask reclassification. (**a**) Result of bone recognition; (**b**) automated segmentation; (**c**) mask after reclassification.

2.3. Elbow Reconstruction

The elbow bone is reconstructed using the marching cube algorithm based on the elbow segmentation results. Then, the image acquisition coordinate system is transformed into the spatial coordinate system, facilitating precise intraoperative localization in clinical applications. Therefore, the coordinate transformation can be expressed as follows:

$$\begin{bmatrix} P' \\ 1 \end{bmatrix} = \begin{bmatrix} & & 0 & \\ O_{image}\Delta S & 0 & & S \\ & & \Delta h & \\ 0 & 0 & 0 & 1 \end{bmatrix} \begin{bmatrix} P \\ 1 \end{bmatrix} \quad (7)$$

where P and P' denote the vertices of the 3D bone model before and after the coordinate transformation, O_{image} and ΔS represent image orientation and pixel spacing obtained from original CT data, Δh denotes slice distance, and S is the spatial position of the point in the upper left corner of the first CT image. Following that, a Laplace smoothing operation is conducted to reduce high-frequency noise while preserving the whole structure of the bone shape. During this process, the smoothed positions are determined based on their adjacent points for all vertices:

$$U(p) = \frac{1}{n}\sum_{i=1}^{n-1} Adj_i(P) \quad (8)$$

where P and $U(p)$ are points before and after smoothing.

3. Results and Discussion

3.1. Dataset

The study received approval from the Beijing Jishuitan Hospital Ethics Committee, and informed consent was waived by the ethics committee. A total of 41 patients treated in the Department of Trauma at Beijing Jishuitan Hospital were enrolled in the study. Inclusion criteria were no history of congenital malformations and arthritis. All patients underwent unilateral elbow CT scans (UIH/uCT500) with voltage and current settings of 100 kV and 65 mA. The resolution of the CT image was 512 × 512, with a pixel spacing ranging from

0.29 to 0.57 mm, and a slice distance of 0.8 mm. Among 41 series of unilateral elbow joint CT scans, there are 17 left elbow joints and 24 right elbow joints. The range of elbow flexion angles was between 82.10° and 170.11°. The entire dataset was divided into training and validation datasets, with 11 series of CT images randomly selected for the transfer learning of MedSAM, and the other 30 series of CT images applied for verification. Three senior orthopedic experts collaborated to label the bones using Mimics 20.0, including the humerus, ulna, and radius, which were regarded as the "ground truth (GT)".

The transfer learning of MedSAM was conducted using Python 3.10 and PyTorch 2.0. The checkpoint of MedSAM is used as the initial value. Adam optimizer was utilized to train the network; the training spanned 100 epochs, utilizing a weight decay of 0.01, and a learning rate set of 0.0001. The loss function is defined by the sum of cross-entropy loss and dice loss, which is reduced from 0.022 to 0.007 for the training dataset.

3.2. Verification of Elbow Bone Recognition

3.2.1. Result of Elbow Bone Recognition

Figure 8 shows the bone recognition results and generated prompt boxes, demonstrating that elbow bone recognition can automatically identify multiple targets in the entire CT images based on the bone position relationships. This facilitated more efficient and intuitive classification of bones in elbow CT images. However, the recognized masks only depicted the general shape of the bones. Specifically, only pixels with high CT intensity values can be captured. The marrow cavity was omitted due to the low bone mineral de BMD, which can be supplemented if the cortical bone is completely segmented. In addition, the BMD at the bone end is relatively low due to the need for flexibility and shock absorption, resulting in missing pixels, which typically require pixel-by-pixel segmentation refinement. Despite these limitations, the recognized masks still provided an approximate indication of the bone's position, guiding the generation of prompt boxes. It helps enhance the automation of subsequent segmentation and 3D reconstruction. To ensure complete segmentation of elbow bones, expanding the prompt boxes is conducted. In our studies, the expansion size was set to 20 pixels at the bone ends and 10 pixels at the bone shafts.

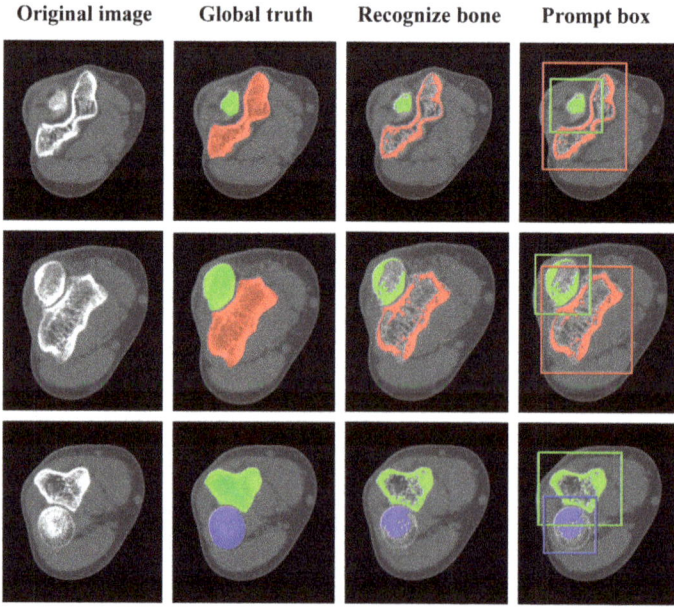

Figure 8. Results of elbow bone recognition and generated prompt boxes. In the figure, the humerus, ulna, and radius are shown in red, green, and blue, corresponding to the prompt boxes.

3.2.2. Impact of Automatic Prompt Box Generation

An ablation study was performed to verify the significance of prompt box generation, which is shown in Figure 9. Obviously, without a prompt box, the segmented mask covered almost the entire image, as shown in Figure 9b, resulting in an invalid segmentation. In addition, for an inappropriate prompt box (such as an overly large prompt box, as shown in Figure 9c), effective segmentation results could not be obtained. This indicates that appropriate prompt boxes are crucial for achieving precise segmentation.

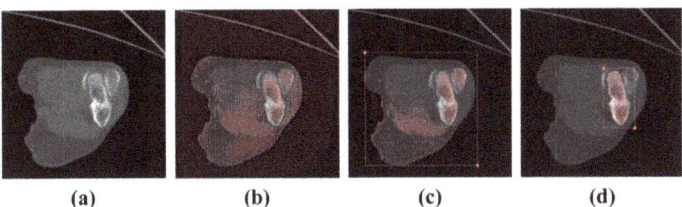

Figure 9. The contrasting impact of prompt boxes on elbow bone segmentation. (**a**) Original image; (**b**) segmentation masks without a prompt box; (**c**) segmentation masks using an inappropriate prompt box; (**d**) segmentation masks using an appropriate prompt box.

Furthermore, it is crucial to recognize elbow bones and generate prompt boxes automatically, as manually drawing prompt boxes by surgeons is time-consuming and labor-intensive. Additionally, identifying the specific elbow joint bone within the prompt box is challenging when only a single image is available, particularly in scenarios where one bone appears as two separate domains in the CT image.

3.3. Impact of Mask Correction and Reclassification

The comparison for a series of CT images before and after mask correction and reclassification was performed, and the intersection over union (IoU) was utilized to gauge segmentation accuracy, which is defined by the following:

$$\text{IoU} = \frac{A \cap B}{A \cup B} \tag{9}$$

where A and B represent two masks being compared, and the numerator means the area of the overlap regions of two masks, and the denominator means the total coverage of two masks, as shown in Figure 10. Obviously, IoU is 1 if the two masks completely overlap, while IoU is 0 if the two masks are entirely unrelated.

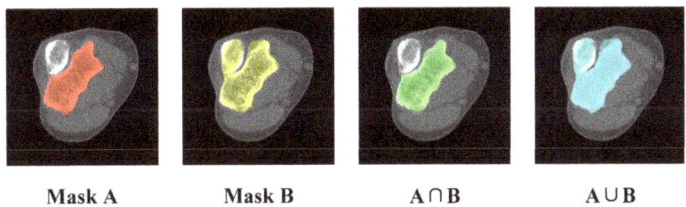

Figure 10. Diagram about the calculation of IoU.

It was noticeable that mask correction had little impact on the IoU at the bone shaft (Figure 11a). The slight differences in IoU values for these layers were primarily due to minor holes attributed to the complex shape and texture of bone tissue. The IoU values were improved after these tiny holes were filled, as shown in Figure 11b.

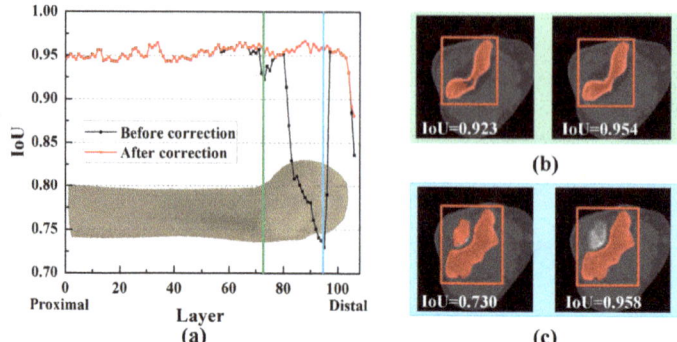

Figure 11. Comparison of the effect of mask correction and reclassification. (**a**) Comparison of IoU values for all CT layers; (**b**,**c**) mask comparison for two specified layers.

However, in the region of the humeral trochlea and capitellum, the difference in IoU values was substantial, with a maximum difference of 0.228 (Figure 11c). Due to the narrow joint space at the humeroulnar joint, there was an overlap between the prompt boxes for the ulna and humerus. As a result, segmented masks within the humerus prompt box might not entirely belong to the humerus (Figure 11c). However, through mask reclassification based on the initial bone recognition, pixels corresponding to the ulna were excluded, resulting in a significant increase in IoU, highlighting the importance of reclassifying segmentation targets.

3.4. Result of Elbow CT Segmentation

3.4.1. Qualitative Evaluation of Segmentation Results

Figure 12 demonstrates the qualitative evaluation of segmentation results using U-Net, origin MedSAM, and elbow MedSAM. There was no significant difference observed for the bone shafts of the humerus (Image 1), ulna, and radius (Image 2) among the three segmentation models. However, the primary segmentation challenge arose near the elbow joint, primarily due to the narrow joint space between the humerus and ulna, resulting in an indistinct boundary between bone and soft tissue. For U-net, it is almost impossible to separate different bones. In the region of the elbow joint, different bones might be segmented as an entire bone (Image 4). Among three segmentation methods, elbow MedSAM demonstrated excellent segmentation results (Images 3 and 4), attributed to the transfer learning, which enhanced the model's effectiveness in distinguishing between joint gaps and bone tissues. Obviously, transfer learning resulted in a sharper feature recognition capability. Even when confronted with small joint spaces, elbow MedSAM demonstrated a reduced likelihood of identifying different bones as a single target, thereby ensuring the relative integrity of bone segmentation. Furthermore, our investigation revealed that even in cases where the elbow joint is in a flexed posture, meaning elbow bones are not fully displayed in cross-section, elbow MedSAM can accurately identify and segment the forearm bones (Image 5).

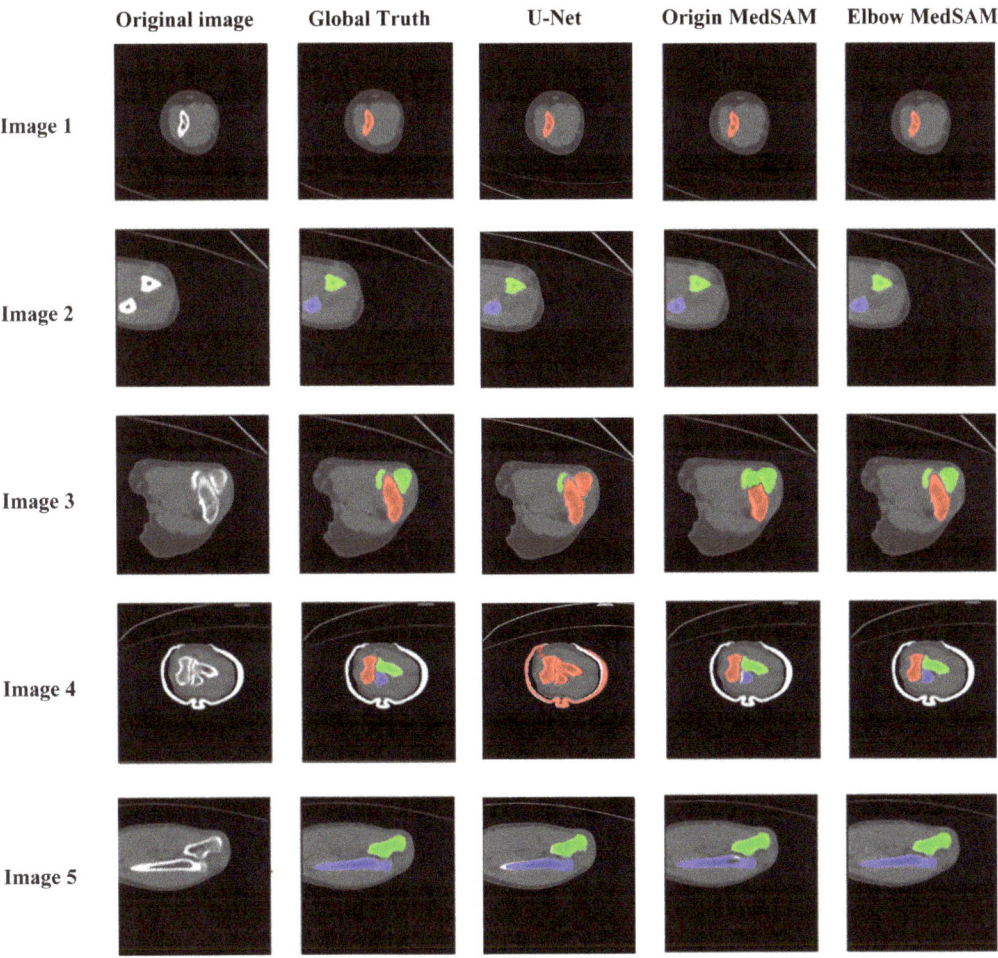

Figure 12. Comparison of segmentation results between origin MedSAM and elbow MedSAM. In the figure, the humerus, ulna, and radius are shown in red, green, and blue, respectively.

3.4.2. Quantitative Evaluation of Segmentation Accuracy

The U-Net exhibited extremely limited segmentation performance at the elbow joint regions. Moreover, it has also been documented in the literature that its segmentation effectiveness is inferior to that of MedSAM for medical images [36]. Therefore, only the origin MedSAM and elbow MedSAM were compared for quantitative evaluation.

The quantitative evaluation was first conducted on a series of CT scans, as shown in Figure 13. It was clear that the elbow MedSAM exhibited stability, with the IoU values ranging from 0.911 to 0.983, 0.894 to 0.981, and 0.865 to 0.980 for the humerus, ulna, and radius, respectively. Compared to the origin MedSAM, the IoU values were significantly improved.

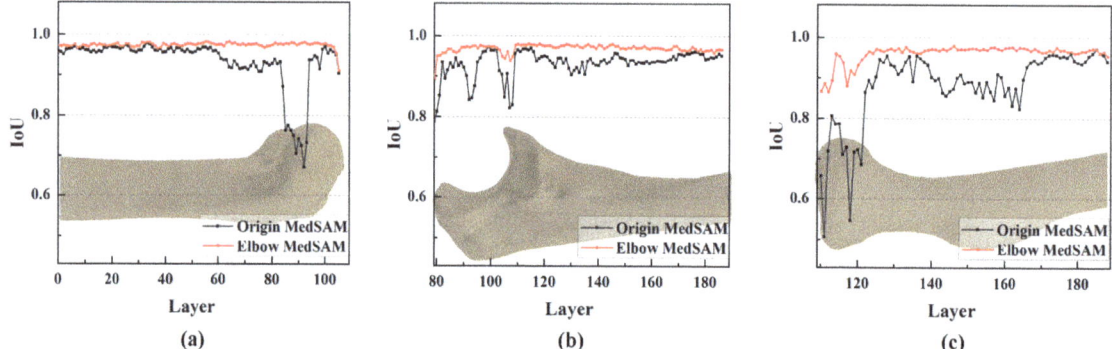

Figure 13. IoU plot of a series of elbow joint CT segmentation results. (a) Humerus; (b) ulna; (c) radius.

In fact, the bone shaft exhibits high BMD and clear bone texture, resulting in relatively good segmentation results for both models. However, near the elbow joint, compared to elbow MedSAM, the origin MedSAM experienced a maximum decrease in IoU values of 0.31, 0.138, and 0.38 for the humerus, ulna, and radius, respectively.

The CT images displaying the most significant disparities in IoU values were utilized for further analysis. For the humerus and ulna, origin MedSAM exhibited the lowest segmentation accuracy at the humeroulnar joint, mainly attributed to the extremely narrow joint space and blurred edges, as shown in Figure 14. Without transfer learning, the segmentation model might erroneously identify the humerus and ulna as a single bone, and the accuracy was hardly enhanced through mask reclassification. For the radius, lower segmentation accuracy was observed near the radial head due to physiological reasons. The radial head requires flexibility and cushioning during elbow joint movement, thus resulting in relatively lower BMD. Additionally, the BMD naturally decreases with increasing age.

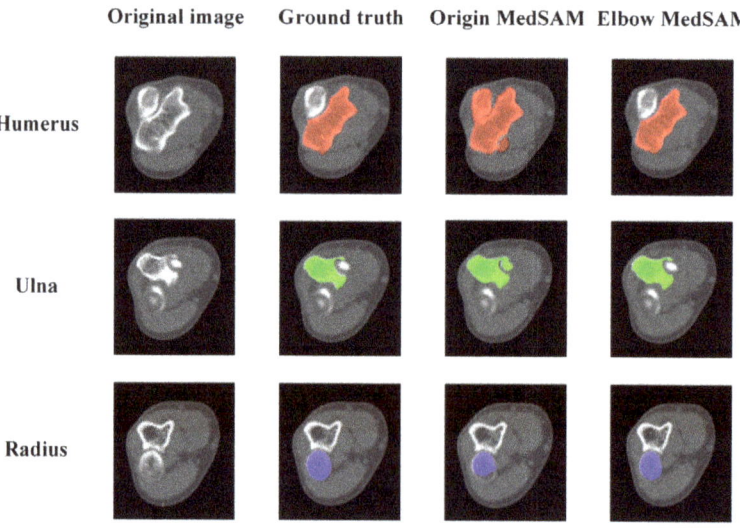

Figure 14. Instances of segmentation result for each bone. In the figure, the humerus, ulna, and radius are shown in red, green, and blue, respectively.

Lower IoU values mainly existed in several layers near the bone end. However, the phenomenon did not indicate poor segmentation quality. In these regions of the CT images, the target bones occupied a very small proportion (just a few dozen pixels). As a result, even a small number of pixels inconsistent with the ground truth can lead to a significantly low IoU value. Moreover, influenced by the partial volume effect, the bone edges appeared quite blurred. This implies that even senior surgeons might not provide completely accurate segmentation results when labeling the ground truth. Furthermore, we noted that despite the partial volume effect causing blurred bone edges, our model demonstrated a certain degree of robustness. Overall, the results indicated that despite encountering some challenges, the elbow MedSAM has achieved satisfactory results in elbow joint CT image segmentation tasks.

Statistics analysis was conducted on the segmentation results among the entire test dataset, and the results were visualized in a box plot, as depicted in Figure 15. The boxes encompassed IoU values from 10% to 90%. From the figure, the median IoU values were 0.963, 0.959, and 0.950 for the humerus, ulna, and radius, respectively. For 90% of the elbow CT images, the segmentation IoU values exceeded 0.939, 0.927, and 0.917, significantly higher than the IoU values achieved by origin MedSAM, which were 0.77, 0.773, and 0.779, respectively. This indicated that elbow MedSAM demonstrated higher stability and can be applied to a broader range of elbow CT images. In contrast, the origin MedSAM showed weaker specificity toward the elbow joint, resulting in slightly inferior performance.

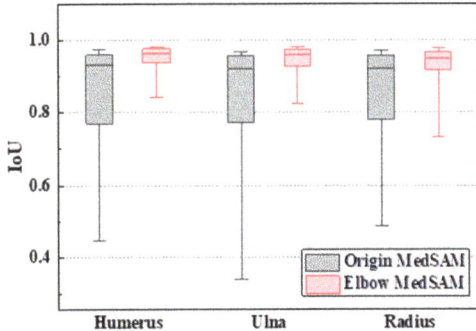

Figure 15. Boxplot of segmentation IoU values of elbow CT images.

3.4.3. Objectivity Discussion of Elbow Joint Segmentation

In clinical practice, the segmentation results of elbow joint bones by senior surgeons exhibit a certain degree of subjective judgment, highlighting the reliance on the clinician's experience for manual segmentation. This subjectivity can lead to inconsistencies and variability in the results. To address this issue, we developed Elbow-MedSAM through transfer learning based on MedSAM, which has been pre-trained on a large dataset of medical images. This model leverages the extensive segmentation experience embedded in these images, thereby eliminating the need for manual adjustments and human intervention. By automating the segmentation process, Elbow-MedSAM ensures a more objective and consistent approach, reducing the impact of individual biases and potentially improving the accuracy and reproducibility of the segmentation outcomes.

3.5. Reliability and Accuracy of 3D Elbow Reconstruction

3.5.1. Result of 3D Elbow Reconstruction

The comparison between the 3D elbow bone surface models reconstructed by the automatic method and ground truth was conducted, and the reconstruction results are depicted in Figure 16. It is evident that the 3D elbow joint model obtained through the automatic method is essentially consistent with the ground truth, showing no obvious holes or bone adhesions.

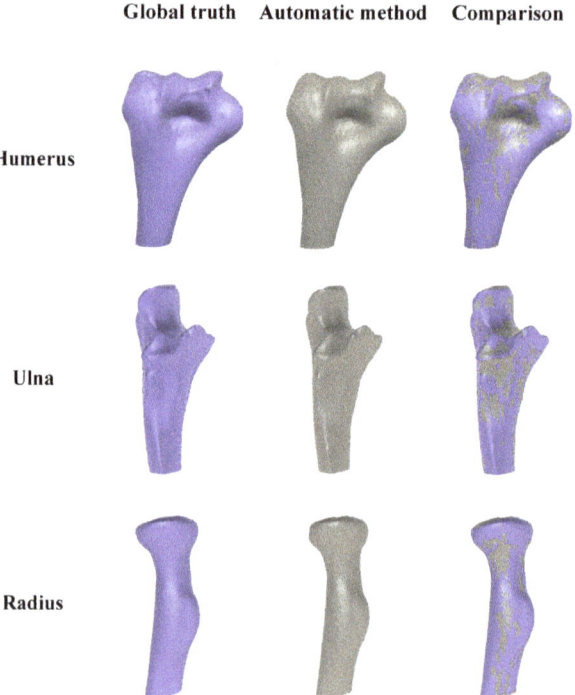

Figure 16. Comparison of reconstruction elbow bone model between ground truth (in purple) and automatic method (in grey).

The surface error of the reconstructed model was calculated, with the statistical results shown in Figure 17. The maximum surface error values were 0.877 mm, 1.246 mm, and 1.274 mm for the three elbow bones. These results underscored the reliability and accuracy of our automatic reconstruction method in capturing the intricate details of the elbow joint morphology, indicating a high degree of consistency with the ground truth.

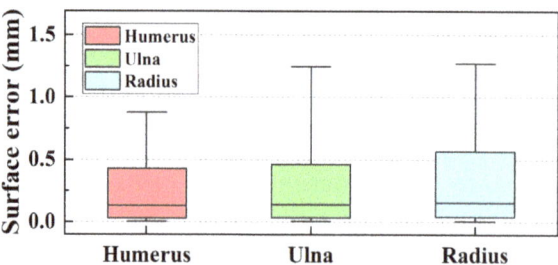

Figure 17. Boxplot of surface error values of reconstructed models between automatic method and ground truth.

3.5.2. Reliability Analysis of Elbow Joint Reconstruction

Further analysis of the reconstruction results of 30 elbow joints was conducted, and the maximum surface errors are shown in Table 1 and Figure 18. Two main reasons accounted for the occurrence of large surface errors at individual points: (1) incomplete segmentation near the radius head in some cases, as depicted in Figure 18a. (2) For the bone shaft positioned at the edge of the CT image, the edges of the rectangular box coincide with the image edge. This situation primarily occurred in the ulnar and radial shaft regions, resulting in part bone marrow cavities in the reconstructed bone models, introducing surface errors from the ideally reconstructed bones, as shown in Figure 18b. Since the main functions of the elbow joint are flexion and extension, these errors are not expected to significantly affect the analysis of elbow joint morphology and kinematic simulation.

Table 1. Max surface error for test CT series.

	Mean (mm)	Quartiles (Q1 to Q3) (mm)	Range (Min. to Max.) (mm)
Humerus	1.127	0.654 to 1.433	0.262 to 2.247
Ulna	1.523	0.976 to 1.906	0.737 to 2.695
Radius	2.062	1.299 to 2.711	0.582 to 3.388

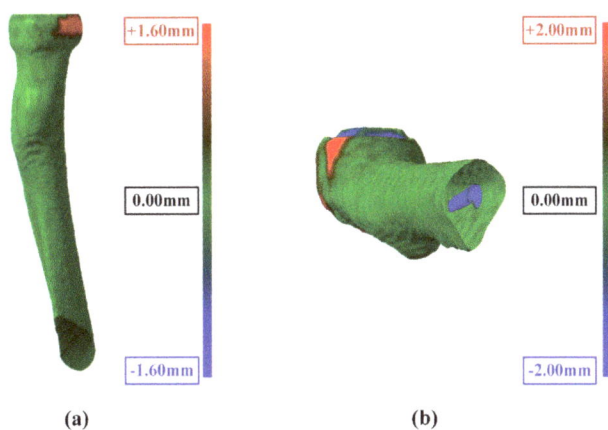

Figure 18. Main error in surface model reconstruction. (a) Incomplete segmentation near the radius head; (b) interference of marrow cavity.

The automatic reconstruction was successfully implemented in all 30 cases of elbow joint CT images, indicating its robustness and general applicability for elbow joint CT images. As the checkpoint is fixed after transfer learning, the segmentation results are repeatable. The reconstructed surface models generally align well with the ground truth except for some special positions, and the reconstruction error is within an acceptable range, further validating the reliability of the proposed method.

4. Conclusions

In this study, an automatic method for elbow joint recognition, segmentation, and reconstruction is proposed. Prompt boxes are generated by automatic elbow bone recognition. Transfer learning is utilized to improve the segmentation accuracy, and hole-filling and mask reclassification are applied to refine the segmented masks. The elbow bone reconstruction can be conducted seamlessly. By comparing the segmentation and reconstruction results with manually labeled images, this automatic method has proved to be reliable, objective, and accurate. The main conclusions are as follows:

(1) This study employs an interpretable algorithm to automatically recognize the humerus, ulna, and radius from elbow joint CT images. The algorithm exhibits stability and effectiveness for elbow joints from flexion (82.10°) to extension (170.11°) postures.

(2) The IoU values near the joint are significantly increased by mask correction and reclassification, with a maximum improvement of 0.028, conclusively boosting segmentation accuracy.

(3) The segmentation accuracy is enhanced by the MedSAM after transfer learning, allowing for more precise capture of bone edges and reducing instances of mistaking

multiple bones as a single target. The median IoU values are 0.963, 0.959, and 0.950 for the humerus, ulna, and radius, respectively, notably surpassing the predictions of the origin MedSAM.

(4) The maximum surface errors for the bone surface model reconstructed by the marching cube algorithm are 1.127, 1.523, and 2.062 mm for the humerus, ulna, and radius, respectively.

Author Contributions: Conceptualization, T.Z. and Y.Z. (Yejun Zha); methodology, Y.C.; software, Y.C.; validation, Y.C.; formal analysis, X.Z. and Y.Z. (Yichuan Zhang); investigation, X.Z.; resources, Y.Z. (Yejun Zha); data curation, S.J.; writing—original draft preparation, Y.C.; writing—review and editing, T.Z. and S.J.; visualization: Y.C.; supervision, X.Z.; project administration, S.J.; funding acquisition, Y.Z. (Yejun Zha). All authors have read and agreed to the published version of the manuscript.

Funding: Supported by Beijing Natural Science Foundation (L192049).

Institutional Review Board Statement: All procedures performed in studies involving human participants were in accordance with the ethical standards of the National Research Committee and with the 1964 Helsinki Declaration and its later amendments or comparable ethical standards. The study was approved by the Beijing Jishuitan Hospital (K2022-197).

Informed Consent Statement: Patient consent was waived due to the retrospective design of this study.

Data Availability Statement: The data used to support the findings of this study are available from the corresponding author upon request. The data are not publicly available due to privacy concerns. We are actively working with hospitals to obtain public permission for the dataset.

Acknowledgments: We are sincerely grateful to all novices and surgeons in the Beijing Jishuitan Hospital who participated in the manual methods. We also thank all the patients who provided elbow CT images in this study.

Conflicts of Interest: The authors declare no conflicts of interest.

References

1. Facchini, G.; Bazzocchi, A.; Spinnato, P.; Albisinni, U. CT and 3D CT of the Elbow. In *The Elbow*; Springer: Cham, Switzerland, 2018; pp. 91–96, ISBN 978-3-319-27805-6.
2. Jackowski, J.R.; Wellings, E.P.; Cancio-Bello, A.; Nieboer, M.J.; Barlow, J.D.; Hidden, K.A.; Yuan, B.J. Computed Tomography Provides Effective Detection of Traumatic Arthrotomy of the Elbow. *J. Shoulder Elb. Surg.* **2023**, *32*, 1280–1284. [CrossRef]
3. Giannicola, G.; Sacchetti, F.M.; Greco, A.; Cinotti, G.; Postacchini, F. Management of Complex Elbow Instability. *Musculoskelet. Surg.* **2010**, *94*, 25–36. [CrossRef] [PubMed]
4. Zubler, V.; Saupe, N.; Jost, B.; Pfirrmann, C.W.A.; Hodler, J.; Zanetti, M. Elbow Stiffness: Effectiveness of Conventional Radiography and CT to Explain Osseous Causes. *Am. J. Roentgenol.* **2010**, *194*, W515–W520. [CrossRef] [PubMed]
5. Acar, K.; Aksay, E.; Oray, D.; Imamoğlu, T.; Gunay, E. Utility of Computed Tomography in Elbow Trauma Patients with Normal X-Ray Study and Positive Elbow Extension Test. *J. Emerg. Med.* **2016**, *50*, 444–448. [CrossRef] [PubMed]
6. Haapamaki, V.V.; Kiuru, M.J.; Koskinen, S.K. Multidetector Computed Tomography Diagnosis of Adult Elbow Fractures. *Acta Radiol.* **2004**, *45*, 65–70. [CrossRef]
7. Hamoodi, Z.; Singh, J.; Elvey, M.H.; Watts, A.C. Reliability and Validity of the Wrightington Classification of Elbow Fracture-Dislocation. *Bone Jt. J.* **2020**, *102-B*, 1041–1047. [CrossRef]
8. Waldt, S.; Bruegel, M.; Ganter, K.; Kuhn, V.; Link, T.M.; Rummeny, E.J.; Woertler, K. Comparison of Multislice CT Arthrography and MR Arthrography for the Detection of Articular Cartilage Lesions of the Elbow. *Eur. Radiol.* **2005**, *15*, 784–791. [CrossRef]
9. Alnusif, N.S.; Matache, B.A.; AlQahtani, S.M.; Isa, D.; Athwal, G.S.; King, G.J.W.; MacDermid, J.C.; Faber, K.J. Effectiveness of Radiographs and Computed Tomography in Evaluating Primary Elbow Osteoarthritis. *J. Shoulder Elb. Surg.* **2021**, *30*, S8–S13. [CrossRef] [PubMed]
10. Kwak, J.-M.; Kholinne, E.; Sun, Y.; Alhazmi, A.M.; Koh, K.-H.; Jeon, I.-H. Intraobserver and Interobserver Reliability of the Computed Tomography-Based Radiographic Classification of Primary Elbow Osteoarthritis: Comparison with Plain Radiograph-Based Classification and Clinical Assessment. *Osteoarthr. Cartil.* **2019**, *27*, 1057–1063. [CrossRef] [PubMed]
11. Sabo, M.T.; Athwal, G.S.; King, G.J.W. Landmarks for Rotational Alignment of the Humeral Component During Elbow Arthroplasty. *J. Bone Jt. Surg.* **2012**, *94*, 1794–1800. [CrossRef]
12. Willing, R.T.; Nishiwaki, M.; Johnson, J.A.; King, G.J.W.; Athwal, G.S. Evaluation of a Computational Model to Predict Elbow Range of Motion. *Comput. Aided Surg.* **2014**, *19*, 57–63. [CrossRef]

13. Yang, F.; Weng, X.; Miao, Y.; Wu, Y.; Xie, H.; Lei, P. Deep Learning Approach for Automatic Segmentation of Ulna and Radius in Dual-Energy X-ray Imaging. *Insights Imaging* **2021**, *12*, 191. [CrossRef]
14. Schnetzke, M.; Fuchs, J.; Vetter, S.Y.; Beisemann, N.; Keil, H.; Grützner, P.-A.; Franke, J. Intraoperative 3D Imaging in the Treatment of Elbow Fractures—A Retrospective Analysis of Indications, Intraoperative Revision Rates, and Implications in 36 Cases. *BMC Med. Imaging* **2016**, *16*, 24. [CrossRef]
15. Iwamoto, T.; Suzuki, T.; Oki, S.; Matsumura, N.; Nakamura, M.; Matsumoto, M.; Sato, K. Computed Tomography–Based 3-Dimensional Preoperative Planning for Unlinked Total Elbow Arthroplasty. *J. Shoulder Elb. Surg.* **2018**, *27*, 1792–1799. [CrossRef] [PubMed]
16. Tarniță, D.; Boborelu, C.; Popa, D.; Tarniță, C.; Rusu, L. The Three-Dimensional Modeling of the Complex Virtual Human Elbow Joint. *Rom. J. Morphol. Embryol.* **2010**, *51*, 489–495. [PubMed]
17. Bizzotto, N.; Tami, I.; Santucci, A.; Adani, R.; Poggi, P.; Romani, D.; Carpeggiani, G.; Ferraro, F.; Festa, S.; Magnan, B. 3D Printed Replica of Articular Fractures for Surgical Planning and Patient Consent: A Two Years Multi-Centric Experience. *3D Print. Med.* **2016**, *2*, 2. [CrossRef] [PubMed]
18. Antoniac, I.V.; Stoia, D.I.; Ghiban, B.; Tecu, C.; Miculescu, F.; Vigaru, C.; Saceleanu, V. Failure Analysis of a Humeral Shaft Locking Compression Plate—Surface Investigation and Simulation by Finite Element Method. *Materials* **2019**, *12*, 1128. [CrossRef] [PubMed]
19. Savic, S.P.; Ristic, B.; Jovanovic, Z.; Matic, A.; Prodanovic, N.; Anwer, N.; Qiao, L.; Devedzic, G. Parametric Model Variability of the Proximal Femoral Sculptural Shape. *Int. J. Precis. Eng. Manuf.* **2018**, *19*, 1047–1054. [CrossRef]
20. Grunert, R.; Winkler, D.; Frank, F.; Moebius, R.; Kropla, F.; Meixensberger, J.; Hepp, P.; Elze, M. 3D-Printing of the Elbow in Complex Posttraumatic Elbow-Stiffness for Preoperative Planning, Surgery-Simulation and Postoperative Control. *3D Printing in Medicine* **2023**, *9*, 28. [CrossRef]
21. Klein, A.; Warszawski, J.; Hillengaß, J.; Maier-Hein, K.H. Automatic Bone Segmentation in Whole-Body CT Images. *Int. J. Comput. Assist. Radiol. Surg.* **2019**, *14*, 21–29. [CrossRef]
22. Rathnayaka, K.; Sahama, T.; Schuetz, M.A.; Schmutz, B. Effects of CT Image Segmentation Methods on the Accuracy of Long Bone 3D Reconstructions. *Med. Eng. Phys.* **2011**, *33*, 226–233. [CrossRef] [PubMed]
23. Liu, X.; Song, L.; Liu, S.; Zhang, Y. A Review of Deep-Learning-Based Medical Image Segmentation Methods. *Sustainability* **2021**, *13*, 1224. [CrossRef]
24. Qureshi, I.; Yan, J.; Abbas, Q.; Shaheed, K.; Riaz, A.B.; Wahid, A.; Khan, M.W.J.; Szczuko, P. Medical Image Segmentation Using Deep Semantic-Based Methods: A Review of Techniques, Applications and Emerging Trends. *Inf. Fusion* **2023**, *90*, 316–352. [CrossRef]
25. Ronneberger, O.; Fischer, P.; Brox, T. *U-Net: Convolutional Networks for Biomedical Image Segmentation*; Springer International Publishing: Cham, Switzerland, 2015; pp. 234–241.
26. Zhou, Z.; Siddiquee, M.M.R.; Tajbakhsh, N.; Liang, J. UNet++: Redesigning Skip Connections to Exploit Multiscale Features in Image Segmentation. *IEEE Trans. Med. Imaging* **2020**, *39*, 1856–1867. [CrossRef] [PubMed]
27. Huang, H.; Lin, L.; Tong, R.; Hu, H.; Zhang, Q.; Iwamoto, Y.; Han, X.; Chen, Y.-W.; Wu, J. UNet 3+: A Full-Scale Connected UNet for Medical Image Segmentation. In Proceedings of the ICASSP 2020–2020 IEEE International Conference on Acoustics, Speech and Signal Processing (ICASSP), Barcelona, Spain, 4–8 May 2020; pp. 1055–1059.
28. Abdul Rahman, A.; Biswal, B.; Hasan, S.; Sairam, M.V.S. Robust Segmentation of Vascular Network Using Deeply Cascaded AReN-UNet. *Biomed. Signal Process. Control* **2021**, *69*, 102953. [CrossRef]
29. Wang, Z.; Zou, Y.; Liu, P.X. Hybrid Dilation and Attention Residual U-Net for Medical Image Segmentation. *Comput. Biol. Med.* **2021**, *134*, 104449. [CrossRef] [PubMed]
30. Wei, D.; Wu, Q.; Wang, X.; Tian, M.; Li, B. Accurate Instance Segmentation in Pediatric Elbow Radiographs. *Sensors* **2021**, *21*, 7966. [CrossRef] [PubMed]
31. Xu, T.; An, D.; Jia, Y.; Chen, J.; Zhong, H.; Ji, Y.; Wang, Y.; Wang, Z.; Wang, Q.; Pan, Z.; et al. 3D Joints Estimation of Human Body Using Part Segmentation. *Inf. Sci.* **2022**, *603*, 1–15. [CrossRef]
32. Zhang, C.; Liu, L.; Cui, Y.; Huang, G.; Lin, W.; Yang, Y.; Hu, Y. A Comprehensive Survey on Segment Anything Model for Vision and Beyond. *arXiv* **2023**, arXiv:2305.08196. [CrossRef]
33. Kirillov, A.; Mintun, E.; Ravi, N.; Mao, H.; Rolland, C.; Gustafson, L.; Xiao, T.; Whitehead, S.; Berg, A.C.; Lo, W.-Y.; et al. Segment Anything. In Proceedings of the IEEE/CVF International Conference on Computer Vision, Paris, France, 2–6 October 2023; pp. 4015–4026.
34. Mazurowski, M.A.; Dong, H.; Gu, H.; Yang, J.; Konz, N.; Zhang, Y. Segment Anything Model for Medical Image Analysis: An Experimental Study. *Med. Image Anal.* **2023**, *89*, 102918. [CrossRef]
35. Zhang, Y.; Shen, Z.; Jiao, R. Segment Anything Model for Medical Image Segmentation: Current Applications and Future Directions. *Comput. Biol. Med.* **2024**, *171*, 108238. [CrossRef] [PubMed]
36. Ma, J.; He, Y.; Li, F.; Han, L.; You, C.; Wang, B. Segment Anything in Medical Images. *Nat. Commun.* **2024**, *15*, 654. [CrossRef] [PubMed]

Disclaimer/Publisher's Note: The statements, opinions and data contained in all publications are solely those of the individual author(s) and contributor(s) and not of MDPI and/or the editor(s). MDPI and/or the editor(s) disclaim responsibility for any injury to people or property resulting from any ideas, methods, instructions or products referred to in the content.

Article

Deep Learning-Based Multi-Class Segmentation of the Paranasal Sinuses of Sinusitis Patients Based on Computed Tomographic Images

Jongwook Whangbo [1,2], Juhui Lee [2], Young Jae Kim [2,3], Seon Tae Kim [4] and Kwang Gi Kim [2,3,5,*]

1 Department of Computer Science, Wesleyan University, Middletown, CT 06459, USA; jwhangbo@wesleyan.edu
2 Medical Devices R&D Center, Gachon University Gil Medical Center, Incheon 21565, Republic of Korea; juhui05134@gmail.com (J.L.); youngjae@gachon.ac.kr (Y.J.K.)
3 Department of Health Sciences and Technology, Gachon Advanced Institute for Health & Sciences and Technology (GAIHST), Gachon University, Incheon 21565, Republic of Korea
4 Department of Otolaryngology-Head and Neck Surgery, Gachon University Gil Hospital, Incheon 21565, Republic of Korea; kst2383@gilhospital.com
5 Department of Biomedical Engineering, College of IT Convergence, Gachon University, Seongnam-si 13120, Republic of Korea
* Correspondence: kimkg@gachon.ac.kr; Tel.: +82-10-3393-4544

Abstract: Accurate paranasal sinus segmentation is essential for reducing surgical complications through surgical guidance systems. This study introduces a multiclass Convolutional Neural Network (CNN) segmentation model by comparing four 3D U-Net variations—normal, residual, dense, and residual-dense. Data normalization and training were conducted on a 40-patient test set (20 normal, 20 abnormal) using 5-fold cross-validation. The normal 3D U-Net demonstrated superior performance with an F1 score of 84.29% on the normal test set and 79.32% on the abnormal set, exhibiting higher true positive rates for the sphenoid and maxillary sinus in both sets. Despite effective segmentation in clear sinuses, limitations were observed in mucosal inflammation. Nevertheless, the algorithm's enhanced segmentation of abnormal sinuses suggests potential clinical applications, with ongoing refinements expected for broader utility.

Keywords: paranasal sinuses; chronic sinusitis; Convolutional Neural Network (CNN); multiclass segmentation

Citation: Whangbo, J.; Lee, J.; Kim, Y.J.; Kim, S.T.; Kim, K.G. Deep Learning-Based Multi-Class Segmentation of the Paranasal Sinuses of Sinusitis Patients Based on Computed Tomographic Images. *Sensors* **2024**, *24*, 1933. https://doi.org/10.3390/s24061933

Academic Editor: Serena Mattiazzo

Received: 19 January 2024
Revised: 7 March 2024
Accepted: 15 March 2024
Published: 18 March 2024

Copyright: © 2024 by the authors. Licensee MDPI, Basel, Switzerland. This article is an open access article distributed under the terms and conditions of the Creative Commons Attribution (CC BY) license (https:// creativecommons.org/licenses/by/ 4.0/).

1. Introduction

In 1994, around 200,000 sinus surgeries were conducted in the United States [1]. By 1996, 12 percent of Americans under the age of 45 reported symptoms indicative of chronic sinusitis [2]. This widespread condition imposes a substantial societal burden, manifesting in frequent office visits, absenteeism from work, and missed school days [1]. When medicinal treatments fail to alleviate the condition, patients are often referred for sinus surgery. Many physicians refer to Computed Tomography (CT) scans when evaluating patients referred for sinus surgery [1,3]. Radiologists report anatomic variants, that can affect operative techniques, and critical variants, that can complicate surgery [4]. Identification of these anatomical variants affords the opportunity to avoid surgical complications [5]. Segmentation data can be used for the diagnosis, surgical planning, or workspace definition of robot-assisted systems. However, manual and semiautomatic segmentation of the paranasal sinuses has been evaluated as impractical in clinical settings because of the amount of time required for both systems [6,7]. The application of machine learning in this process warrants attention due to its potential to substantially mitigate the time and labor costs associated with manual segmentation. Ultimately, this holds promise for making the segmentation process feasible and practical in clinical settings.

Artificial intelligence is gaining popularity in the medical imaging field for developing models that produce human-interpretable results [8–10]. Because of the clustered arrangement of regions, including the frontal, ethmoid, and sphenoid sinuses, developing models that can produce practical results for the paranasal sinuses is an ongoing challenge. Two published studies focused on processing cone-beam computed tomography images to achieve segmentation of the maxillary sinus. In 2022, Choi et al. [11] trained a U-Net model to segment maxillary sinuses. The segmented results were refined using post-processing techniques to isolate and remove disconnected false positives. The trained model made predictions with a Dice similarity coefficient (DSC) value of 0.90 ± 0.19 before post-processing and 0.90 ± 0.19 after post-processing. Morgan et al. [12] trained two U-Net models to segment the maxillary sinus. The first model suggested crop boxes in the original image of the maxillary sinus, which were used to train the second part of the model to produce high-resolution segmentation results. The final segmentation results achieved a DSC score of 0.98 for the first model and 0.99 for the second model. Kuo et al. [13] proposed a 6-class segmentation model that segmented four different areas of the paranasal sinuses, treating the ethmoid sinus as two different areas: the anterior and posterior ethmoid sinus. A secondary model was trained to generate pseudo-labels on the unlabeled datasets. The model used in this study was an adaptation of the U-Net model [14] with the addition of depth-wise separable convolution, squeeze-and-excitation networks, and residual connections. The model was able to make predictions with a DSC value of 0.90. The approaches proposed by Choi et al. [11] and Morgan et al. [12] exhibited performance adequate for clinical applications. However, the aim of both studies was limited to the binary segmentation of the maxillary sinus.

We proposed a 5-class segmentation model for the four regions of the paranasal sinus: frontal sinus, ethmoid sinus, sphenoid sinus, and maxillary sinus. Training and validation were conducted on clinical-level CT scans sourced from patients exhibiting high degrees of genetic and biological variations. The objective was to develop a model capable of generating clinical data with sufficient accuracy to be practically applicable in clinical settings.

2. Materials and Methods

This study was approved by the Institutional Review Board (IRB) of Gachon University Gil Medical Center (GAIRB2022-182) and was conducted in accordance with the relevant guidelines and ethical regulations.

A total of 39,605 paranasal CT scans were collected from 201 patients with varying degrees of chronic sinusitis, including 3821 images from 20 patients without sinusitis. A total of 40 datasets were randomly selected as the hold-out test set, with 20 datasets originating from the patient group without sinusitis. These subsets were then labeled as "normal" and "abnormal" to reflect the respective patient group characteristics. Training was performed on the remaining 161 datasets with 5-fold cross validation, where 128 datasets were used for training and 33 for validation. In summary, the dataset was divided into sets comprising 128 patients for training, 33 patients for validation, and 40 patients for testing. Demographic information of the participating patients is summarized in Table 1.

Table 1. Patient distribution by age group and gender.

Age Group	Male	Gender Ratio	Female	Gender Ratio	Total	Ratio by Age
10–20	6	40.00%	9	60.00%	15	7.58%
20–30	15	68.18%	7	31.82%	22	11.11%
30–40	21	84.00%	4	16.00%	25	12.63%
40–50	14	63.64%	8	36.36%	22	11.11%
50–60	34	62.96%	20	37.04%	54	27.27%

Table 1. *Cont.*

Age Group	Male	Gender Ratio	Female	Gender Ratio	Total	Ratio by Age
60–70	32	71.11%	13	28.89%	45	22.73%
70–80	8	72.73%	3	27.27%	11	5.56%
80–	2	50.00%	2	50.00%	4	2.02%
Total	132	66.67%	66	33.33%	198	100%

Data collection and storage were performed using Excel (version 16.83, Microsoft, Redmond, WA, USA) and statistical analyses were performed using MedCalc (version 22, MedCalc Software Ltd., Ostend, Belgium). Training was performed on an Ubuntu server (version 20.04.6 LTS) with four Nvidia A100 80Gb GPUs (NVIDIA, Santa Clara, CA, USA), an AMD EPYC 7452 32-Core Processor (AMD, Santa Clara, CA, USA), and 1,031,900 Mb of RAM. The following libraries were used for training: Python (version 3.7), TensorFlow (version 2.6.0), and Keras (version 2.6.0).

Using the collected sinus data, we meticulously curated a ground truth dataset by labeling the sinus region for each patient. The oversight and guidance of two experienced otorhinolaryngologists was integral to this process, ensuring the utmost quality and accuracy of the dataset. The final ground truth data were congregated through a consensus between the two physicians. The ground truth was labeled along the axial, sagittal, and coronal axes, as visually depicted in Figure 1. The volumetric reconstruction (Figure 1d) presents the data in its authentic form, providing insight into how it is inputted into the deep learning model. The axial view (Figure 1a) shows the maxillary and sphenoid sinuses beneath the ethmoid sinuses. The sagittal view (Figure 1b) shows the left maxillary sinus and part of the sphenoid sinus. The coronal view (Figure 1c) shows the frontal and maxillary sinuses surrounding the ocular area.

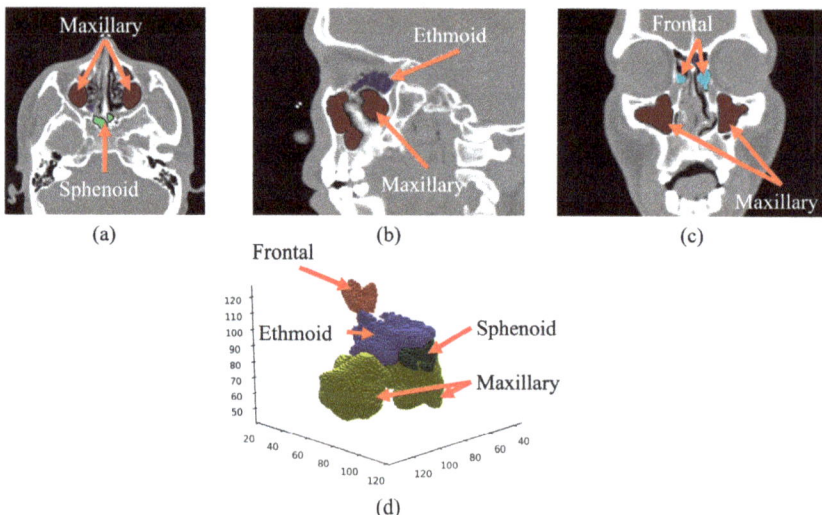

Figure 1. CT image of the paranasal sinuses with ground truth data overlayed. (**a**) Axial view, (**b**) sagittal view, (**c**) coronal view, and (**d**) volumetric reconstruction.

To facilitate the extraction of features within the CT scans, the datasets underwent several enhancements (Figure 2A), including window setting adjustments, isotropic voxel reconstruction, contrast-limited adaptive histogram equalization (CLAHE), and region of interest (ROI) cropping. The preprocessed images were used to train the segmentation model (Figure 2B) to produce segmentation results (Figure 2C). The overall training process is presented in Figure 2. A bone window with a width of 2,000 and a level of 0 was set and converted into 8-bit encoding. This setting has been established as the imaging technique of choice for examining patients before functional endoscopic sinus surgery [15,16].

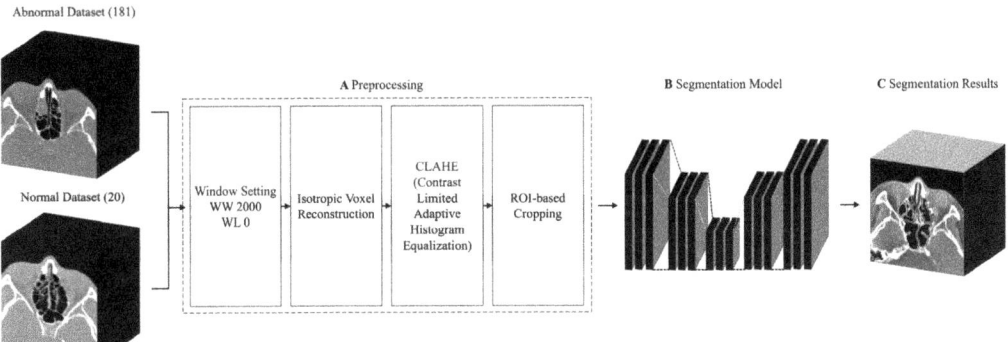

Figure 2. Flowchart of the multiclass sinus segmentation training process.

Depending on the acquisition process, CT images can have varying slice thickness and pixel spacing within the protocol range [17,18]. The acquired images exhibited a consistent 1 mm slice thickness but varying pixel spacings, resulting in the disproportionate volumetric ratio of planar CT images. To eliminate unwanted ratio variations among the dataset, an isotropic voxel reconstruction algorithm was applied across the dataset to equalize the slice thickness to pixel spacing ratio. The ratio of slice thickness to pixel spacing was calculated to downsample the images accordingly using cubic spline interpolation [19] such that the volumetry of the resized images matched real proportions.

Adaptive histogram algorithms are commonly used in medical imaging to create images with equal intensity levels, thereby generating an image with an increased dynamic range, leading to an increase in contrast [20,21]. CLAHE [22,23] was employed in this study to restrict amplification and prevent overamplification of noise in areas with relatively homogeneous contrast.

To equalize the image dimensions for training, a cropping algorithm was used to crop images based on the region of interest. To guarantee the comprehensive inclusion of the region of interest, specific dimensions were set, with a target depth of 192, a height of 128, and a width of 128. The dimensions were chosen based on an analysis of the ground truth data in the entire dataset. The algorithm used in the analysis calculated the 3-dimensional coordinates of the edges for the largest ground truth data. As the voxel reconstruction algorithm resized the CT scans in accordance with the actual proportions of the paranasal sinuses, a greater amount of ground truth data became available along the depth axis.

The U-Net architecture is commonly used for medical image segmentation models because of its reliable performance on medical images [24–26]. Furthermore, its utilization of depth-wise 3D convolution operations allows for the simultaneous extraction of features along the 3 axes: axial, sagittal, and coronal. Three variants of the 3D U-Net architecture, each deeper than the last, were trained and compared: 3D U-Net with residual connections [27], 3D U-Net with dense blocks [28], and 3D U-Net with dense blocks and residual connections [29]. The 3D U-Net architecture, which served as the basis for constructing our model, is presented in Figure 3.

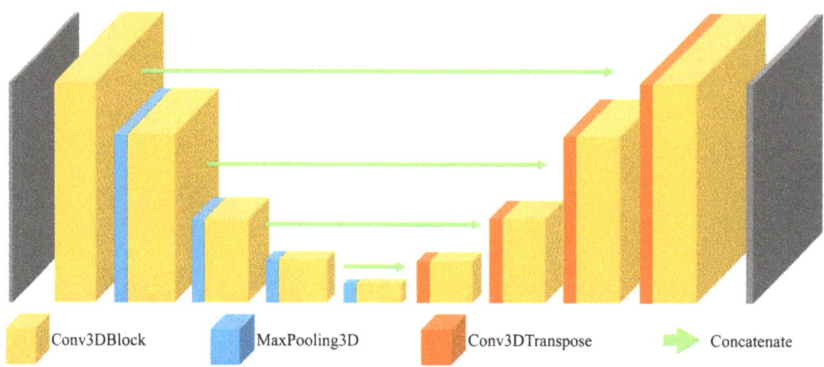

Figure 3. Architecture of the 3D U-Net.

The 3D U-Net used in this study comprised 18 convolutional layers with 5,644,981 trainable parameters. The residual 3D U-Net comprised 63 convolutional layers and 2,350,989 trainable parameters. The dense 3D U-Net comprised 28 convolutional layers and 10,960,437 trainable parameters. The residual dense 3D U-Net comprised 34 convolutional layers and 47,078,117 trainable parameters. A summary of the parameter and layer counts for each model is provided in Table 2, along with the kernel-wise feature map details summarized in Table 3. All models were trained on the same hyperparameters. The Adam [30] optimizer was used with an initial learning rate of 0.0001. Categorical cross-entropy loss was used to monitor validation loss, and accuracy was used as the evaluation metric. Learning rates on plateaus, early stoppers, and model checkpoints were used to prevent issues such as overfitting and plateauing. The tolerance for learning rate reduction was configured to 20 epochs, while the early stopper tolerance was set at 30 epochs.

Table 2. Parameter and layer count by model.

	3D U-Net		Residual		Dense		Residual-Dense	
Count	Parameter	Layer	Parameter	Layer	Parameter	Layer	Parameter	Layer
block 1	7376	4	4456	13	42,352	10	84,480	17
block 2	41,536	4	19,840	13	125,088	9	388,416	17
block 3	166,016	4	78,592	13	499,008	9	1,550,976	17
block 4	663,808	4	312,832	13	1,993,344	9	6,198,528	17
block 5	2,654,720	3	1,248,256	12	2,657,664	8	24,783,360	16
block 6	1,589,632	4	517,632	13	4,117,504	9	10,622,208	18
block 7	397,504	4	129,792	13	1,063,424	9	2,656,896	18
block 8	99,424	4	32,640	13	266,496	9	664,896	18
block 9	24,880	4	8256	13	73,872	9	138,624	18
Output	85	1	165	1	85	1	4325	1
Total	5,644,981	36	2,352,461	117	10,838,837	82	47,092,709	157

Table 3. Layer-by-layer kernel-wise details of each model. The 3D U-Net and dense 3D U-Net models share feature map details, while the residual 3D U-Net and residual-dense 3D U-Net models also share feature map details.

		3D U-Net/Dense 3D U-Net		Residual 3D U-Net/Residual-Dense 3D U-Net	
	Name	Feat Maps (Input)	Feat Maps (Output)	Feat Maps (Input)	Feat Maps (Output)
Encoding path	conv3d_block_1	192 × 128 × 128 × 1	192 × 128 × 128 × 16	192 × 128 × 128 × 1	192 × 128 × 128 × 32
	maxpool3d_1	192 × 128 × 128 × 16	96 × 64 × 64 × 16	192 × 128 × 128 × 32	96 × 64 × 64 × 32
	conv3d_block_2	96 × 64 × 64 × 16	96 × 64 × 64 × 32	96 × 64 × 64 × 32	96 × 64 × 64 × 64
	maxpool3d_2	96 × 64 × 64 × 32	48 × 32 × 32 × 32	96 × 64 × 64 × 64	48 × 32 × 32 × 64
	conv3d_block_3	48 × 32 × 32 × 32	48 × 32 × 32 × 64	48 × 32 × 32 × 64	48 × 32 × 32 × 128
	maxpool3d_3	48 × 32 × 32 × 64	24 × 16 × 16 × 64	48 × 32 × 32 × 128	24 × 16 × 16 × 128
	conv3d_block_4	24 × 16 × 16 × 64	24 × 16 × 16 × 128	24 × 16 × 16 × 128	24 × 16 × 16 × 256
	maxpool3d_4	24 × 16 × 16 × 128	12 × 8 × 8 × 128	24 × 16 × 16 × 256	12 × 8 × 8 × 256
Bridge		12 × 8 × 8 × 128	12 × 8 × 8 × 256	12 × 8 × 8 × 256	12 × 8 × 8 × 512
Decoding path	conv3d_trans_1	12 × 8 × 8 × 256	24 × 16 × 16 × 128	12 × 8 × 8 × 512	24 × 16 × 16 × 256
	conv3d_block_5	24 × 16 × 16 × 128	24 × 16 × 16 × 128	24 × 16 × 16 × 256	24 × 16 × 16 × 256
	conv3d_trans_2	24 × 16 × 16 × 128	48 × 32 × 32 × 64	24 × 16 × 16 × 256	48 × 32 × 32 × 128
	conv3d_block_6	48 × 32 × 32 × 64	48 × 32 × 32 × 64	48 × 32 × 32 × 128	48 × 32 × 32 × 128
	conv3d_trans_3	48 × 32 × 32 × 64	96 × 64 × 64 × 32	48 × 32 × 32 × 128	96 × 64 × 64 × 64
	conv3d_block_7	96 × 64 × 64 × 32	96 × 64 × 64 × 32	96 × 64 × 64 × 64	96 × 64 × 64 × 64
	conv3d_trans_4	96 × 64 × 64 × 32	192 × 128 × 128 × 16	96 × 64 × 64 × 64	192 × 128 × 128 × 32
	conv3d_block_8	192 × 128 × 128 × 16	192 × 128 × 128 × 5	192 × 128 × 128 × 32	192 × 128 × 128 × 5

3. Results

Each model was tested against the hold-out test set to generate segmentation results. The segmentation results were evaluated using the following five performance metrics: intersection over union (IoU), accuracy, recall, precision, and F1 score. The results are expressed as the mean ± 95% confidence interval, with statistical significance set at $p < 0.05$.

The segmentation results from the normal test set were evaluated using the performance metrics and summarized in Table 4. Overall, the models were able to make predictions with an F1 score in the range of 0.843–0.785, of which the 3D U-Net model achieved the highest F1 score with a value of 0.843. Conversely, the residual 3D U-Net model recorded the lowest F1 score, standing at 0.785.

Table 4. Prediction results obtained on the normal test set, reported in performance metrics per model.

Metrics	Base	Residual	Dense	Residual-Dense
F1 score	**0.843 ± 0.699**	0.785 ± 0.066	0.790 ± 0.073	0.802 ± 0.093
Accuracy	**0.995 ± 0.003**	0.992 ± 0.001	0.993 ± 0.002	0.993 ± 0.003
Precision	**0.857 ± 0.056**	0.789 ± 0.059	0.801 ± 0.060	0.822 ± 0.073
Recall	**0.854 ± 0.064**	0.821 ± 0.060	0.822 ± 0.068	0.836 ± 0.078
Mean IoU	**0.787 ± 0.071**	0.703 ± 0.067	0.714 ± 0.074	0.742 ± 0.092

The segmentation results from the abnormal test set are summarized in Table 5. In the abnormal test set, the segmentation results were evaluated to record a lower overall F1 score in the range of 0.793–0.740. The 3D U-Net model made predictions with the highest F1 score of 0.793, whereas the predictions made by the residual-dense 3D U-Net model recorded the lowest F1 score of 0.741.

A comparative plot of IoU values across the models in the normal and abnormal test set is presented in Figure 4. The average IoU difference across the models was 0.082 ± 0.034 (mean ± 95% confidence interval). Paired t-tests of the IoU across the models showed statistically insignificant differences in IoU values between the four models ($p < 0.05$). The average F1 score difference, encompassing both test sets, between the 3D U-Net and the other three models were as follows: 0.067 ± 0.016 for the residual model, 0.069 ± 0.028

for the dense model, and 0.082 ± 0.037 for the residual-dense 3D U-Net. Paired *t*-tests of the F1 scores between the models showed statistically insignificant F1 score variation across the models ($p < 0.05$). The average differences in F1 scores between the two test sets (normal and abnormal) were as follows: 0.170 ± 0.067 for the 3D U-Net, 0.188 ± 0.064 for the residual 3D U-Net, 0.206 ± 0.072 for the dense 3D U-Net, and 0.257 ± 0.099 for the residual-dense 3D U-Net. Statistical analysis using paired *t*-tests showed a statistically significant difference in the F1 scores between the normal and abnormal test sets ($p > 0.05$).

Table 5. Prediction results obtained on the abnormal test set, reported in performance metrics per model.

Metrics	Base	Residual	Dense	Residual-Dense
F1 score	**0.793 ± 0.063**	0.741 ± 0.069	0.747 ± 0.074	0.740 ± 0.095
Accuracy	**0.994 ± 0.002**	0.991 ± 0.002	0.992 ± 0.002	0.991 ± 0.003
Precision	**0.839 ± 0.057**	0.779 ± 0.067	0.785 ± 0.071	0.793 ± 0.089
Recall	**0.785 ± 0.067**	0.755 ± 0.076	0.756 ± 0.068	0.745 ± 0.092
Mean IoU	**0.717 ± 0.061**	0.653 ± 0.063	0.666 ± 0.074	0.670 ± 0.089

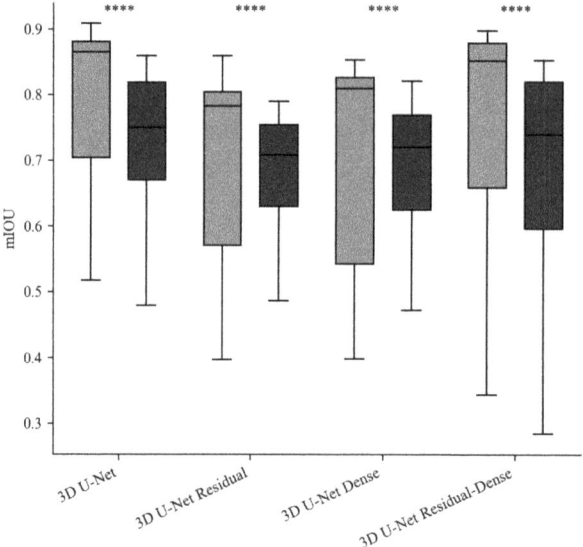

Figure 4. mIoU comparison of each model in the normal and abnormal test set (gray: normal dataset results, black: abnormal dataset results, ****: statistical significance ($p < 0.05$)).

Visual overviews of the segmentation results for the normal and abnormal test sets are shown in Figures 5 and 6, respectively. The figures show the segmentation results for the ethmoid sinus, maxillary sinus, and sphenoid sinus; each area is color-coded for better visual representation. The images were chosen randomly from the fold with the best mIoU score. Each row represents predictions from different models. From left to right, the three columns represent the ground truth, prediction, and overlay comparison of the ground truth and prediction.

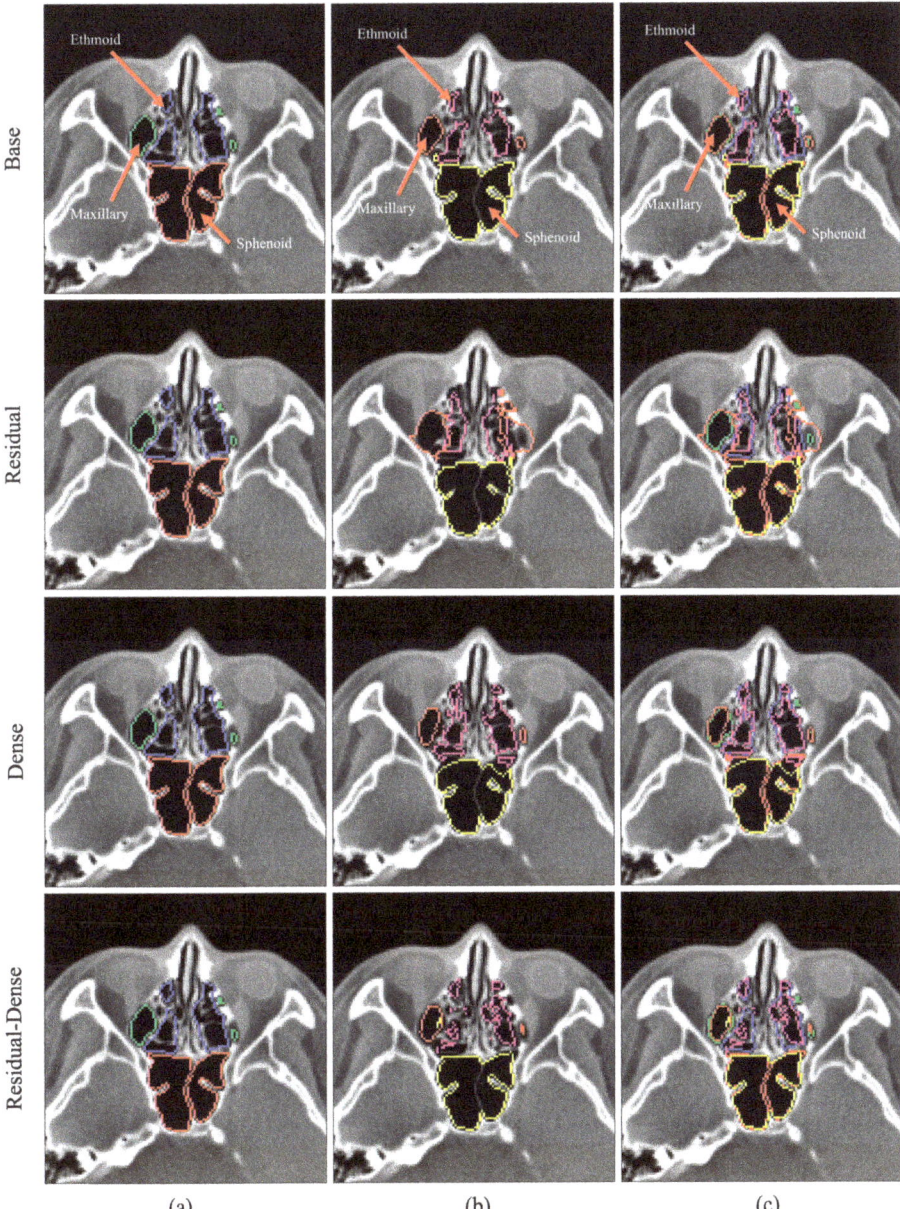

Figure 5. 3D U-Net segmentation comparison for the normal test case. Color legend: green, orange—maxillary sinus; blue, pink—ethmoid sinus; red, yellow—sphenoid sinus; (**a**) ground truth data, (**b**) model prediction, (**c**) overlay comparison of ground truth and prediction.

Normalized true positive (TP) distribution per class as a heatmap for the 3D U-Net is shown in Figure 6. For the normal dataset, the sphenoid sinus showed the highest TP rate of 0.95, whereas the ethmoid sinus showed the lowest at 0.82. For the abnormal dataset, the sphenoid sinus reported the highest TP rate at 0.88, and the lowest for the frontal sinus at 0.67.

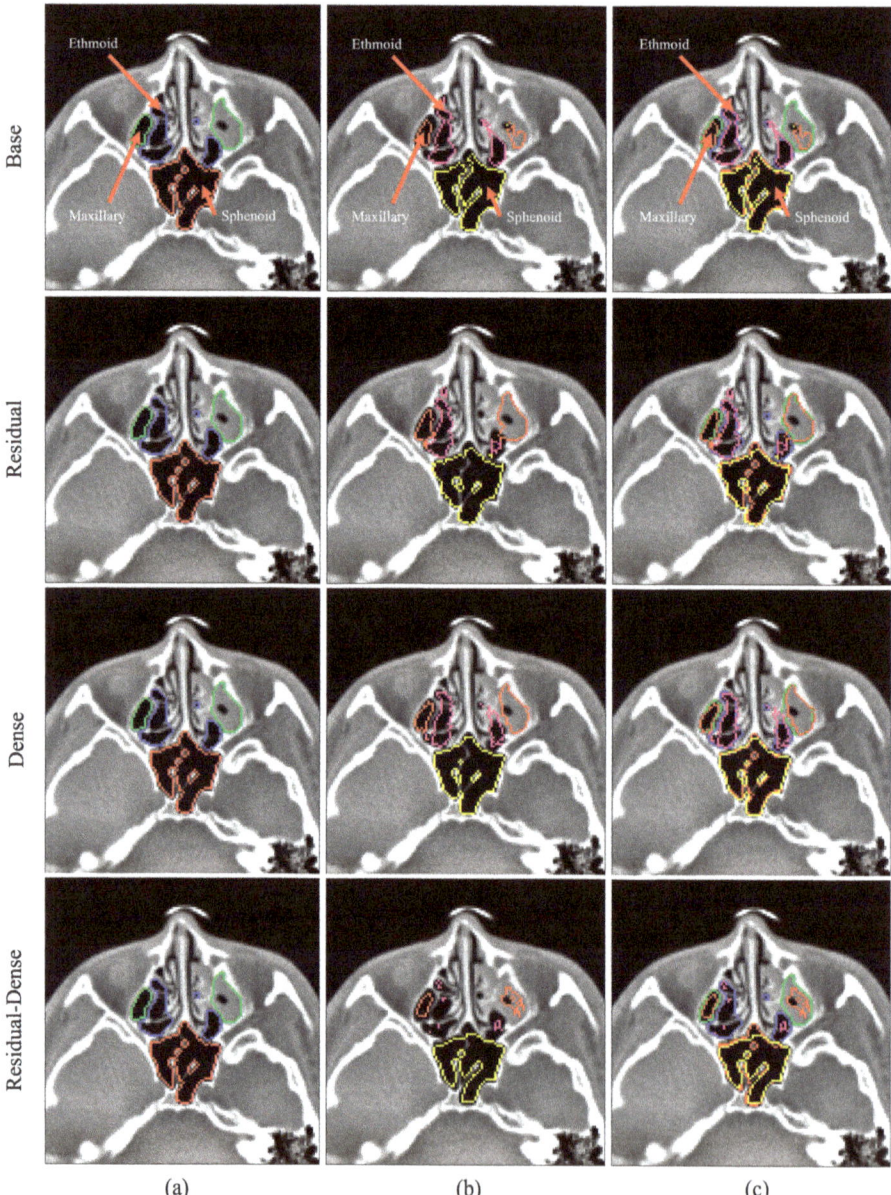

Figure 6. 3D U-Net segmentation comparison for the abnormal test case. Color legend: green, orange—maxillary sinus; blue, pink—ethmoid sinus; red, yellow—sphenoid sinus; (**a**) ground truth data, (**b**) model prediction, (**c**) overlay comparison of ground truth and prediction.

4. Discussion

In this study, a 3D segmentation model for the four areas of the paranasal sinus based on CT images was developed and evaluated. Four models based on the 3D U-Net were trained and evaluated on a hold-out test set of 40 datasets, comprising 20 datasets from patients without sinusitis and 20 datasets from patients with sinusitis. Prediction results were further validated using 5-fold cross validation. In the normal test set, the models showed performances in the range of 0.843–0.785 with an average F1 score of 0.805. In the abnormal test set, the models performed in the range of 0.793–0.740 with an average F1 score of 0.755. In both test sets, the base 3D U-Net was able to make predictions with the highest F1 score of 0.843, and 0.793, respectively, in the normal test set and the abnormal test set. Statistical analysis of performance metrics was performed across the four models between normal and abnormal test sets with statistical significance set at $p = 0.05$. Performance metrics across the models exhibited statistically insignificant variations. However, mucosal inflammation had a greater impact on the performance metrics across the models.

The method proposed by Choi et al. [11] reported an F1 score of 0.972 in normal sinuses and 0.912 in sinuses with mucosal inflammation. Morgan et al. [12] reported an F1 score of 0.984 and 0.996, respectively, for normal and abnormal sinuses. Note that these studies were limited to binary segmentation of the maxillary sinus, manifesting in the higher F1 score. The study by Kuo et al. [13] trained multiple models with the aim of multi-class segmentation of the sinus, in which the U-Net model reported an average F1 score of 0.896. This is within 6.2% of the highest performing model in our study, the base 3D U-Net.

We performed a thorough analysis of prediction accuracy for the 3D U-Net model across the four main sinus regions, focusing on true positive rates. The outcomes underscored notable limitations in the precise prediction of the frontal and ethmoid sinus regions. The abnormal test set showed lower prediction metrics, overall, in comparison to the normal test set. The frontal and ethmoid sinuses showed particularly lower TP rates in the abnormal test set, at 0.67 and 0.75, respectively. The frontal and ethmoid sinuses are anatomically adjacent structures, and both have smaller volumes than the sphenoid and maxillary sinuses [31]. In sinus cavities with mucosal inflammation, the cavities of the ethmoid and frontal sinuses had much less pronounced features compared to other areas of the paranasal sinuses. This limitation is evident in Figure 7 of the right ethmoid sinus, where the contrast between the sinus bone and cavity appears less pronounced compared to the left ethmoid sinus.

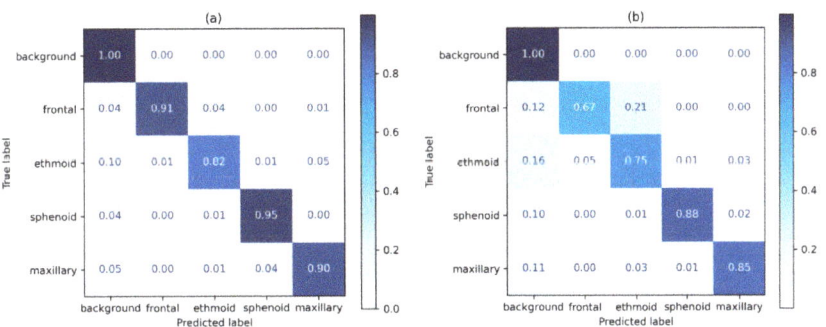

Figure 7. 3D U-Net prediction heatmap; (**a**) prediction results for the normal test set, (**b**) prediction results for the abnormal test set.

Despite the substantial size of the dataset collected for this study, the clinical nature of the CT scans led to an uneven distribution of data between patients with sinusitis and those without the condition. Moreover, training data was obtained solely from a single institution, suggesting the possibility that the trained models could exhibit limited generalization capabilities on external datasets. A comprehensive follow-up study should encompass a well-balanced dataset, including an equal distribution of data from patients with sinusitis and those without the condition. It would be advantageous to source this data from multiple institutes to enable internal and external validations.

Accurate segmentation of the paranasal sinuses is crucial for the preoperative evaluation of patients undergoing sinus surgery. To this end, this study aimed to evaluate the segmentation efficacy in patients with mucosal inflammation. While limitations do exist in the segmentation of paranasal sinuses with mucosal inflammation, the proposed method exhibited promising results. With minor refinements, our segmentation model has the potential to enhance surgical accuracy when integrated into guidance systems. Such integration can aid surgeons in avoiding healthy mucosal tissue, thereby reducing the risk of complications.

Author Contributions: J.W., J.L., Y.J.K., S.T.K. and K.G.K. contributed to the conception and design of this study. Material preparation, data collection, and analyses were performed by J.L. The first draft of this manuscript was written by J.W., and all authors commented on the previous versions of this manuscript. All authors have read and agreed to the published version of the manuscript.

Funding: This work was supported by the Technology Innovation Program (K_G012001187801, "Development of Diagnostic Medical Devices with Artificial intelligence Based Image Analysis Technology") funded By the Ministry of Trade, Industry & Energy (MOTIE, Korea), and by Gachon University (GCU-202205980001).

Institutional Review Board Statement: The study was conducted in accordance with the Declaration of Helsinki, approved by the Institutional Review Board (IRB) of Gachon University Gil Medical Center (GAIRB2022-182), and was conducted in accordance with relevant guidelines and ethical regulations.

Informed Consent Statement: As this study is retrospective in nature, the Institutional Review Board (IRB) of Gachon University Gil Medical Center has granted a waiver for the requirement of patient informed consent. The retrospective design involves the analysis of existing data and poses minimal risk to the participants. This waiver ensures the ethical conduct of the study while respecting the rights and privacy of the individuals involved.

Data Availability Statement: The raw data supporting the conclusions of this article will be made available by the authors on request.

Conflicts of Interest: The authors declare no conflicts of interest.

References

1. Hamilos, D.L. Chronic Sinusitis. *J. Allergy Clin. Immunol.* **2000**, *106*, 213–227. [CrossRef] [PubMed]
2. Kaliner, M.; Osguthorpe, J.; Fireman, P.; Anon, J.; Georgitis, J.; Davis, M.; Naclerio, R.; Kennedy, D. Sinusitis: Bench to BedsideCurrent Findings, Future Directions. *J. Allergy Clin. Immunol.* **1997**, *99*, S829–S847. [CrossRef] [PubMed]
3. Bhattacharyya, N. Clinical and Symptom Criteria for the Accurate Diagnosis of Chronic Rhinosinusitis. *Laryngoscope* **2006**, *116*, 1–22. [CrossRef] [PubMed]
4. Hoang, J.K.; Eastwood, J.D.; Tebbit, C.L.; Glastonbury, C.M. Multiplanar Sinus CT: A Systematic Approach to Imaging Before Functional Endoscopic Sinus Surgery. *Am. J. Roentgenol.* **2010**, *194*, W527–W536. [CrossRef] [PubMed]
5. O'Brien, W.T.; Hamelin, S.; Weitzel, E.K. The Preoperative Sinus CT: Avoiding a "CLOSE" Call with Surgical Complications. *Radiology* **2016**, *281*, 10–21. [CrossRef]
6. Tingelhoff, K.; Moral, A.I.; Kunkel, M.E.; Rilk, M.; Wagner, I.; Eichhorn, K.W.G.; Wahl, F.M.; Bootz, F. Comparison between Manual and Semi-Automatic Segmentation of Nasal Cavity and Paranasal Sinuses from CT Images. In Proceedings of the 2007 29th Annual International Conference of the IEEE Engineering in Medicine and Biology Society, Lyon, France, 22–26 August 2007; pp. 5505–5508.
7. Pirner, S.; Tingelhoff, K.; Wagner, I.; Westphal, R.; Rilk, M.; Wahl, F.M.; Bootz, F.; Eichhorn, K.W.G. CT-Based Manual Segmentation and Evaluation of Paranasal Sinuses. *Eur. Arch. Oto-Rhino-Laryngol.* **2009**, *266*, 507–518. [CrossRef]
8. Varoquaux, G.; Cheplygina, V. Machine Learning for Medical Imaging: Methodological Failures and Recommendations for the Future. *NPJ Digit. Med.* **2022**, *5*, 48. [CrossRef]
9. Litjens, G.; Kooi, T.; Bejnordi, B.E.; Setio, A.A.A.; Ciompi, F.; Ghafoorian, M.; van der Laak, J.A.W.M.; van Ginneken, B.; Sánchez, C.I. A Survey on Deep Learning in Medical Image Analysis. *Med. Image Anal.* **2017**, *42*, 60–88. [CrossRef]
10. Cheplygina, V.; de Bruijne, M.; Pluim, J.P.W. Not-so-Supervised: A Survey of Semi-Supervised, Multi-Instance, and Transfer Learning in Medical Image Analysis. *Med. Image Anal.* **2019**, *54*, 280–296. [CrossRef] [PubMed]
11. Choi, H.; Jeon, K.J.; Kim, Y.H.; Ha, E.-G.; Lee, C.; Han, S.-S. Deep Learning-Based Fully Automatic Segmentation of the Maxillary Sinus on Cone-Beam Computed Tomographic Images. *Sci. Rep.* **2022**, *12*, 14009. [CrossRef] [PubMed]
12. Morgan, N.; Van Gerven, A.; Smolders, A.; de Faria Vasconcelos, K.; Willems, H.; Jacobs, R. Convolutional Neural Network for Automatic Maxillary Sinus Segmentation on Cone-Beam Computed Tomographic Images. *Sci. Rep.* **2022**, *12*, 7523. [CrossRef]
13. Kuo, C.-F.J.; Liu, S.-C. Fully Automatic Segmentation, Identification and Preoperative Planning for Nasal Surgery of Sinuses Using Semi-Supervised Learning and Volumetric Reconstruction. *Mathematics* **2022**, *10*, 1189. [CrossRef]
14. Çiçek, Ö.; Abdulkadir, A.; Lienkamp, S.S.; Brox, T.; Ronneberger, O. 3D U-Net: Learning Dense Volumetric Segmentation from Sparse Annotation. In *Medical Image Computing and Computer-Assisted Intervention–MICCAI 2016: 19th International Conference,*

Athens, Greece, October 17–21, 2016, Proceedings, Part II 19; Springer International Publishing: Berlin/Heidelberg, Germany, 2016; pp. 424–432.
15. Som, P.M. CT of the Paranasal Sinuses. *Neuroradiology* **1985**, *27*, 189–201. [CrossRef]
16. Melhem, E.R.; Oliverio, P.J.; Benson, M.L.; Leopold, D.A.; Zinreich, S.J. Optimal CT Evaluation for Functional Endoscopic Sinus Surgery. *AJNR Am. J. Neuroradiol.* **1996**, *17*, 181–188.
17. Huang, K.; Rhee, D.J.; Ger, R.; Layman, R.; Yang, J.; Cardenas, C.E.; Court, L.E. Impact of Slice Thickness, Pixel Size, and CT Dose on the Performance of Automatic Contouring Algorithms. *J. Appl. Clin. Med. Phys.* **2021**, *22*, 168–174. [CrossRef]
18. Cantatore, A.; Müller, P. *Introduction to Computed Tomography*; DTU Mechanical Engineering: Lyngby, Denmark, 2011.
19. Dyer, S.A.; Dyer, J.S. Cubic-Spline Interpolation. 1. *IEEE Instrum. Meas. Mag.* **2001**, *4*, 44–46. [CrossRef]
20. Salem, N.; Malik, H.; Shams, A. Medical Image Enhancement Based on Histogram Algorithms. *Procedia Comput. Sci.* **2019**, *163*, 300–311. [CrossRef]
21. Rahman, T.; Khandakar, A.; Qiblawey, Y.; Tahir, A.; Kiranyaz, S.; Abul Kashem, S.B.; Islam, M.T.; Al Maadeed, S.; Zughaier, S.M.; Khan, M.S.; et al. Exploring the Effect of Image Enhancement Techniques on COVID-19 Detection Using Chest X-Ray Images. *Comput. Biol. Med.* **2021**, *132*, 104319. [CrossRef] [PubMed]
22. Pizer, S.M.; Amburn, E.P.; Austin, J.D.; Cromartie, R.; Geselowitz, A.; Greer, T.; ter Haar Romeny, B.; Zimmerman, J.B.; Zuiderveld, K. Adaptive Histogram Equalization and Its Variations. *Comput. Vis. Graph. Image Process.* **1987**, *39*, 355–368. [CrossRef]
23. Sørensen, T.; Sørensen, T.; Biering-Sørensen, T.; Sørensen, T.; Sorensen, J.T. A Method of Establishing Group of Equal Amplitude in Plant Sociobiology Based on Similarity of Species Content and Its Application to Analyses of the Vegetation on Danish Commons. *Biol. Skr.* **1948**, *5*, 1–34.
24. Ronneberger, O.; Fischer, P.; Brox, T. U-Net: Convolutional Networks for Biomedical Image Segmentation. In *Medical Image Computing and Computer-Assisted Intervention–MICCAI 2015: 18th International Conference, Munich, Germany, 5–9 October 2015, Proceedings, Part III 18*; Springer International Publishing: Berlin/Heidelberg, Germany, 2015; pp. 234–241.
25. Azad, R.; Aghdam, E.K.; Rauland, A.; Jia, Y.; Avval, A.H.; Bozorgpour, A.; Karimijafarbigloo, S.; Cohen, J.P.; Adeli, E.; Merhof, D. Medical Image Segmentation Review: The Success of U-Net. *arXiv* **2022**, arXiv:2211.14830.
26. Siddique, N.; Paheding, S.; Elkin, C.P.; Devabhaktuni, V. U-Net and Its Variants for Medical Image Segmentation: A Review of Theory and Applications. *IEEE Access* **2021**, *9*, 82031–82057. [CrossRef]
27. Lee, K.; Zung, J.; Li, P.H.; Jain, V.; Seung, H.S. Superhuman Accuracy on the SNEMI3D Connectomics Challenge. *arXiv* **2017**, arXiv:1706.00120.
28. Kolařík, M.; Burget, R.; Uher, V.; Dutta, M.K. 3D Dense-U-Net for MRI Brain Tissue Segmentation. In Proceedings of the 2018 41st International Conference on Telecommunications and Signal Processing (TSP), Athens, Greece, 4–6 July 2018; pp. 1–4.
29. Sarica, B.; Seker, D.Z.; Bayram, B. A Dense Residual U-Net for Multiple Sclerosis Lesions Segmentation from Multi-Sequence 3D MR Images. *Int. J. Med. Inform.* **2023**, *170*, 104965. [CrossRef] [PubMed]
30. Kingma, D.P.; Ba, J. Adam: A Method for Stochastic Optimization. *arXiv* **2014**, arXiv:1412.6980.
31. Mynatt, R.G.; Sindwani, R. Surgical Anatomy of the Paranasal Sinuses. In *Rhinology and Facial Plastic Surgery*; Springer: Berlin/Heidelberg, Germany, 2009; pp. 13–33.

Disclaimer/Publisher's Note: The statements, opinions and data contained in all publications are solely those of the individual author(s) and contributor(s) and not of MDPI and/or the editor(s). MDPI and/or the editor(s) disclaim responsibility for any injury to people or property resulting from any ideas, methods, instructions or products referred to in the content.

Article

Metric Learning in Histopathological Image Classification: Opening the Black Box

Domenico Amato [1], Salvatore Calderaro [1], Giosué Lo Bosco [1], Riccardo Rizzo [2,*] and Filippo Vella [2]

[1] Department of Mathematics and Computer Science, University of Palermo, 90123 Palermo, Italy; domenico.amato01@unipa.it (D.A.); salvatore.calderaro01@unipa.it (S.C.); giosue.lobosco@unipa.it (G.L.B.)
[2] Institute for High-Performance Computing and Networking, National Research Council of Italy, 90146 Palermo, Italy; filippo.vella@icar.cnr.it
* Correspondence: riccardo.rizzo@icar.cnr.it; Tel.: +39-091-680-9714

Abstract: The application of machine learning techniques to histopathology images enables advances in the field, providing valuable tools that can speed up and facilitate the diagnosis process. The classification of these images is a relevant aid for physicians who have to process a large number of images in long and repetitive tasks. This work proposes the adoption of metric learning that, beyond the task of classifying images, can provide additional information able to support the decision of the classification system. In particular, triplet networks have been employed to create a representation in the embedding space that gathers together images of the same class while tending to separate images with different labels. The obtained representation shows an evident separation of the classes with the possibility of evaluating the similarity and the dissimilarity among input images according to distance criteria. The model has been tested on the BreakHis dataset, a reference and largely used dataset that collects breast cancer images with eight pathology labels and four magnification levels. Our proposed classification model achieves relevant performance on the patient level, with the advantage of providing interpretable information for the obtained results, which represent a specific feature missed by the all the recent methodologies proposed for the same purpose.

Keywords: metric learning; triplet networks; embedding; BreakHis; patient level accuracy; breast cancer imaging; WSI; classification interpretability; visualization

Citation: Amato, D.; Calderaro, S.; Lo Bosco, G.; Rizzo, R.; Vella, F. Metric Learning in Histopathological Image Classification: Opening the Black Box. *Sensors* **2023**, *23*, 6003. https://doi.org/10.3390/s23136003

Academic Editors: Sergiu Nedevschi and Mitrea Delia-Alexandrina

Received: 9 May 2023
Revised: 15 June 2023
Accepted: 22 June 2023
Published: 28 June 2023

Copyright: © 2023 by the authors. Licensee MDPI, Basel, Switzerland. This article is an open access article distributed under the terms and conditions of the Creative Commons Attribution (CC BY) license (https://creativecommons.org/licenses/by/4.0/).

1. Introduction

Van der Laak et al. in [1] point out that the digitisation of patient tissue samples, usually called Whole Slide Images (WSI), enabled the development of a set of techniques in the field of biomedical image analysis under the name of computational pathology. In the histopathology field, deep learning algorithms perform similarly to trained pathologists, but only very few of these have reached a clinical implementation.

The resolution of the WSI image can reach $10,000 \times 10,000$ pixels and may present high morphological variance and various artifacts [2]. Due to the general complexity of such kinds of images, the analysis of WSI requires a high degree of expertise and can be very time-consuming. In addition, the complexity of this task is further increased by the need to explore the samples at different magnification scales. As a consequence, a complete diagnosis is often obtained through a discussion among specialised physicians that compare the outcomes of different medical analyses (not only images).

Due to their dimensions, WSI images are challenging to process as a whole. For this reason, they are broken into patches or tiles and given as input to the machine learning (ML) algorithms for classification purposes. The attribution of the class to the WSI images is then obtained by combining the labels predicted for the related patches [3]. This task is not simple due to the morphological variance inside the WSI; however, patch analysis remains the most-used technique for WSI processing, and the development of patch processing and classification systems is an active research field.

To measure and compare the contribution of such approaches, some datasets have been proposed in the literature. One of them, considered as a standard, is the BreakHis dataset (BH) [4]. This dataset comprises 7909 histopathological images of eight different kinds of breast cancer collected from 82 patients. For a fair comparison among the proposed approaches, the authors provided a specific split into a 5-fold structure. A deeper description of BH is reported in Section 3. Many deep neural networks (DNN) have been proposed for histopathological image classification [1], due to their state-of-the-art performances in the generic problem of image classification. In many cases, the architecture used to classify BH images is very close to the ones proposed for other classification tasks, such as the case of ImageNet Large Scale Visual Recognition Challenge [5], that are based on convolutional layers [6]. These kinds of networks have a common architecture constituted by two parts: a first one devoted to features extraction and a second combining the extracted features for classification.

Despite the relevant performance that deep neural networks (DNNs) can achieve in the image classification task, their adoption in the medical domain is not straightforward because they are *black-box* methods, making it difficult to understand the logic behind specific decisions.

Today explainability and interpretability are an important part of the discussion as well as research work about deep neural systems and their performances [7]. What we need are systems that make consistent decisions using an explainable mechanism and provide information that users can understand and find meaningful, particularly in the medical domain. An example of an interpretability mechanism giving information that is not fully significant is CAM or Grad-CAM [8], which can highlight an image region crucial for the classification result; however, this mechanism can not explain what, in that region, is truly important or whether it fits with the shared medical knowledge [9].

The paper is structured as follows: the following section reports the related works in the field of classification of the BreakHis dataset, and in the interpretable deep ANN, the Section 3 reports the details on the BreakHis dataset, the proposed architecture and the training method; the obtained results and the discussions are reported in Sections 4 and 5; finally, in Section 6 some conclusions are drawn.

2. Related Works

We already proposed some studies on image classification supporting tools based on metric learning with fuzzy techniques [10] or convolutional deep networks [11,12], or X ray image classification [13]. The issues related to the classification of histological images are due to the variability of the different image acquisition equipment. For example, authors in [14] indicate stain variability as one of the challenges in classification; stain normalisation was also addressed in [15].

In this paper, we focus on the BreakHis dataset [4], which is widely adopted as a workbench and used in all the papers discussed in this section. The structure of the dataset will be discussed in the Section 3.1, but it is necessary to highlight here some characteristics related to the image classes, the magnification level, and the train/test split. The dataset has a hierarchical organisation with two super-classes (Benign and Malignant) and eight sub-classes (four for each super-class). The images in the dataset have four different magnification levels $40\times, 100\times, 200\times$, and $400\times$, which are hierarchically related. For example, images at $200\times$ will contain details of images at $100\times$ and $40\times$.

The availability of images at different magnification levels leads to two possible set-ups: the first is the study of magnification-specific (MS) classification when a classifier is trained and tested for each magnification level and the second is the magnification-independent (MI) classification when the classifier is trained and tested using all the images regardless of their magnification level.

The developers of the BreakHis dataset released a 5-fold train/test split that can be used to facilitate the comparison of the results among different approaches [4]. The proposed split is created at *patient level* so that images of the same patient are never in the training and testing set of the same fold run.

This split is often overridden by many authors for various reasons, but in order to make fair and reliable comparisons, its use is highly recommended, as in the case of this manuscript.

The availability of patient information means that the assessment of the results can be performed considering the images related to a single patient (patient-level assessment or PLA) or the single images (image-level assessment or ILA). In this work, we have decided to adopt only the PLA paradigm since it is undoubtedly the most reliable in terms of closeness to the real case of histopathological image classification. The specific formulas of PLA will be discussed in Section 3.

In the following subsections, we survey most of the methodologies proposed so far for the histopathology image classification problem that use the BH dataset, differentiating them in terms of image magnification, whether specific or independent. Special attention will be devoted to the contributions that follow the folds separation suggested by the BreakHis dataset developers, whose results can be directly compared with the methodology we present in this manuscript.

2.1. Magnification Independent Methods

Magnification Independent Binary (MIB) classifiers are trained to separate benign from malignant images using all the available images, regardless of magnification factors. Only a few works fall into this category, and only two use the 5-fold split suggested by Spanhol et al. in [4]. In the following, we will first introduce the two works that use this split, while the latter uses a different split.For each examined work, the PLA value is reported. PLA is an interesting comparison measure since it summarizes the accuracy for a patient and is the most realistic measure of the system performance when a new patient is examined. Other metrics, even if not present in the literature works, are computed for the proposed approach and are shown in Section 4.

Bayramoglu et al. [16] propose two convolutional neural networks to classify breast cancer histopathology images for the MIB case. These models are used to predict the malignancy of the input sample (single-task fashion) or both the malignancy and magnification factor of the input (multi-task fashion). For the single task, a PLA of 82% is obtained.

Another model, proposed by Sun et al. [17], uses an approach similar to the one presented in this paper with both MIB and MSB tasks. Authors adopt a siamese and a triplet network to create a new representation of the images, and they use a different loss function that also considers the samples' imbalance. The images, without any pre-processing, are fed to the network that implements the classification, and the obtained PLA accuracy is about 88%, which is slightly lower than ours.

The last work, proposed by Gupta et al. [18], adopts selected colour and texture descriptors, baseline classifiers such as SVM, kNN, Decision Trees and Discriminant Analysis, fused together with majority voting. For the model validation, a different train/test split was used. A total of 58 patients (70%) were randomly chosen for the training set and the remaining 24 (30%) for the test set and the process was repeated for five trials. For the magnification-independent classification, the model achieves an average PLA of 87.53%.

2.2. Magnification Specific Methods

Several methodologies have been proposed for the case of Magnification Specific Binary classification (MSB). We have decided to filter them on the basis of the adoption of the training-test split proposed in [4] since this choice provides a solid comparison process. As a consequence, all the accuracy values of the methodology reported in the following works use the same training-test split. This is the case for the methodology proposed by Sun et al. [17] described in the previous paragraph, which has also been used for the MSB case. The results show an average PLA ranging from $\sim$87% (400$\times$) to $\sim$91% (200$\times$) (see Table 1 for details).

Apart from the provided split, the proposers Spanhol et al. have also investigated two approaches based on patches extraction [19].

The patches for training were obtained by two strategies: using a sliding window with 50% overlap or a random extraction of 1000 patches with no overlap check. These strategies were repeated twice to obtain patches of 32×32 pixels and patches of 64×64 pixels. For the classification task, the authors proposed a variant of the AlexNet architecture. The images of the test set were obtained by using patches extracted with a sliding window and no overlap. The final classification of an image was the result of a combination of classification results, each one computed on a single patch. The best results were obtained with the training of 1000 patches of the largest dimension and using the Max Fusion Rule as a combination paradigm. The patient accuracy values for each magnification level range from 84% ($200\times$) to 90% ($200\times$) for PLA (see Table 1 for details).

Spanhol et al. have proposed another solution [20] exploiting the adoption of specific image features, named DeCAF. They are obtained by extracting the outputs of the top 3 layers of a pre-trained AlexNet-like network and using them as the input of a CNN classifier. The experiments were organised considering a patch-based recognition, using a different number of patches ranging from 1 to 16. The achieved accuracy ranges from 82% ($400\times$) to 86% ($200\times$) for PLA (see Table 1 for details).

The idea of DeCAF features has also been investigated by Benhammou et al. [21] using a pre-trained Inception v3 [22]. To obtain the features, during the fine-tuning step, only the weights of the last fully connected layer (situated before the softmax layer) are retrained, while the other net layers are frozen. A pre-processing step on the images based on the mean-pixel subtraction is used. The achieved accuracy ranges from $\sim$80% ($400\times$) to 87% ($40\times$) for PLA (see Table 1 for details).

Another couple of studies that use pre-defined features was proposed by Song et al. [23]. The two approaches exploit Fisher Vectors encoding [24] as feature representation. In their first approach [23], the authors used the descriptors together with a linear support vector machine. First, the features of VGG-D network, pre-trained on the ImageNet dataset, were extracted and then represented with Fisher vector encoding. Next, an adaptation layer formed by two locally connected layers was adopted, and an additional classification layer was also added. The descriptors obtained from the adaptation layer were used to train linear-kernel support vector machines. The obtained average PLA is bounded by $\sim$86% ($400\times$) and $\sim$90% ($40\times$) (see Table 1). The second approach [25] provides a supervised intra-embedding algorithm that uses a neural network to transform the Fisher Vector encoding into more discriminating feature representations. The input images are re-scaled to multiple sizes, and for each re-scaled image, the VGG-VD ConvNet pre-trained on ImageNet is applied so that the last layer produces a feature vector of size 512. The features collected from all the re-scaled images are pooled together to generate the so-called ConvNet-based Fisher Vectors (CVF) encoding of the image. To perform the classification, a Support Vector Machine is used. The average PLA values are bounded by $\sim$87% ($400\times$) and $\sim$90% ($40\times$) (see Table 1 for details).

Sudharshan et al. adopted a Multiple Instance Learning (MIL) framework for convolutional neural networks [26]. MIL is concerned with learning from sets (bags) of objects (instances), where the classification label is assigned to the bag, not to the single instance. For the training phase, 1000 patches of size 64×64, are randomly extracted from each image, while for the testing phase, a grid of non-overlapping patches is extracted, yielding around 100 patches per image. Each patch is represented by a specific 162-long feature vector of Parametric-Free Threshold Adjacency Statistics (PFTAS) features. The work presents an evaluation of twelve different MIL variants, both parametric and non-parametric.

The results are provided for two different settings: one where each patient is considered as a bag, considering multiple images for each of them and the second where each image is a bag composed of patches. The best results are obtained at Patient Level using a non-parametric MIL. They range from $\sim$87% ($200\times$) to $\sim$92% ($40\times$) and are listed in Table 1 for details.

Table 1. Triplet net embedding: MSB classification (PLA) results and comparison with the other methods.

Method	40×	100×	200×	400×
Sun et al. [17]	87.51 ± 4.07	89.12 ± 2.86	90.82 ± 3.31	87.10 ± 3.80
Spanhol et al. [19]	90.0 ± 6.7	88.4 ± 4.8	84.6 ± 4.2	86.1 ± 6.2
Spanhol et al. [20]	84.0 ± 6.9	83.9 ± 5.9	86.3 ± 3.5	82.1 ± 2.4
Benhammou et al. [21]	87.6 ± 3.9	82.4 ± 2.7	86.1 ± 0.7	79.7 ± 3.2
Song et al. [23]	90.0 ± 3.2	88.9 ± 5.0	86.9 ± 5.2	86.3 ± 7.0
Song et al. [25]	90.2 ± 3.2	91.2 ± 4.4	87.8 ± 5.3	87.4 ± 7.2
Sudharshan et al. [26]	92.1 ± 5.9	89.1 ± 5.2	87.2 ± 4.3	87.8 ± 5.6
Proposed method (k-NN)	87.60 ± 3.92	88.17 ± 3.86	89.37 ± 3.26	85.98 ± 2.78

The papers reported and discussed so far are the ones that can be directly compared with our proposal since all of them share the same dataset split for computing classifier performances. Other train-test splits at the patient level are available, such as the one proposed by Kumar and Rao [27], but it is not actually considered a benchmark dataset. Interested readers could decide to adopt it in their benchmark studies.

As said before, it is very important that the train and test sets must not contain images from the same patients; otherwise, the samples from the same patient could occur in both train and test sets at different magnification levels. This point makes the accuracy artificially higher than the other approaches. Several methodologies overlooking this issue have been proposed. An example is the work in Wei et al. [28], which proposes a CNN architecture trained from scratch with ImageNet first and fine-tuned with the BH dataset, or in Bardou et al. [29], which compared the use of two sets of features, handcrafted and automatically extracted. Other works use GoogleNet [30], different versions of VGG networks [31] and Restricted Boltzmann Machine [32]. Finally, some contributions restrict the study to images at specific magnification sizes, such as [33,34], which used only 40× images.

3. Materials and Methods

3.1. The BreakHis Dataset

The BreakHis dataset comprises 7909 images at a resolution of 700 × 460 pixels obtained from tissue samples of 82 breast cancer patients. These samples are divided into two classes, Benign and Malignant, and each of them is separated into four sub-classes, according to the structure in Table 2. The images were acquired at different magnification levels (40×, 100×, 200×, 400×); Figure 1 reports sample images from the dataset.

In this paper, we are only interested in the dichotomy of Benign vs. Malignant, and Table 2 shows that the number of images in the malignant class is double that of the number of benign images.

Although many authors used pre-processing techniques on images, such as stain-normalisation or whitening, we do not consider any of these techniques because our preliminary experiments show that they provide no performance increase. Moreover, these pre-processing techniques require an additional processing time, sometimes very long, and add a new set of parameters: for example, in stain normalisation a reference image must be selected from the training dataset; or in a whitening procedure, color mean and variance must be calculated on available training images.

As said before, the original paper presenting the dataset [4] also proposes a train/test split for the training of a classifier. This split is based on a 70–30% proportion that also considers the patient set to avoid the presence of images of the same patient both in the training and test set.

Table 2. The BreakHis dataset: for each patient the set of images is divided in magnifications 40×, 100×, 200×, 400×.

Benign	Cancer Types	No. of Images	No. of Patients
	Adenosis	444	4
	Fibroadenoma	1014	10
	Tubular Adenoma	453	7
	Phyllodes Tumour	569	3
Total		2480	24
Malignant	**Cancer Types**	**No. of Images**	**No. of Patients**
	Ductal Carcinoma	3451	38
	Lobular Carcinoma	626	5
	Mucinous Carcinoma	792	9
	Papillary Carcinoma	560	6
Total		5429	58

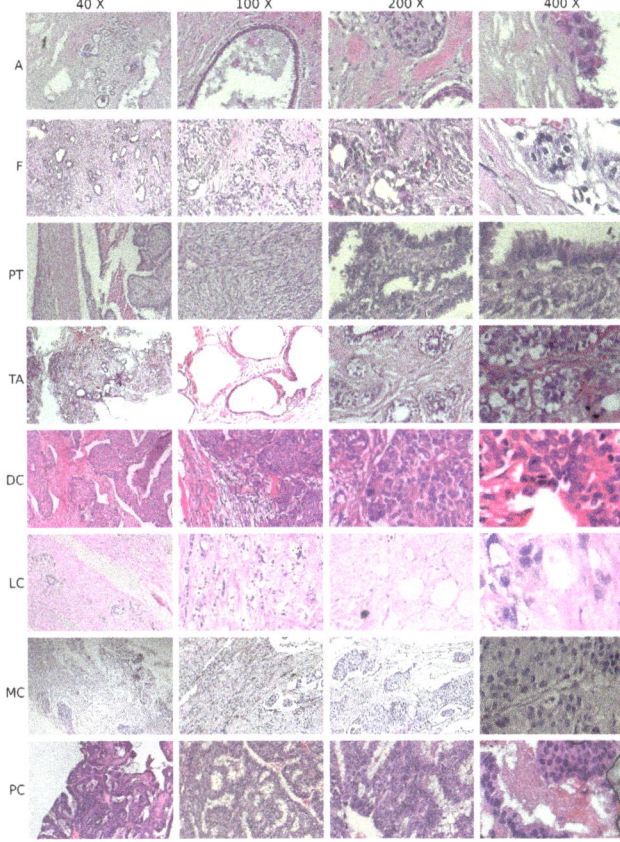

Figure 1. Some images from BreakHis dataset. There is a column for each magnification factor and a row for each subclass: A (Adenosis), F (Fibroadenoma), TA (Tubular Adenoma) and PT (Phillodes Tumor) are benign, DC (Ductal Carcinoma), LC (Lobular Carcinoma), MC (Mucinous Carcinoma) and PC (Papillary Carcinoma) are malignant.

The availability of this specific partitioning is of great advantage for the assessment of the overall quality and potential limitations of the results, which, in the general case, must be carefully performed by the adoption of specific statistical tools [35].

Using this method, five training/test splits were generated and can be downloaded with the dataset allowing a complete comparison of the classification results.

The classification results will be reported using the so-called PLA [4], i.e., a performance index that takes into account the results on a patient level in the following way:

$$Patient\ Score = \frac{N_{rec}}{N_P} \qquad (1)$$

where N_P is the number of images available of the patient P, and N_{rec} is the fraction of N_P images correctly classified. The Patient Level Accuracy (PLA) is defined as follows:

$$PLA = \frac{\sum Patient\ Score}{Total\ Number\ of\ patients}. \qquad (2)$$

Considering the images of the benign class as positive and the images of the malignant class as negative, the Patient Score is defined as the accuracy for a single patient. PLA is an average of the Patient Scores, evaluated as an average among the patients.

3.2. The Proposed Neural Network Architecture

The general architecture of a deep neural network image classifier can be viewed as a stack comprising a feature extraction part, typically composed of pre-trained convolutional layers, followed by a set of fully connected (FC) layers that implement a Multi-Layer Perceptron (MLP) architecture and serve as a classifier. Among all the layers, only the final layer is trained specifically for network specialisation.

Sometimes in these deep architectures, the last convolutional layer is followed by a "squeeze" operation that precedes the fully connected part of the architecture. The MLP layers have decreasing dimensions from thousands of units to the number of classes. Connecting a layer with dimension H to a layer with dimension D units in a fully connected architecture generates an $H \times D$ weight matrix. As explained in [36], introducing a linear layer K, with $K \ll D$ and $K \ll H$, results in two matrices $H \times K$ and $K \times D$ and in a number of weights $K \times (H + D)$. The reduced number of weights result in a faster training and can beneficial.

In our architecture, we used a linear layer as a filter for the "signals" generated by the feature extraction lower layers. The training of this linear layer is aimed at separating the input of different classes and can significantly improve the classification results. We perform this training using the metric learning technique implemented with the triplet network learning paradigm; separating the two classes allowed us to remove the MLP layers and use a simpler classifier.

The network proposed in this work uses a ResNet152 network [37], pre-trained on the ImageNet dataset [5] as feature extractor (indicated in Figure 2). These features are linearly projected onto a lower-dimensional space embedding layer (see the Embedding block in Figure 2a). The number of features generated by the ResNet152 is 2048, and the size of the embedding layer is 512. The resulting embedding provides a data representation that is so effective that the MLP layers can be substituted by a simpler k-NN classifier, as shown in Figure 2a. In the ablation study (see Figure 2b), we will demonstrate that without the linear embedding, the results are noticeably worse.

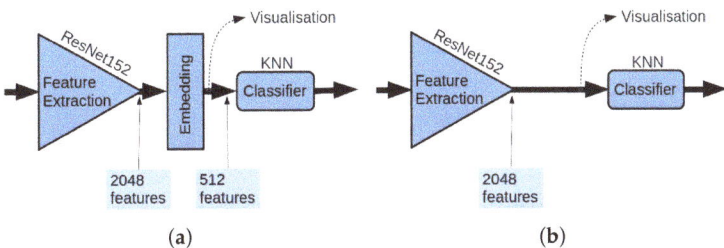

Figure 2. A representation of the classification network proposed in this work: (**a**) The architecture with the embedding layer. (**b**) The architecture without the embedding layer used in the ablation study.

The feature extraction of the proposed model is not interpretable or explainable, since ResNet is a very complex model and is already trained via transfer learning.

The training procedure of the linear layer is aimed to transform the feature space, obtained from the feature extraction layer, in a lower dimension space where images of the same class are near each other and images of different classes are taken apart. In this training mechanism, we are trying to confirm underlying information in the training set: images of the same class should have similar characteristics and they should not share features with images of another class.

3.3. Metric Learning

Deep metric learning aims to derive effective embeddings from the input data using one or more neural networks and an optimisation strategy based on a chosen distance. In this field, the most-used architectures are the siamese networks [38] and the triplet networks [39]. Both models are constituted by neural networks with shared weights: siamese networks use two neural networks while the triplet network, which we used in this work, has three neural networks.

Figure 3 reports a representation of the training phase; in this figure three deep networks ResNet152 produce the representations x_a, x_p, x_n of the three inputs: x_a corresponds to the anchor example, x_p is the representation of the positive example, an input of the same class of x_a, and x_n is the negative example, an input of a different class. The embedding layer will transform the representations $x_* \in \Re^n$ to the embeddings $r_* \in \Re^{n'}$.

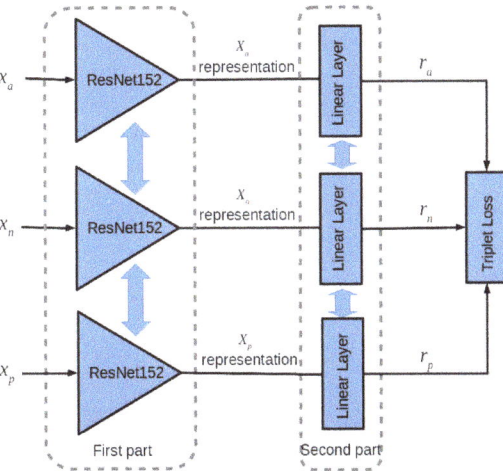

Figure 3. A representation of the network; the vertical double arrow indicates the weights sharing.

Given a distance metric d, during training the shared weights of the embedding layer are adjusted so that the value $d(r_a, r_n)$ is greater than a prefixed margin m w.r.t. the distance $d(r_a, r_p)$. The distance d and the margin m are parameters of the model. The most-used distances are defined using the cosine function or euclidean distance. The representation with three networks and shared weights, like the one in Figure 3, is a common way to indicate the training procedure; however, it is important to point out that the implementation uses a single network receiving the three inputs organised into a sequence, then collecting and storing the values to calculate the global result. The loss function used for training the model is the so-called triplet margin loss [40], defined as:

$$l_{triplet} = \max\left\{0, d(r_a, r_p) - d(r_a, r_n) + m\right\}. \tag{3}$$

During the training phase a mining strategy searches—inside the input mini batch—the most effective triplets to update the model. There are different mining strategies [40]; here we adopt the so-called semi-hard mining strategy, defined as:

$$d(r_a, r_p) < d(r_a, r_n) < d(r_a, r_p) + m. \tag{4}$$

This strategy chooses the negative sample to be farther away from the anchor, with respect to the positive, but always bounded by the margin m. As a consequence, the network's loss is bounded by the margin m.

3.4. Interpretability of the Proposed Model

The interpretability that distinguishes our proposed paradigm belongs to the category of interpretable methodology based on visual analytics [41].

The deep metric learning network finds a proper embedding, where the mappings of all the items to classify can be visualised in two or three dimensions using a dimensionality reduction algorithm. One possibility that we have adopted in this paper is the use of Uniform Manifold Approximation and Projection (UMAP) [42], which accomplishes dimensionality reduction using Riemannian geometry and algebraic topology. Metric learning and UMAP introduce a first level of interpretability to the model since they enable the visualisation of the training images that are close to the test ones, and consequently perform neighbour detection (see Figure 4a for an example).

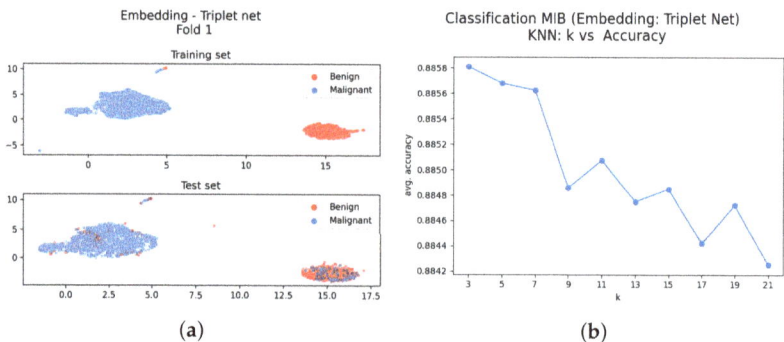

Figure 4. Classification MIB: (**a**) Bi-dimensional representation of the input dataset both training (upper part of Figure 4a) and test set (bottom part of Figure 4b); (**b**) Number of neighbours vs. average accuracy.

The visualisation of the embedding could also be beneficial to study the distribution of the images related to a specific chosen patient (see Figure 5).

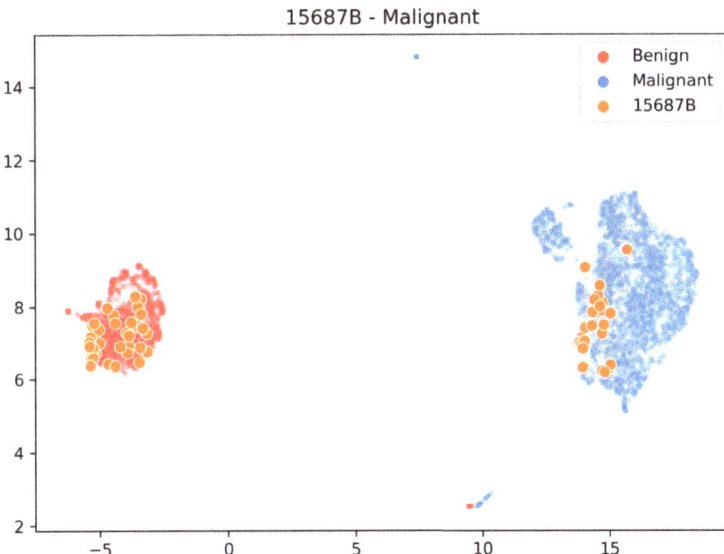

Figure 5. Position of the embedded images of the test patient 15687B in the embedding space w.r.t. the training set embedded images. Our system does not correctly classify most images relating to this patient.

The second level of interpretability is introduced by a classifier that clearly maps the input to the output, such as a linear classifier, a Support Vector Machine, or a simple k-NN, like the one we adopted here. Unlike the others, with k-NN, the nearest neighbors can be used to visually estimate the support for a prediction, providing human-interpretable explanations of predictions. Furthermore, it is always possible for each classified image to display the training images that led to the class labeling (see Figure 6). We find this option very useful as part of a decision support system, where analysts can have interpretable information about the suggested output.

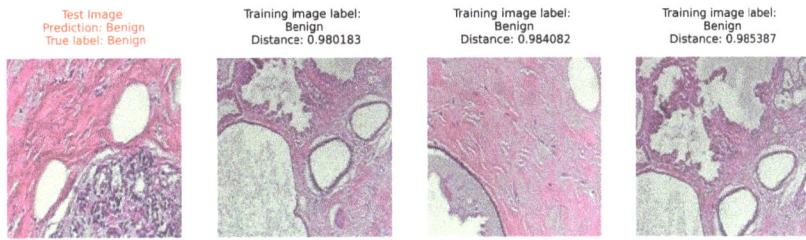

Figure 6. An example of correct classification of a benign image. The test image is on the (**left**); its nearest neighbours with the associated distance are on the (**right**).

The performances of the proposed model are similar, if not better, to those of other classification systems of the same kind, confirming that the interpretability can be obtained without compromising the performances [7].

3.5. Experimental Setup

Experiments have been performed using a workstation equipped with a 12th Gen Intel® Core™ i9-12900KS and an Nvidia 3090 Ti GPU. The total amount of system memory is 32 Gbyte of DDR5. The GPU is also supplied with 24 Gbyte of DDR5 memory and adopts a CUDA parallel computing platform. The operating system is Ubuntu 22.10. We used the TensorFlow Python library [43] to carry out the experiments. The training time for the MIB configuration is about 11 min, while for the MSB configuration about 12 min (3 min for each magnification). We have trained the network for 20 epochs with a mini-batch size of 32 and using the Adam optimisation algorithm [44] with a learning rate of 1×10^{-5} and weight decay factor of 1×10^{-4}. All the images were resized to 224×224.

4. Results

The classification performances obtained with the proposed architecture in both MIB and MSB tasks are reported using the PLA approach. In order to justify the Embedding Layer, an ablation study was carried out, which is described in Section 4.2, using the same classification tasks with the architecture in Figure 2b. Finally, an example of the efficacy of the projection is reported in Section 4.3.

4.1. Results with the Proposed Architecture: Triplet Net Embedding

The proposed architecture has been used to classify the BH images in the Magnification Independent (MIB) and Magnification Specific (MSB) fashion. The train/test split used is the 5-fold split proposed by the authors of the dataset.

The training with the MIB procedure uses all the available images in the training set regardless of the magnification factor. After the training of the embedding system, the UMAP procedure was used to visualise the images as points in the embedding space. Then a set of trials was carried out to optimise the k-nearest neighbourhood value. The UMAP visualisation and the plot of the accuracy vs. k-values are in Figure 4; the lower section of Figure 4a shows the embedding of the test images. The upper part of Figure 4a shows that the two clusters of the embedding points obtained from the training images are well-separated. The left cluster collects all the malignant training image projections, and the right cluster collects the benign ones. The same configuration can be observed at the bottom of the sub-figure that reports the test image embeddings. It can be noticed that both sets present two well-separated clusters, and no image is mapped outside these specific areas. Even the misclassified test images are represented as dots in the wrong cluster; there is only a small detached cluster of images on the upper part of the figure.

The plot in Figure 4b shows that the value of k has very little influence on the accuracy results: the accuracy change from $k = 3$ to $k = 21$ is less than $1/100$.

The PLA accuracy values are in Table 3, compared with the results obtained by Bayramoglu et al. [16] and Sun et al. [17]. The proposed method shows a better performance while requiring a simpler network.

Table 3. Triplet net embedding: MIB classification results and comparison with the other methods.

Method	PLA
Bayramoglu et al. [16]	82.13
Sun et al. [17]	88.40 ± 4.10
Proposed method (k-NN)	88.90 ± 2.41

The confusion matrix is averaged over five folds and is reported in Figure 7. It shows that the number of malignant images misclassified as benign is less than that of benign images misclassified as malignant. This point is crucial in the medical domain since a malignant tumor will receive more attention and probably require more exams than a benign one, and the error has a significant probability of being corrected.

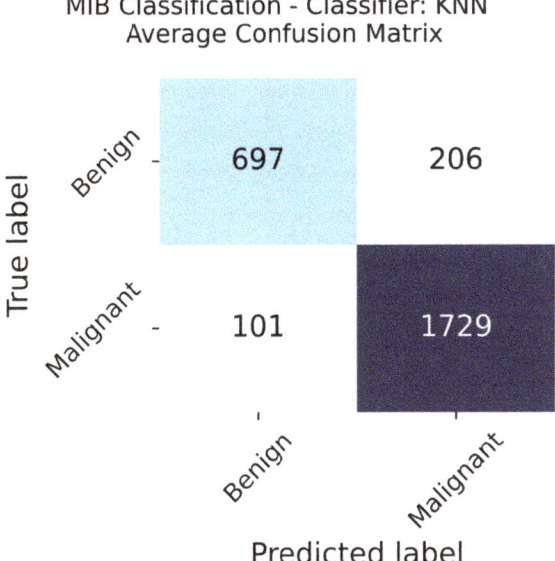

Figure 7. MIB Classification, average confusion matrix.

Figure 8 reports the classification results for each patient of the five folds. Using a blue scale, the ratio between the images labelled as malignant (benign) and the total number of images for that patient is presented. The second level of interpretability is introduced by a classifier that clearly maps the input to the output, such as a linear classifier, a Support Vector Machine, or a simple k-NN, like the one we adopted here. Unlike the others, with k-NN, the nearest neighbors can be used to visually estimate the support for a prediction, providing human-interpretable explanations of predictions. Furthermore, it is always possible for each classified image to display the training images that led to the class labeling (see Figure 6). We find this option very useful as part of a decision support system, where analysts can have interpretable information about the suggested output.

The assigned class is represented with a cross. The ground truth is presented according to the rows. The first nine rows are from the benign class and the remaining are from the malignant class. In the figure, we can see that 3 patients, 19854C, 9146, and 15687B, are classified differently in the five folds. These samples show that the correct or incorrect classification is heavily influenced by the composition of the network's training set, which requires further investigation. The images of these patients are distributed in both clusters (for example, see Figure 5 for patient 15687B), and the patient classification changes according to the fraction of images correctly classified.

Finally, we notice that each fold has a different error rate: Fold 5 has no errors, while Fold 2 and 3 have one error, Fold 1 has two errors and Fold 4 has 4 errors.

MSB classification is obtained by training four classifiers, one for each magnification factor. Also, in this case, we obtained two well-separated clusters for each training, and the k value has little influence (less than 2/100) on the accuracy values, figures are not reported in this case; we selected $k = 3$ for classification.

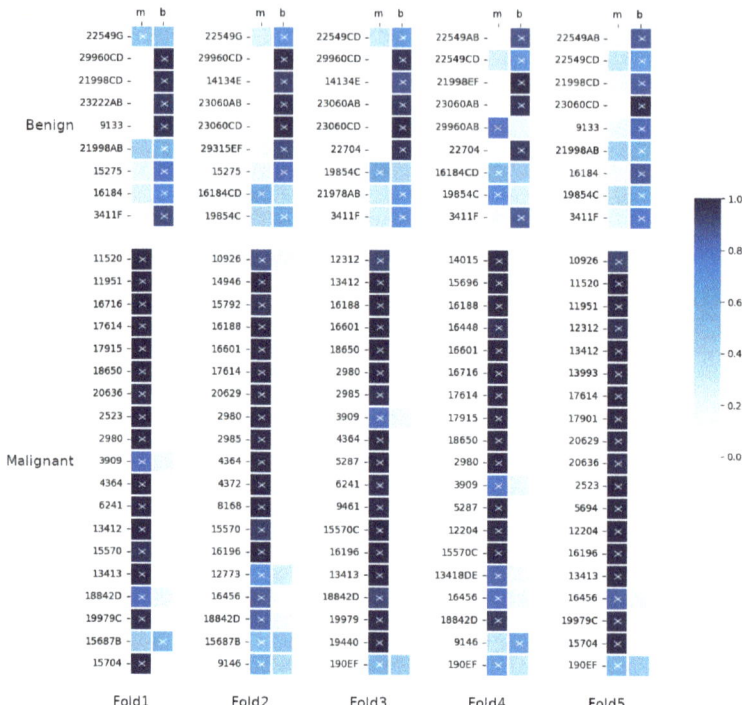

Figure 8. Classification results for each test fold. The colour inside each cell represents the ratio between the image labelled as malignant (benign) and the total number of images of the relative patient. The assigned label is presented as a cross. The ground truth is represented according to the rows.

The PLA is reported in Table 1, and compared with the other methods cited in Section 2.2.

Note that to evaluate the performance of an image classification system, the widely used metrics are accuracy and F1-Score. According to us, in histopathological image classification, the metric that best simulates reality is patient-based accuracy (PLA); indeed, the pathologist makes the diagnosis by evaluating images of different patients. For completeness, we have reported in Tables 4 and 5 the values of the other metrics we calculated for the proposed approach: precision, recall, F1-score and the area under the ROC curve (AUC).

Table 4. Triplet net embedding: MIB classification metrics.

Accuracy	Precision	Recall	F1-Score	AUC
88.71 ± 2.10	88.24 ± 2.78	85.87 ± 2.02	86.84 ± 2.28	85.87 ± 2.02

Table 5. Triplet net embedding: MSB classification metrics.

Metric	40×	100×	200×	400×
Accuracy	84.60 ± 3.53	86.34 ± 3.13	88.70 ± 3.22	84.98 ± 3.17
Precision	85.48 ± 4.45	87.16 ± 3.40	88.36 ± 3.20	84.86 ± 2.78
Recall	94.42 ± 2.36	95.24 ± 3.13	94.79 ± 4.42	94.04 ± 4.50
F1-Score	89.15 ± 2.51	90.31 ± 2.25	91.90 ± 2.34	89.23 ± 2.36
AUC	82.36 ± 3.67	83.14 ± 3.70	87.68 ± 4.12	81.67 ± 2.94

4.2. Ablation Study: ResNet152 Embedding

The ablation study was conducted to motivate the use of the embedding layer. The new considered model is depicted in Figure 2b. Cancelling the linear layer after the ResNet152 model was used as a feature extractor, embeddings in the metric space were not generated, and the extracted features were used to train the *k*-NN classifier. The visual representation of the images in the 2048-dimensions space is obtained using the UMAP method [42].

Figure 9 reports the visualisation of the embeddings in the MIB case, together with the plot of accuracy vs. k value. We can notice that the embedded training or test points are not separated into two clusters, and the k value still has little influence on the accuracy value. The same results can be observed for the MSB classification (results not shown), but in order to obtain the best performance, the k value must be varied for each magnification; the chosen values were 3, 7, 5 and 9 for the magnification values 40×, 100×, 200× and 400×, respectively. The obtained PLA results are reported in Table 6 for MIB and MSB cases.

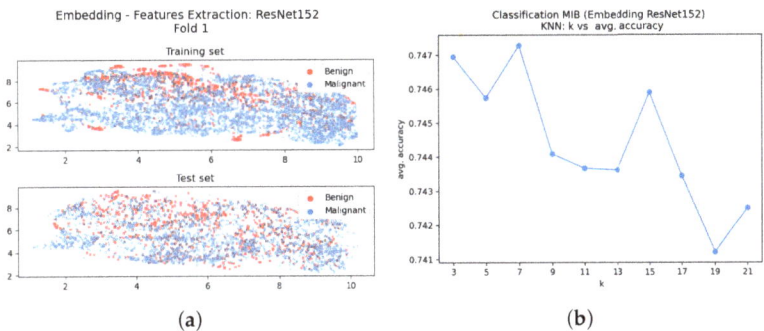

Figure 9. Ablation study with classification MIB: (**a**) Bi-dimensional representation of the input dataset; (**b**) number of neighbours vs. average accuracy.

Table 6. Ablation study: Magnification Independent (MIB) and Magnification Specific (MSB) classification with *k*-NN.

		PLA-*k*-NN
MIB		75%
MSB	40×	79%
	100×	78%
	200×	86%
	400×	80%

4.3. Mapping Other Set Images in the Embedded Space

The purpose of the embedding layer is to map the input information into a space that makes simple and effective the classification process and, to support this point, we used a very simple classifier, still obtaining a very good classification accuracy. If we input a new histopathology image, we expect that the system will correctly classify it as malignant or benign, using, at its best, the image features obtained from the lower layers.

The embedding layer was trained with breast cancer histopathology images, and the whole system has knowledge about this kind of images. According to this consideration, the system should not recognise any other kind of image.

Of course, any image can be mapped in the embedding space and the *k*-NN will always provide an output according to the k neighbourhoods. At the same time, we can check the cluster, learned from the training process, where the image is placed.

If we input a random image into the system, it would be desirable that the system answer that the features were not appropriate to classify it meaningfully. We expect that a random image should be in an area of the embedding space far from the embedding of the

BH images. In order to test this hypothesis, we visualise the position where the system, trained with BreakHis dataset, maps the images of two different histological image datasets: the MHIST dataset [45] (Minimalist HISTopathology contains images of colon-rectal polyps) and the Epistroma dataset [46,47] (that contains histopathology images belonging to two tissue types: epithelium and stroma). These image sets were selected because they are of the same kind (histology images) but from a different origin. Figure 10a,b show the results: the embedding point clouds are far away from the BH image projection.

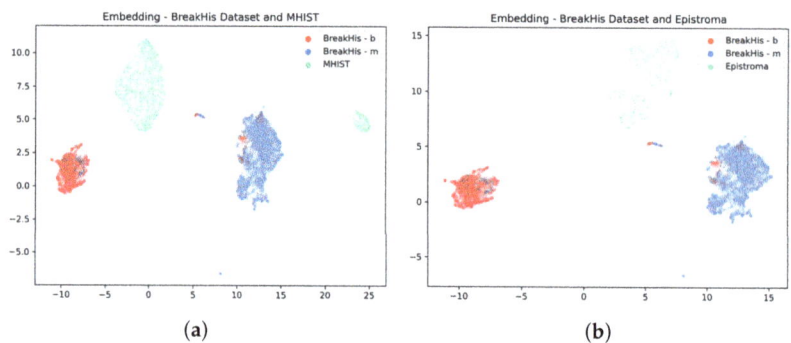

Figure 10. UMAP representation of the embeddings of the BreakHis with other histology image datasets; (**a**) The embedding of the MHIST dataset obtained using the triplet net trained on the BreakHis dataset; (**b**) The embedding of the Epistroma dataset obtained using the triplet net trained on the BreakHis dataset.

5. Discussion

There is a common idea that interpretability should be paid with a substantial loss of performance. This claim can be found in [48]; however, C. Rudin argued in [7] that there is not evidence supporting this statement. In the last section, we show that this is not necessarily true; it is possible to support interpretability without a significant loss of performance. In the following, we will discuss the result comparison with the state of the art and the interpretability of the proposed framework.

5.1. The Performances of the System

There are many works on the classification of BH dataset, those reported in Section 2 are only the one that contains enough information for a comparison using the PLA metrics, the only metrics that make sense in this kind of studies. According to these premises, the presented work can be compared with very few works on the MIB classification, while there are more works for the MSB classification. Our results for MIB are better than the state-of-the-art results, as reported in Table 3, and we notice that the accuracy values in the MSB task are somewhat lower than the ones in the state-of-the-art works. In Table 1, we report the results from seven different works; the comparison shows that only in $40\times$ magnification our results are less than 5% below the work of Sudharshan et al. [26]; for other magnification values, the accuracy values are just 2 or 3% below. In the MSB classification, it is necessary to train one neural network for each magnification value, and separating the training images into the different magnification sub-classes results in a smaller number of training images for each neural network. However, in our experiments, this effect is not recovered using augmentation, and this requires more investigations. All the methods that perform better than our proposal involve complex learning paradigms such as classifier combination (for the case of Spanhol et al. [20] or Multiple Instance Learning (Sudharshan et al. [26]) or some kind of specific pre-processing (re-scaling for the case of Song et al. [23]). While benefits can be observed from the performances, they remain not very interpretable.

The proposed method is based on a clear and simple mechanism, the ablation study shows that the cluster separation is only due to the presence of the linear layer and the metric learning training, and this is clear comparing Figures 4a and 9a. Table 6 also shows a significant performance drop both in MIB and MSB classification.

5.2. The Interpretability Aspects

The interpretability of the proposed method is based on the set of information that can be obtained at the end of the embedding layer and from the *k*-NN classification module. The metric learning layer rearranges the embedding space while reducing the number of dimensions. In the new embedding space, the training images are meaningfully clustered, and the visualisation allows the user to understand the system response. This visualisation constitutes the first set of information that can help the user to understand the first processing steps. Test images are clustered in the same areas of the training ones; this happens for both classification methods MIB and MSB. The visualization of the training image in the embedding space in Figure 4a shows that metric learning creates a sort of "areas" where input images should be projected. The visualisation of the embedding space is also useful during the classification of a new input. Observing the position of the images of a new patient in the embedding space, such as the ones reported in Figure 5, the user can check if this input is near the embeddings of the training images set, meaning that the new input came from the same distribution of the training images.

Finally, the same visualisation can spot when the network is not able to process an input image. Figure 10 shows what happens if the input is an out-of-distribution image set. In this case, the images are embedded in points outside the two clusters obtained from the BH training images. In cases like this, the *k*-NN classifier will assign a class to the input based on the nearest images, but the visualisation shows that the features of the inputs are different from the ones of the training images. This consideration allows an examiner to say that the system classification output will have no meaning since the images are outside what the system "knows".

Focusing on the classification algorithm, notice that *k*-NN is considered one of the naturally interpretable classification algorithm. For each classification result is always possible to visualise the train images that produce the result and, thanks to the application of metric learning, the number of neighbourhoods is very low, less than the one obtained in the ablation study.

This organisation of the embedding space allows using a *k*-NN classifier with a very low k, so that an user can visualise the neighbourhood of a test image and easily spot the similarity and the differences with its neighbourhood. An example of classification in MIB is reported in Figure 6, with the neighbour images. This figure shows the three nearest neighbourhood images used to label the test image (left image). As pointed out in the book [49], the interpretation of the classification results is translated to the interpretation of the neighbourhood images, and a domain expert can give the right meaning to the images similarity.

6. Conclusions

The high performances obtained with the DNN in the medical domain are confirmed by many studies and applications, and histopathology image classification is one of these applications. However, there are two main problems connected with many of these studies: the first is connected with the train and test set; not all the works consider that the focus is on the patient, not on the images, and the patient should be completely unknown for the system undergoing testing. The second problem is related to the interpretability of the supporting systems; in the medical domain the answer of a supporting system should always be followed with an explanation of the decisions steps.

In this paper, we have presented a classification model based on a deep neural network and an interpretable classifier, which take into account the above-mentioned considerations. In particular, we used a triplet network based on a ResNet152 as a deep network, which allows us to map the input samples into an embedding space learned to represent images of different classes as separate clusters. The representation in this embedding space is viable for the comparison of a test image with its neighbourhood, providing an explanation of the classification as compared with images of the same kind. The final classification is performed with a simple classifier (k-NN) since the input information is represented through the neural network. Interestingly, only a few layers of the model are trained, while the majority of the network is pre-trained and does not need to be updated.

The results obtained with this technique are better than those obtained with the ad-hoc full-trained deep neural networks and allow the user to visualise the representation of the input image together with the most similar training image. The proposed technique can be used in many classifier architectures based on representation/classification schema. Future research will be focused on the analysis of the neighbourhood images in order to find the details that can be considered characteristic of a specific class.

Author Contributions: Conceptualization, S.C., G.L.B., R.R. and F.V.; methodology, S.C., G.L.B., R.R. and F.V.; software, S.C.; validation, D.A., S.C., G.L.B., R.R. and F.V.; writing—original draft preparation, D.A., S.C., G.L.B., R.R. and F.V.; writing—review and editing, D.A., S.C., G.L.B., R.R. and F.V. All authors have read and agreed to the published version of the manuscript.

Funding: This research has been supported by Piano Nazionale per gli investimenti Complementari al PNRR, project DARE-Digital Lifelong Prevention, CUP B53C22006460001, Decreto Direttoriale (Direzione Generale Ricerca) Ministero Università e Ricerca n. 1511 del 30/09/2022. Additional support to G.L.B. has been granted by the University of Palermo FFR (Fondo Finalizzato alla ricerca di Ateneo) year 2023.

Institutional Review Board Statement: This study has not involved any human subject.

Informed Consent Statement: This study has not involved any human subject.

Data Availability Statement: The BreakHis dataset is publicly available at https://web.inf.ufpr.br/vri/databases/breast-cancer-histopathological-database-breakhis/ (accessed on 1 June 2022). The MHIST dataset is available at https://bmirds.github.io/MHIST/ (accessed on 1 June 2022). The Epistroma dataset is publicly available at http://fimm.webmicroscope.net/Research/Supplements/epistroma (accessed on 1 June 2022). The source code is available at https://github.com/Calder10/BreakHis-MIB-MSB-Classification-Metric-Learning (accessed on 1 June 2022).

Conflicts of Interest: The authors declare no conflict of interest.

Abbreviations

The following abbreviations are used in this manuscript:

BH	BreakHis (dataset)
CNN	Convolution Neural Network
CVF	ConvNet-based Fisher Vectors
DNN	Deep Neural Network
FC	Fully Connected
ILA	Image Level Accuracy
k-NN	k Nearest Neighbour
MI	Magnification Independent
MIB	Magnification Independent Binary (classification)
MLP	Multi Layer Perceptron
MS	Magnification Specific
MSB	Magnification Specific Binary (classification)
PFTAS	Parametric-Free Threshold Adjacency Statistics
PLA	Patient Level Accuracy
AUC	Area under the Roc Curve
UMAP	Uniform Manifold Approximation and Projection

References

1. Van der Laak, J.; Litjens, G.; Ciompi, F. Deep learning in histopathology: The path to the clinic. *Nat. Med.* **2021**, *27*, 775–784. [CrossRef]

2. Dimitriou, N.; Arandjelović, O.; Caie, P.D. Deep Learning for Whole Slide Image Analysis: An Overview. *Front. Med.* **2019**, *6*, 264. [CrossRef]
3. Campanella, G.; Hanna, M.G.; Geneslaw, L.; Miraflor, A.; Werneck Krauss Silva, V.; Busam, K.J.; Fuchs, T.J. Clinical-Grade Computational Pathology Using Weakly Supervised Deep Learning on Whole Slide Images. *Nat. Med.* **2019**, *25*, 1301–1309. [CrossRef]
4. Spanhol, F.A.; Oliveira, L.S.; Petitjean, C.; Heutte, L. A Dataset for Breast Cancer Histopathological Image Classification. *IEEE Trans. Biomed. Eng.* **2016**, *63*, 1455–1462. [CrossRef]
5. Deng, J.; Dong, W.; Socher, R.; Li, L.J.; Li, K.; Fei-Fei, L. ImageNet: A large-scale hierarchical image database. In Proceedings of the 2009 IEEE Conference on Computer Vision and Pattern Recognition, Miami, FL, USA, 20–25 June 2009; pp. 248–255. [CrossRef]
6. Lecun, Y.; Bottou, L.; Bengio, Y.; Haffner, P. Gradient-Based Learning Applied to Document Recognition. *Proc. IEEE* **1998**, *86*, 2278–2324. [CrossRef]
7. Rudin, C. Stop explaining black box machine learning models for high stakes decisions and use interpretable models instead. *Nat. Mach. Intell.* **2019**, *1*, 206–215. [CrossRef]
8. Selvaraju, R.R.; Cogswell, M.; Das, A.; Vedantam, R.; Parikh, D.; Batra, D. Grad-cam: Visual explanations from deep networks via gradient-based localization. In Proceedings of the IEEE International Conference on Computer Vision, Venice, Italy, 22–29 October 2017; pp. 618–626.
9. Ghassemi, M.; Oakden-Rayner, L.; Beam, A.L. The false hope of current approaches to explainable artificial intelligence in health care. *Lancet Digit. Health* **2021**, *3*, e745–e750. [CrossRef]
10. Calderaro, S.; Lo Bosco, G.; Rizzo, R.; Vella, F. Fuzzy Clustering of Histopathological Images Using Deep Learning Embeddings. *CEUR Workshop Proc.* **2021**, *3074*, 18.
11. Calderaro, S.; Lo Bosco, G.; Rizzo, R.; Vella, F. Deep Metric Learning for Histopathological Image Classification. In Proceedings of the 2022 IEEE Eighth International Conference on Multimedia Big Data (BigMM), Naples, Italy, 5–7 December 2022; pp. 57–64.
12. Calderaro, S.; Lo Bosco, G.; Vella, F.; Rizzo, R. Breast Cancer Histologic Grade Identification by Graph Neural Network Embeddings. In Proceedings of the 2023 10th International Work-Conference on Bioinformatics and Biomedical Engineering (IWBBIO), Gran Canaria, Spain, 12–14 July 2023.
13. Calderaro, S.; Lo Bosco, G.; Rizzo, R.; Vella, F. Deep Metric Learning for Transparent Classification of COVID-19 X-Ray Images. In Proceedings of the 2022 16TH International Conference on Signal Image Technology & Internet Based Systems (SITIS), Dijon, France, 19–21 October 2022.
14. Khened, M.; Kori, A.; Rajkumar, H.; Krishnamurthi, G.; Srinivasan, B. A Generalized Deep Learning Framework for Whole-Slide Image Segmentation and Analysis. *Sci. Rep.* **2021**, *11*, 11579. [CrossRef]
15. Gandomkar, Z.; Brennan, P.C.; Mello-Thoms, C. MuDeRN: Multi-Category Classification of Breast Histopathological Image Using Deep Residual Networks. *Artif. Intell. Med.* **2018**, *88*, 14–24. [CrossRef]
16. Bayramoglu, N.; Kannala, J.; Heikkilä, J. Deep learning for magnification independent breast cancer histopathology image classification. In Proceedings of the 2016 23rd International Conference on Pattern Recognition (ICPR), Cancun, Mexico, 4–8 December 2016; pp. 2440–2445. [CrossRef]
17. Sun, Y.; Huang, X.; Wang, Y.; Zhou, H.; Zhang, Q. Magnification-independent Histopathological Image Classification with Similarity-based Multi-scale Embeddings. *arXiv* **2021**, arXiv:2107.01063.
18. Gupta, V.; Bhavsar, A. Breast Cancer Histopathological Image Classification: Is Magnification Important? In Proceedings of the 2017 IEEE Conference on Computer Vision and Pattern Recognition Workshops (CVPRW), Honolulu, HI, USA, 21–26 July 2017; pp. 769–776. ISSN 2160-7516. [CrossRef]
19. Spanhol, F.A.; Oliveira, L.S.; Petitjean, C.; Heutte, L. Breast cancer histopathological image classification using Convolutional Neural Networks. In Proceedings of the 2016 International Joint Conference on Neural Networks (IJCNN), Vancouver, BC, Canada, 24–29 July 2016; pp. 2560–2567. ISSN 2161-4407. [CrossRef]
20. Spanhol, F.A.; Oliveira, L.S.; Cavalin, P.R.; Petitjean, C.; Heutte, L. Deep features for breast cancer histopathological image classification. In Proceedings of the 2017 IEEE International Conference on Systems, Man, and Cybernetics (SMC), Banff, AL, Canada, 5–8 July 2017; pp. 1868–1873. [CrossRef]
21. Benhammou, Y.; Tabik, S.; Achchab, B.; Herrera, F. A First Study Exploring the Performance of the State-of-the Art CNN Model in the Problem of Breast Cancer. In Proceedings of the International Conference on Learning and Optimization Algorithms: Theory and Applications, Rabat, Morocco, 2–5 May 2018. [CrossRef]
22. Szegedy, C.; Vanhoucke, V.; Ioffe, S.; Shlens, J.; Wojna, Z. Rethinking the Inception Architecture for Computer Vision. *arXiv* **2015**, arXiv:1512.00567. [CrossRef]
23. Song, Y.; Zou, J.J.; Chang, H.; Cai, W. Adapting fisher vectors for histopathology image classification. In Proceedings of the 2017 IEEE 14th International Symposium on Biomedical Imaging (ISBI 2017), Melbourne, Australia, 18–21 June 2017; pp. 600–603. ISSN: 1945-8452. [CrossRef]
24. Sanchez, J.; Perronnin, F.; Mensink, T.; Verbeek, J. Image Classification with the Fisher Vector: Theory and Practice. *Int J. Comput. Vision* **2013**, *105*, 222–245. [CrossRef]

25. Song, Y.; Chang, H.; Huang, H.; Cai, W. Supervised Intra-Embedding of Fisher Vectors for Histopathology Image Classification, Proceedings of the Medical Image Computing and Computer Assisted Intervention-MICCAI 2017, Quebec City, QC, Canada, 10–14 June 2017; Descoteaux, M., Maier-Hein, L., Franz, A., Jannin, P., Collins, D.L., Duchesne, S., Eds.; Lecture Notes in Computer Science; Springer International Publishing: Cham, Switzerland, 2017; pp. 99–106. [CrossRef]
26. Sudharshan, P.J.; Petitjean, C.; Spanhol, F.; Oliveira, L.E.; Heutte, L.; Honeine, P. Multiple instance learning for histopathological breast cancer image classification. *Expert Syst. Appl.* **2019**, *117*, 103–111. [CrossRef]
27. Kumar, K.; Rao, A.C.S. Breast cancer classification of image using convolutional neural network. In Proceedings of the 2018 4th International Conference on Recent Advances in Information Technology (RAIT), Dhanbad, India, 15–17 March 2018; pp. 1–6. [CrossRef]
28. Wei, B.; Han, Z.; He, X.; Yin, Y. Deep learning model based breast cancer histopathological image classification. In Proceedings of the 2017 IEEE 2nd International Conference on Cloud Computing and Big Data Analysis (ICCCBDA), Chengdu, China, 28–30 April 2017; pp. 348–353. [CrossRef]
29. Bardou, D.; Zhang, K.; Ahmad, S.M. Classification of Breast Cancer Based on Histology Images Using Convolutional Neural Networks. *IEEE Access* **2018**, *6*, 24680–24693. [CrossRef]
30. Das, K.; Karri, S.P.K.; Guha Roy, A.; Chatterjee, J.; Sheet, D. Classifying histopathology whole-slides using fusion of decisions from deep convolutional network on a collection of random multi-views at multi-magnification. In Proceedings of the 2017 IEEE 14th International Symposium on Biomedical Imaging (ISBI 2017), Melbourne, Australia, 18–21 April 2017. [CrossRef]
31. Cascianelli, S.; Bello-Cerezo, R.; Bianconi, F.; Fravolini, M.L.; Belal, M.; Palumbo, B.; Kather, J.N. Dimensionality Reduction Strategies for CNN-Based Classification of Histopathological Images. In *Intelligent Interactive Multimedia Systems and Services*; De Pietro, G., Gallo, L., Howlett, R.J., Jain, L.C., Eds.; Springer International Publishing: Cham, Switzerland, 2018; Smart Innovation, Systems and Technologies; pp. 21–30. [CrossRef]
32. Nahid, A.; Kong, Y. Histopathological breast-image classification with restricted Boltzmann machine along with backpropagation. *Biomed. Res.* **2018**, *29*, 2068–2077. [CrossRef]
33. Akbar, S.; Peikari, M.; Salama, S.; Nofech-Mozes, S.; Martel, A. Transitioning Between Convolutional and Fully Connected Layers in Neural Networks. In *Deep Learning in Medical Image Analysis and Multimodal Learning for Clinical Decision Support*; Cardoso, M.J., Arbel, T., Carneiro, G., Syeda-Mahmood, T., Tavares, J.M.R., Moradi, M., Bradley, A., Greenspan, H., Papa, J.P., Madabhushi, A., et al., Eds.; Lecture Notes in Computer Science; Springer International Publishing: Cham, Switzerland, 2017; pp. 143–150. [CrossRef]
34. Nejad, E.M.; Affendey, L.S.; Latip, R.B.; Bin Ishak, I. Classification of Histopathology Images of Breast into Benign and Malignant using a Single-layer Convolutional Neural Network. In Proceedings of the International Conference on Imaging, Signal Processing and Communication, Penang, Malaysia, 26–28 July 2017; Association for Computing Machinery: New York, NY, USA, 2017; pp. 50–53. [CrossRef]
35. Halacli, B.; Yildirim, M.; Kaya, E.K.; Ulusoydan, E.; Ersoy, E.O.; Topeli, A. Chronic critical illness in critically ill COVID-19 patients. *Chronic Illn.* **2023**. [CrossRef]
36. Ba, J.; Caruana, R. Do deep nets really need to be deep? In Proceedings of the 27th International Conference on Neural Information Processing Systems-Volume 2, Montreal, QC, Canada, 8–13 December 2014; MIT Press: Cambridge, MA, USA, 2014; Volume 27, pp. 2654–2662.
37. He, K.; Zhang, X.; Ren, S.; Sun, J. Deep Residual Learning for Image Recognition. In Proceedings of the 2016 IEEE Conference on Computer Vision and Pattern Recognition (CVPR), Las Vegas, NV, USA, 26 June–1 July 2016; pp. 770–778. [CrossRef]
38. Bromley, J.; Guyon, I.; LeCun, Y.; Säckinger, E.; Shah, R. Signature Verification Using a "Siamese" Time Delay Neural Network. In *Advances in Neural Information Processing Systems*; Morgan Kaufmann: Burlington, MA, USA, 1993; Volume 6.
39. Hoffer, E.; Ailon, N. Deep Metric Learning Using Triplet Network. In *Similarity-Based Pattern Recognition*; Feragen, A., Pelillo, M., Loog, M., Eds.; Lecture Notes in Computer Science; Springer International Publishing: Cham, Switzerland, 2015; pp. 84–92. [CrossRef]
40. Kaya, M.; Sakir, B.H. Deep Metric Learning: A Survey. *Symmetry* **2019**, *11*, 1066. [CrossRef]
41. Alicioglu, G.; Sun, B. A survey of visual analytics for Explainable Artificial Intelligence methods. *Comput. Graph.* **2022**, *102*, 502–520. [CrossRef]
42. McInnes, L.; Healy, J. UMAP: Uniform Manifold Approximation and Projection for Dimension Reduction. *arXiv* **2018**, arXiv:1802.03426.
43. Abadi, M.; Agarwal, A.; Barham, P.; Brevdo, E.; Chen, Z.; Citro, C.; Corrado, G.S.; Davis, A.; Dean, J.; Devin, M.; et al. TensorFlow: Large-Scale Machine Learning on Heterogeneous Systems. 2015. Available online: tensorflow.org (accessed on 1 June 2022).
44. Kingma, D.P.; Ba, J. Adam: A Method for Stochastic Optimization. In Proceedings of the 3rd International Conference on Learning Representations, ICLR 2015, San Diego, CA, USA, 7–9 May 2015.
45. Wei, J.; Suriawinata, A.; Ren, B.; Liu, X.; Lisovsky, M.; Vaickus, L.; Brown, C.; Baker, M.; Tomita, N.; Torresani, L.; et al. A Petri Dish for Histopathology Image Analysis. In *Artificial Intelligence in Medicine*; Tucker, A., Henriques Abreu, P., Cardoso, J., Pereira Rodrigues, P., Riaño, D., Eds.; Springer International Publishing: Cham, Switzerland, 2021; pp. 11–24.
46. Linder, N.; Konsti, J.; Turkki, R.; Rahtu, E.; Lundin, M.; Nordling, S.; Haglund, C.; Ahonen, T.; Pietikäinen, M.; Lundin, J. Identification of tumor epithelium and stroma in tissue microarrays using texture analysis. *Diagn. Pathol.* **2012**, *7*, 22. [CrossRef] [PubMed]

47. Linder, N.; Martelin, E.; Lundin, M.; Louhimo, J.; Nordling, S.; Haglund, C.; Lundin, J. Xanthine oxidoreductase—Clinical significance in colorectal cancer and in vitro expression of the protein in human colon cancer cells. *Eur. J. Cancer* **2009**, *45*, 648–655. [CrossRef] [PubMed]
48. DARPA. Broad Agency Announcement, Explainable Artificial Intelligence (XAI). DARPA-BAA-16-53. 2016. Available online: https://www.darpa.mil/attachments/DARPA-BAA-16-53.pdf (accessed on 1 June 2022).
49. Molnar, C. Interpretable Machine Learning; Lulu. com. 2020. Available online: https://christophm.github.io/interpretable-ml-book (accessed on 1 June 2022).

Disclaimer/Publisher's Note: The statements, opinions and data contained in all publications are solely those of the individual author(s) and contributor(s) and not of MDPI and/or the editor(s). MDPI and/or the editor(s) disclaim responsibility for any injury to people or property resulting from any ideas, methods, instructions or products referred to in the content.

Article

Hexagonal-Grid-Layout Image Segmentation Using Shock Filters: Computational Complexity Case Study for Microarray Image Analysis Related to Machine Learning Approaches

Aurel Baloi [1,2,†], Carmen Costea [3,†], Robert Gutt [4,†], Ovidiu Balacescu [5], Flaviu Turcu [4,6] and Bogdan Belean [4,*]

1. Research Center for Integrated Analysis and Territorial Management, University of Bucharest, 4-12 Regina Elisabeta, 030018 Bucharest, Romania
2. Faculty of Administration and Business, University of Bucharest, 030018 Bucharest, Romania
3. Department of Mathematics, Faculty of Automation and Computer Science, Technical University of Cluj-Napoca, 400114 Cluj-Napoca, Romania
4. Center of Advanced Research and Technologies for Alternative Energies, National Institute for Research and Development of Isotopic and Molecular Technologies, 400293 Cluj-Napoca, Romania
5. Department of Genetics, Genomics and Experimental Pathology, The Oncology Institute, Prof. Dr. Ion Chiricuta, 400015 Cluj-Napoca, Romania
6. Faculty of Physics, Babes-Bolyai University, 400084 Cluj-Napoca, Romania
* Correspondence: bogdan.belean@itim-cj.ro
† These authors contributed equally to this work.

Abstract: Hexagonal grid layouts are advantageous in microarray technology; however, hexagonal grids appear in many fields, especially given the rise of new nanostructures and metamaterials, leading to the need for image analysis on such structures. This work proposes a shock-filter-based approach driven by mathematical morphology for the segmentation of image objects disposed in a hexagonal grid. The original image is decomposed into a pair of rectangular grids, such that their superposition generates the initial image. Within each rectangular grid, the shock-filters are once again used to confine the foreground information for each image object into an area of interest. The proposed methodology was successfully applied for microarray spot segmentation, whereas its character of generality is underlined by the segmentation results obtained for two other types of hexagonal grid layouts. Considering the segmentation accuracy through specific quality measures for microarray images, such as the mean absolute error and the coefficient of variation, high correlations of our computed spot intensity features with the annotated reference values were found, indicating the reliability of the proposed approach. Moreover, taking into account that the shock-filter PDE formalism is targeting the one-dimensional luminance profile function, the computational complexity to determine the grid is minimized. The order of growth for the computational complexity of our approach is at least one order of magnitude lower when compared with state-of-the-art microarray segmentation approaches, ranging from classical to machine learning ones.

Keywords: hexagonal grids; shock-filter; machine learning; image segmentation; computational complexity; gene expression; microarray

Citation: Baloi, A.; Costea, C.; Gutt, R.; Balacescu, O.; Turcu, F.; Belean, B. Hexagonal-Grid-Layout Image Segmentation Using Shock Filters: Computational Complexity Case Study for Microarray Image Analysis Related to Machine Learning Approaches. *Sensors* **2023**, *23*, 2582. https://doi.org/10.3390/s23052582

Academic Editors: Stefano Berretti and Kuo-Liang Chung

Received: 2 February 2023
Revised: 17 February 2023
Accepted: 21 February 2023
Published: 26 February 2023

Copyright: © 2023 by the authors. Licensee MDPI, Basel, Switzerland. This article is an open access article distributed under the terms and conditions of the Creative Commons Attribution (CC BY) license (https://creativecommons.org/licenses/by/4.0/).

1. Introduction

In digital-image processing and computer vision, image segmentation represents the process of dividing an image into multiple segments, representing non-overlapping pixel areas with homogeneous features. The resulting image segments are meaningful for defining objects according to human visual perception within the image under analysis. In biomedical and material science applications, when digital images are used to characterize either multiple biological samples or material structural patterns, the image segments (objects) are often disposed using a grid layout. By the grid layout, one can understand a network of lines that cross each other to form a series of geometrical figures which confine all image objects according to their pattern. Hexagonal grid layouts are used when printing space needs to be efficiently managed. An eloquent example is the microarray technology, where the hexagonal grid is considered advantageous compared to the rectangular grid, since it allows more DNA specific probes to be printed onto the same surface [1]. Moreover, images illustrating the hexagonal grid layout of the material structure are registered in cases of different applications. In cell-cluster-array fabrication, self-assembled hexagonal superparamagnetic cone structures induce a local magnetic field gradient which inhibits the cancer cells' migration [2]. In material science applications, benefits such as increased optical performance or material resistance are added by hexagonal grid structures. The performances of the pixelated CsI(Tl) scintillation screens in X-ray imaging are enhanced by using a hexagonal array structure for the micro-columns' shapes [3]. Microlens arrays consisting of circular nanostepped pyramids disposed in hexagonal arrangements have shown efficient bidirectional light focusing and maximal numerical apertures [4]. Considering the above cases, imaging techniques such as grid alignment and registration can be employed to determine the locations of objects in images. After targeting the resulting locations, further analysis by means of image segmentation is performed in order to extract the features of the image objects. Much research effort has been devoted to the development of image segmentation methods, and a wide range of applications exist in the field of image analysis and understanding. In medical image analysis for example, segmentation plays an important role in tasks such as visualization, measurement and reconstruction of shapes and volumes [5–7]; medical diagnosing [8,9]; and even image guided-surgery [10]. Recent research has proposed a large variety of techniques for image segmentation, which can be mainly classified as region-based segmentation, feature-based clustering or machine learning ML-based segmentation. Clustering-based techniques divide the image pixels based on their intensities into homogenous clusters while ignoring the spatial information, which makes them sensitive to image artifacts [11]. Considering its efficiency among the clustering-based algorithms, fuzzy C-means (FCM) has been widely used for image segmentation [12]. Improved variants which make use of the spatial information have been proposed to overcome the aforementioned limitations [13]. Regarding the machine learning approaches for image segmentation, both supervised and unsupervised ones are available. Unsupervised learning has the advantage of automatic segmentation without any prior knowledge of the object features within the training dataset [14]. Computationally expensive tools such as support vector machines, and probabilistic models such as Markov-random fields or Gaussian mixtures [15,16], are nevertheless used. The supervised ML techniques for image segmentation are more accurate and reliable, mainly since the input data are labeled and well known. Despite their computational complexity, deep learning algorithms, decision trees and Bayesian networks are broadly used in applied research [17]. Thus, computer-aided medical diagnosis is carried out on the basis of deep learning algorithms [18–21]. Bayesian networks [22–24] successfully conduct the detection of different geometries related to objects of interest in medical images, such as coronary arteries and retinal vasculature. A decision tree classifier can be used to obtain an adaptive threshold for the optic disc segmentation [25]. The advantages of both decision trees and conditional random fields have also been exploited [26,27]. In order to overcome the disadvantages of the supervised training, implicit deep supervision is assured by the hyper-densely connected convolutional neural network (CNN) proposed for natural image classification tasks [28,29], whereas level-set segmentation leads to semi-supervised CNN segmentation [30]. Considering various imaging technologies, there are cases when image objects are disposed of using a specific grid layout within the same image. In these cases, prior to image segmentation, a grid alignment or image registration procedure is mandatory. Thus, we focus on the registration and segmentation of hexagonal-grid-layout images.

As referred to the aforementioned image processing tasks (i.e., grid alignment and segmentation), when taking into account the large variety of state-of-the-art segmentation approaches, the main challenge is to choose the appropriate image-processing methods for feature extractions while considering both accuracy and computational complexity. In this context, the main findings are presented as follows.

1.1. Main Findings

The present paper proposes a set of image-transformation methods based on shock-filters applied on hexagonal-grid-layout images, aiming for both grid alignment and segmentation of the image objects. The PDE formalism of the shock filter, together with mathematical morphology, is used to evolve image profiles in order to determine the grid layout, by identifying a pair of sub-images, each containing a rectangular grid. The superposition of the two sub-images generates the initial image. Within each rectangular-grid-layout image, another procedure of profile evolution based on the same shock-filter formalism is used to confine the foreground information for each image object into an area of interest. For accurate segmentation of non-homogenous or irregular image objects, pixel intensity refinement classifies pixels to foreground or background, to better fit the true shape of the object. The proposed image-processing workflow was successfully applied to hexagonal-grid-layout microarray images, the hexagonal array structure of pixelated scintillation screens and hexagonal nanodisk–nanohole structure arrays. For microarray images classified as having a hexagonal grid layout, in spite of their advantages and intensive use [31,32], relatively few image-processing methods have been proposed. In [1], a spot-indexing algorithm successfully located microarray spots for hexagonal grids with different spacing and rotation. Giannakeas et al. also proposed a growing concentric hexagon algorithm [33], which detects spots in microarray images with a hexagonal grid layout. As compared with existing approaches, the main benefits of the proposed work are underlined as follows:

- The image-processing workflow represents a general solution for both rectangular and hexagonal grid alignment, which has been successfully applied to both medical images and images of material structures.
- The shock-filter-based grid alignment also delivers segmentation information, and guided by an autocorrelation procedure, it estimates the locations of missing objects within the hexagonal grid layout.
- The computational complexity required to determine the grid layout is minimized, taking into account that the PDEs are targeting the one-dimensional luminance function profiles,
- The segmentation accuracy was evaluated by computing the means and standard deviations of distances between the annotated and detected centers and showed improved results compared with state-of-the-art research.

In order to underline the main findings, the paper is organized as follows. Firstly, in the introductory section, the shock filters in the context of image segmentation and grid alignment are shortly summarized. Section 2 describes the shock-filter-segmentation approach applied for hexagonal-grid-layout microarray images. The results are shown in Section 3, in terms of segmentation accuracy, and the same section underlines the results obtained using the proposed methodology for two other types of hexagonal-layout images. In addition, the computational complexity of our approach is evaluated in the context of existing classical and machine learning solutions for grid alignment. Finally, the Conclusions section summarizes the main results.

1.2. Shock-Filter Fundamentals

An important task in image processing is to separate image areas containing background from foreground information. A shock-filter-based approach involves a process of selectively applying erosion or dilation in a localized manner in order to create a "shock" between two image areas, one belonging to a maximum and the other to a minimum. By iterating this process according to time increments, the resultant image reveals discontinuities only at the edges of the initial image. Moreover, the image areas delineated by the underlined edges become uniform in terms of pixel intensity values, delivering image segmentation information. Commonly, image enhancement processes, such as the one described before, are modeled through a partial differential equation (PDE).

Taking account of the importance of total variations in TV principles which appear for shock calculations in fluid dynamics, Osher and Rudin [34] have applied these ideas to image processing. This was revealed to be a useful method to restore discontinuities in images, such as edges. Their method relies on total variation techniques subject to a certain nonlinear and time-dependent partial differential equation:

$$\partial_t u = -|\nabla u| F(L(u)), \quad (1)$$

where $L(u)$ is a second-order, nonlinear elliptic operator whose zero-crossings correspond to edges. The filtering process (the edge enhancement process) is represented by the evolution of the initial image data $u_0(x)$ into a steady-state solution $u_\infty(x)$ as $t \to \infty$, through $u(x,t), t > 0$. The total variation of the solution,

$$TV(u) := \int_D |\nabla u| dx, \quad (2)$$

at any given state, is preserved and satisfies a maximum principle.

The steady state solution is achieved relatively fast, making it a good candidate for microarray image segmentation. As mentioned in [34], it is an $O(kN)$ method, where N is the number of points and k the number of time iterations. It was pointed out by [35] that the one-dimensional Equation of (1) with $F(u) := sgn(u)$, i.e.,

$$u_t = -sign(u_{xx})|u_x|, \quad (3)$$

is based on the image-enhancement algorithm of Kramer and Bruckner [36], which was proved to converge after a finite number of iterations.

From a morphological perspective, such a filter aims to produce a flow field which is directed from the interior of a region towards its edges, where it develops shock, generating a piecewise constant solution with discontinuities only at the edges of the original image. However, TV preserving methods suffer from fluctuations due to noise, which also create shocks. Therefore, Alvarez and Mazorra [37] considered the operator $L(u) = u_{xx}$ in (1) to be the Gaussian-smoothed version $L(G * u) = G * u_{xx}$, which supplemented the evolution with a noise-eliminating mean-curvature process, for which they proved that the discrete scheme is well-posed and satisfies a maximum—minimum principle. Smoothed morphological operators (dilations, erosions) for shock filters were also employed in [38] to enhance contours through smoothed ruptures, while preserving homogeneous regions.

2. Shock-Filter-Based Approach for Microarray Image Segmentation

Genes represent DNA sequences which determine particular characteristics in living organism, as follows: the genetic information is transmitted from nucleus to cytoplasm by an intermediate molecule called mRNA, which is further on translated into functional gene products known as proteins. Genes' expression levels are reflected in the amounts of respective mRNA present in each cell, providing information on the cell's biochemical pathways and its functions. By measuring mRNA levels for fully sequenced genomes printed on a solid surface, microarray technology is known to be a valuable tool for determining genes' functionality and expression levels in different conditions [39].

The workflow of a microarray experiment aiming at gene expression estimation starts with labeling mRNA samples with different fluorescent markers and hybridized onto the same solid surface. Depending on researchers' needs, gene expression analysis is performed by a one-color or a two-color experiment [40]. After hybridization, laser scanning is performed using one or two light sources with different wavelengths, one for each marker. The fluorescence induced by each light source is captured, and a composite image is produced. The microarray image thus obtained represents a collection of microarray spots, each spot corresponding to a specific gene.

Current technologies allow accurate fluorescence quantification [41], considering different numbers of spots at different densities printed onto a microarray slide, offering a broad view that represents all known genes and their transcripts in the human genome. Two spot layouts can be distinguished: the rectangular grid layout and the hexagonal grid layout, corresponding to the single-density and double-density microarrays, respectively. Commonly, microarray manufacturers use single-density microarrays, where spots are disposed in a rectangular grid. Nevertheless, taking into account that no matter the grid format, sensitivity and performance are preserved, the hexagonal grid is considered advantageous compared to the rectangular grid, since it allows more probes to be printed onto the same surface. Later-stage image-processing techniques, including object registration and segmentation, are used to estimate gene expression. Logical coordinates are determined for each spot of the microarray image, and the segmentation classifies pixels either as foreground, representing the DNA spots, or as background. A great deal of research has been conducted for processing microarray images having a rectangular grid layout. Bariamis et al. [42] used a SVM approach for automatic grid alignment. That, and an approach consisting of optimal multilevel thresholding, followed by a refinement procedure and hill climbing [43,44], lead to accurate grid detection. For spot segmentation, adaptive pixel clustering [45,46], snake fisher models [47,48], 3D spot modeling [49], bio-inspired algorithms [50] and Markov random field modeling [51] were proposed by state-of-the-art research. Nevertheless, considering the reduced publications tackling the hexagonal-grid-layout images [1,33,52], as underlined in the main findings sub-section, we propose a general approach for hexagonal- and rectangular-grid-layout microarray images.

2.1. Materials and Methods

The microarray scanning process delivers $16x$ bit gray-scale images, in TIFF format, in which spot fluorescence levels are captured as intensities of the image pixels which fall within the microarray spot. To identify the position, intensity and background intensity values of each microarray spot, preprocessing techniques, image registration and image segmentation approaches are applied. The preprocessing methods aim at image enhancement based on logarithmic and top-hat image transforms to further improve the accuracy of spot detection. Further on, using the shock-filtered image profiles, each spot line is detected, and making use of a refinement procedure based on morphological filtering, the original image is decomposed into two sub-images, each of them containing a rectangular grid of spots. Next, the segmentation classifies pixels as belonging to the microarray spot or to the image background using the same PDE formalism specific to the shock filters, and the segmentation accuracy metrics are computed. The entire workflow can be depicted in Figure 1. The subsequent sub-sections detail the proposed image-processing techniques for automatic hexagonal-grid-layout microarray image analysis.

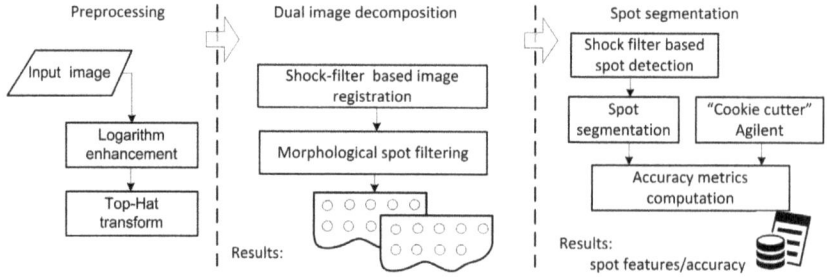

Figure 1. Image-processing workflow for hexagonal-grid-layout image registration and segmentation.

2.2. Preprocessing

Weakly expressed spots and image rotation are common characteristics of the microarray images delivered by the scanning process. Thus, to enhance weekly expressed spots, a logarithm point-wise transform was applied on the image, followed by an intensity adjustment procedure so that the intensity histogram would fit the full dynamic range of the image (the dynamic range was from 1 to 2^{16}). Moreover, a top hat transform was used to reduce the background influence on the microarray spots [53]. In case of misaligned input image, a rotation detection and correction algorithm (the Radon transform) was employed [54]. Figure 2 shows the results of the aforementioned preprocessing techniques for the *US218398* microarray image.

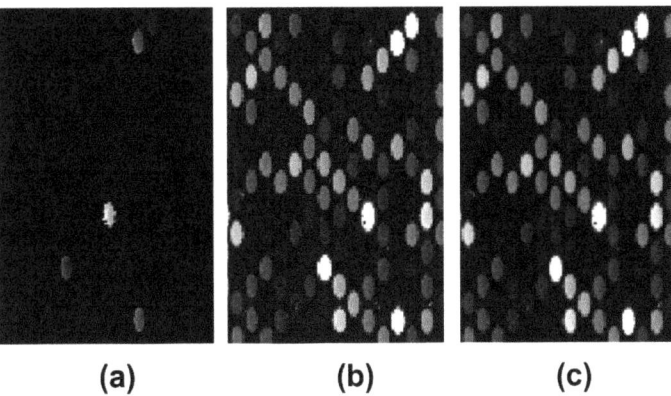

Figure 2. Image preprocessing techniques applied to the *AT218398* microarray image: (**a**) original image, (**b**) logarithmically transformed and normalized image, (**c**) top-hat transformed image.

2.3. Grid-Line Detection for Image Registration

Let $I_P = p_{i,j}$ be the preprocessed, $M \times N$-pixel microarray image, with $p_{i,j}$ being the 16-bit intensity of the pixel found on row i and column j within the microarray image. The vertical image profile was computed as described by the equation

$$V(i) = \frac{1}{N} \sum_{j=0}^{N-1} p_{i,j}, \quad i = 0, \ldots, M-1, \tag{4}$$

whereas the horizontal profile is described by

$$H(j) = \frac{1}{M} \sum_{i=0}^{M-1} p_{i,j}, \quad j = 0, \ldots, N-1. \tag{5}$$

The vertical profile is evolved further on using the shock-filter partial differential Equation (3) given by

$$u_t = -sign(u_{xx})|u_x|, \tag{6}$$

where u_x and u_{xx} are the first- and the second-order derivatives of the image profile. The initial value of u at time $t = 0$ is the image's luminance function profile $V(i)$.

Let the shock-filtered profile of the preprocessed microarray image vertical profile be denoted by $SFP = V(i)$. The inflexions points are marked within the SFP, and their locations respect a specific pattern which reveals the borderlines for the separation of lines of spots. Figure 3 shows how the spots' line separation is performed. The total number of lines of spots (see the line presented in Figure 3d), within the overall microarray image is considered to be n. The positions of all inflexion points detected along the profiles are stored in an uni-dimensional vector *pos*, and the resultant vector size is $2n$. To define each line of spots, the positions of four inflexions points are considered. Thus, as shown in Figure 3a, the uneven lines of spots are defined as the positions $4u - 1$, $4u$, $4(u + 1) - 1$ and $4(u + 1)$ within the *pos* vector. u ranges from 1 to $n/2 - 1$. In a similar manner, the positions $4u - 3$, $4u - 2$, $4u + 1$ and $4u + 2$ define the even lines of spots. The average position between $4u - 1$ and $4u$ and the average position between $4(u + 1) - 1$ and $4(u + 1)$ mark the positions of the horizontal separation's lines for the uneven line of spots denoted by u. As presented in Figure 3b, the continuous lines over the 90^0-rotated section of the original image are the separation lines for spot line u. Based on the aforementioned separation lines, all even and uneven lines of spots were detected. An example of such a line is presented in Figure 3c. It can be observed that the detected uneven lines of spots also include half of the spots within the neighboring even lines of spots. A similar situation describes the even lines of spots. In order to decompose the microarray image in two sub-images, one including the even lines of spots and the other one with the uneven line of spots, mathematical morphology is applied.

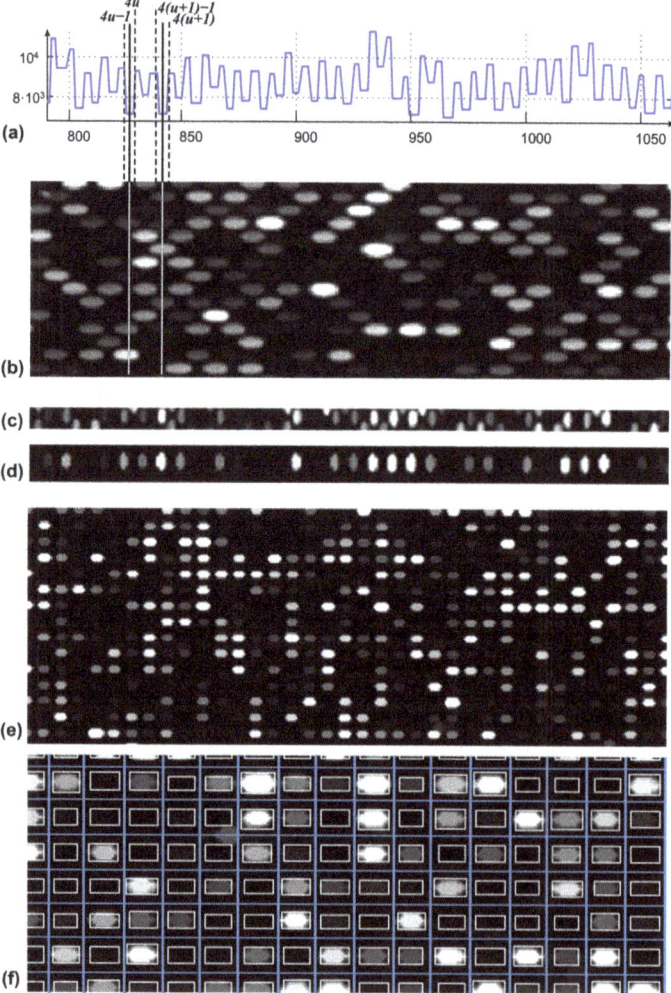

Figure 3. Hexagonal grid segmentation process: (**a**) horizontal profile of (**b**) the preprocessed microarray image, (**c**) selected lines based on separation lines, (**d**) morphological exclusion of neighboring spots, (**e**) final rectangular even spots I_{ev} and (**f**) segmentation of the rectangular image.

As denoted by Figure 3e, each microarray spot is defined by an elliptic shape characterized by the horizontal radius a and the vertical radius b, confined in a rectangular area. In the subsequent step, the average horizontal and vertical radii a and b considering all spots are estimated based on the autocorrelation applied on vertical and horizontal profiles of the image, described by Equations (4) and (5), respectively. A structuring element having an elliptical shape with a and b radii was defined. The upper and lower parts of each line of spots were padded with $b/2$ lines of pixels; each pixel has the lowest intensity value. The resulting image was morphologically opened with respect to the defined structural element. The outcome was similar to the lines of spots from Figure 3d, where the half spots from the neighboring lines were excluded. Further on, the original image was reconstructed once using the even lines of spots and once using the even lines of spots. Each resulted sub-image is characterized by a rectangular grid layout of its microarray spots (see Figure 3e). The sub-images containing the even lines of spots and uneven lines of spots are denoted by I_{ev} and I_{uev}, respectively.

By applying the shock filter to the original vertical profile, the inflexion points at the positions $4u$ and $4(u+1)-1$ within the vertical profile (see Figure 3a,d) can determine the locations on the vertical axes for all the spots from the column of spots denoted by u. Within the sub-image containing the column u, the horizontal profile h, as referred to in Figure 3d, is evolved using shock filters to determine the positions of spots on the x axis. The autocorrelation-based approach applied on the h image profile, described in [55], is used to estimate the positions of missing spots. Consequently, for each microarray spot position within the initial hexagonal grid, an area of interest, denoted by S, which confines the microarray spots, is determined according to Figure 3e. On each area S, image segmentation is applied next to determine the pixels which belong to the microarray spots and which belong to the spot's local background. The aforementioned procedure is consistent with the "cookie cutter" approach used by the software platform Agilent Feature Extraction (FE) and detailed in [56].

2.4. Spot Segmentation

The shock filters deliver segmentation information by identifying simple geometric objects of rectangles for the entire set of microarray spots. For accurate segmentation of spots with spatial non-homogeneous intensity distribution and irregular shapes, a simple threshold procedure is introduced for the S area. As demonstrated in [55], pixels intensity refinement yields a rearrangement of pixels to the foreground and background that better fits the true shape of the spots.

3. Results and Discussions

Our study included a set of four microarray images used for one-color analysis of gene expression data performed using Agilent Technologies (G2505C scanner) on homo sapiens samples. The samples were printed on microarray glass slides formatted with four high-definition 44K, arrays and the images within the dataset have a hexagonal grid layout.

3.1. Microarray Image Registration and Segmentation Accuracy

We evaluate the results obtained using the proposed hexagonal grid alignment procedure compared with state-of-the-art results and with the results delivered by Agilent Feature Extraction Software (FE). Spot centers were annotated by FE for each microarray image from our dataset. The value d_i representing the distance between an annotated spot's center and the one determined using our proposed approach was computed for each spot (i) included in the image under analysis. The mass center's locations (m_i) was determined for each spot I_i, and compared to the mass center's location (m_i^{FE}) determined by FE software. The mean Euclidean distance m_E between the two mass centers for the whole population of spots was used as a metric for the accuracy evaluation. The mean Euclidean distance m_E was measured in pixels, and it is denoted by the equation

$$m_E = \frac{1}{N_S} \sum_{i=1}^{N_S} |m_i - m_i^{FE}|, \quad (7)$$

Table 1 shows the average distance $d = 1.52$ for the whole population of spots included in our dataset and underlines that the proposed approach delivered the lowest standard deviation for the distance d distribution over the whole population of spots compared with state-of-the-art hexagonal grid alignment approaches. The methodology's accuracy given by the percentage of spots correctly positioned by the grid alignment procedure on the selected images was 100%.

Our proposed automatic image processing approach for hexagonal-grid-layout microarray images is evaluated in terms of accuracy and reproducibility, with regard to the whole population of spots within the microarray dataset. The mean spot intensity value I_i was computed by subtracting the mean intensity value of the background pixels and the mean intensity value of pixels which fall within the microarray spots. The range of i is from 1 to N_S, where N_S is the total number of spots within the microarray image. The results are compared with the ones delivered by the Agilent Feature Extraction software (FE) for the same set of images. Consequently, the accuracy estimation of our proposed segmentation method was performed independently on each microarray image from our dataset.

The regression ratio (R) represents an independent measure defined by the slope of the least-squares best-fit regression line of the fluorescence intensity values for each pixel against each other for a given microarray spot. The regression ratio indicates individual spot quality. Considering the regression pixels used to calculate R values, the coefficient of determination R^2 for the least-squares-regression fit of a microarray spot is defined as the square of the correlation coefficient and ranges in value between 0 and 1 [57]. For validating our approach, we correlated the coefficients of determination computed by our approach with the ones determined by FE for the entire population of microarray spots within each microarray image. Let R^2 be the coefficient of determination computed using the proposed approach and R_{FE}^2 be the coefficient of determination annotated by FE. The correlation coefficient, together with the mean difference between our results and the FE results, is described by:

$$r = Pearson(R^2, R_{FE}^2), \quad (8)$$

$$agv_{diff} = \frac{1}{N_s} \sum_{i=1}^{N_s} |R_i - R_{FE}^2|, \quad (9)$$

The Pearson coefficient exceeded values of 0.98, and hence, indicated a high correlation of our data (intensities) with the reference values. Moreover, the reproducibility of the segmentation technique was quantified by means of mean absolute error MAE and coefficient of variation CV, as presented in Equations (10) and (11), according to [58,59], respectively. The lower the MAE and CV values are, the better the performance of the proposed method. $r = 4$ replicates of the microarray experiment were used for evaluation. MAE indicates the spot sameness of the spot's intensities, Equation (10), where I_j is the mean spot intensity over the j experimental replicates and $\underline{I}$ is the overall mean, computed from the means of the spots within all the r replicates.

$$MAE_{spot} = \frac{1}{r} \sum_{j=1}^{r} |I_j - \underline{I}|, \quad (10)$$

Spots intensity variations are expressed by the CV parameter denoted by Equation (11), based on the standard deviation σ of spot intensity with subtracted background and the mean spot intensity v.

$$CV_{spot} = \frac{\sigma}{v}. \quad (11)$$

Table 1. Evaluation of the image registration and segmentation accuracy.

Exp. ID	$r(I_i, I_i^{FE})$	avg_{diff}	avg. MAE	avg. MAE^{FE}	CV	CV^{FE}
FE18398	0.988	0.075			0.420	0.412
FE18399	0.993	0.029			0.395	0.406
FE18400	0.982	0.093	536	524	0.414	0.385
FE18401	0.994	0.046			0.392	0.397

The small CV values correspond to small variation among the pixel intensity values for given microarray spots, showing the reliability of the proposed grid alignment procedure together with the spot segmentation approach. As referred to for the MAE values given in Table 1, smaller values are obtained compared with the full dynamic range of the microarray spot (i.e., spot intensity values range is 1 to 2^{16}).

To evaluate the performance of the proposed methodology for spot detection compared with the one already available, the means and the standard deviations of the distances between the centers of the Agilent annotations and the detected spots centers were computed for the entire datasets included in Table 1 and denoted by FEdata. Moreover, the accuracy of the detection denoted by the ratio of correctly identified microarray spots and the total number of spots was also computed. The results are included in Table 2, together with the results delivered by all approaches referenced in [52], employed for the detection of microarray spots disposed in both rectangular and hexagonal grids. A mean of 1.52 pixels with a standard deviation of less than 1 pixel and a spot detection accuracy of 100% underline the superior performance of our approach.

Table 2. Results of the proposed grid alignment methodology: means and standard deviations of distances between annotated and detected centers, and accuracy.

Reference/ Dataset	Method Description	Image, Grid Type	Image Size (M, N)/ Number of Spots	Spot Diam.	Metric	Value
SMD [42,60]	Gridding based on support vector machines and genetic algorithms	Real, Rectangular grid	1980 × 1917 9196	10	Mean Std Acc	2.52 2.59 96.4
Nycter [61]	K-nearest neighbor	Synthetic, Rectangular grid	3188 × 9552 576,756	14	Mean Std Acc	1.77 1.16 98.9
CNV370 [52]	Voronoi diagrams	Real, Rectangular grid	2800 × 2800 9216	6	Mean Std Acc	1.88 0.82 99.8
Nycter	Gridding based on support vector machines and genetic algorithms	Real, Rectangular grid	2800 × 2800 9216	14	Mean Std Acc	1.91 1.03 99.3
SMD Nycter [52]	Voronoi diagrams	Real, Synthetic with rectangular and hexagonal grids	various sizes	14	Mean Std Acc	1.94 2.32 97.5
FEdata (present)	Shock filter driven by mathematical morphology	Real, Hexagonal	1650 × 4320 9196	14	Mean Std Acc	1.52 0.68 100

3.2. Shock Filters as a General Approach for Hexagonal-Grid-Layout Registration

Hexagonal grid layouts are becoming increasingly popular as the fields of nano- and meta-materials develop. In cell-cluster-array fabrication, self-assembled hexagonal superparamagnetic cone structures induce a local magnetic field gradient which inhibits the cancer cells' migration [2]. For materials science applications, benefits such as increased optical performance and material resistance are added by hexagonal grid structures. Microlens arrays consisting of circular nanostepped pyramids disposed in hexagonal arrangements have showed efficient bidirectional light focusing and a maximal numerical aperture [4]. Considering the above cases, imaging techniques such as grid alignment and registration can be employed to determine the locations of image objects and to analyze and validate the respective structures of the materials in question.

Thus, in pixelated CsI(Tl) scintillation screens for X-ray imaging, the resolution for the pixelated screen with the hexagonal array structure is approximately 8.5% higher than for the screen with the square array structure [3]. Moreover, ultrathin hexagonal nanodisk–nanohole hybrid structure arrays have been employed for developing a novel plasmonic metasurface for subtractive color printing [62]. For the hexagonal-grid-layout image segmentation approaches, a crucial challenge is to develop a robust method which targets various types of hexagonal layout. In order to underline the generality of our proposed approach, both the hexagonal array structure of pixelated CsI(Tl) scintillation screens and the ultrathin hexagonal nanodisk-nanohole hybrid structure were processed using the proposed workflow. The obtained results are presented in Figure 4.

Regarding the main limitations of the proposed approach, the small size of the datasets considered for evaluation is mentioned. Nevertheless, the similarities between the segmentation accuracy metrics delivered by our approach and the ones delivered by the commercial Agilent Feature Extraction Software for over 100,000 microarray spots (Table 1) show the reproducibility of the results. The generality of the approach has also been proven by the results presented in Figure 4 for two other types of hexagonal grid layout. Since the propose approach was designed for microarray images, extensive testing and validation procedures are needed for segmentation procedures applied for other types of hexagonal-grid-layout images (e.g., ultrathin hexagonal nanodisk-nanohole hybrid structure arrays, pixelated CsI(Tl) scintillation screens), which are outside the the scope of current paper. Another drawback of the proposed method is that images with skewed, rotated or irregular hexagonal layouts require special attention. Rotation correction using the Radon transform is included within the proposed workflow, but irregular and skewed layouts are still not addressed. De-skewing algorithms are available [63,64], whereas for the irregular layouts, correction algorithms are to be designed based on the specifics of the irregularities.

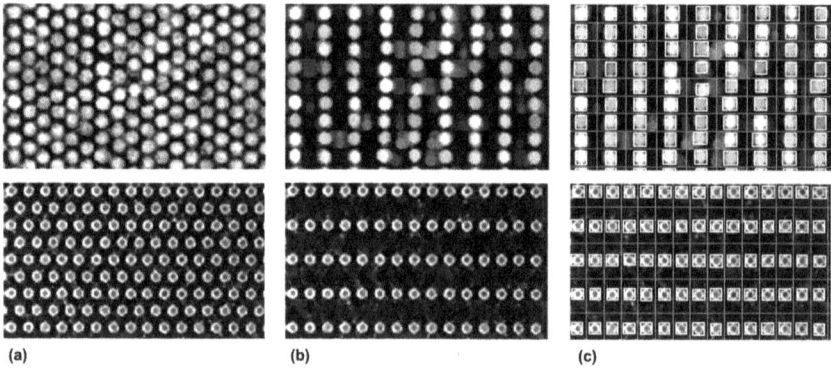

Figure 4. (**a**) Preprocessed images registered from hexagonal-array structure of pixelated CsI(Tl) scintillation screens (**top**) and hexagonal nanodisk–nanohole hybrid structure arrays (**bottom**). (**b**) Dual image decomposition based on shock filters driven by mathematical morphology; (**c**) segmentation.

3.3. Computational Complexity Analysis for the Hexagonal-Grid-Layout Image Segmentation

A large variety of image segmentation approaches are available, from complex ones such as deep learning approaches to reduced complexity ones which perform on image profiles, for example. Moreover, as detailed in [65,66], an interest in reducing the complexity of machine learning algorithms is shown. Thus, the user has to carefully evaluate the image analysis task and choose the appropriate processing approach. For hexagonal-grid-layout image segmentation, the computational complexity is estimated for the state-of-the-art approaches and compared with the proposed approach, in order to offer an overview of available methods from the computational complexity perspective. The comparison is detailed as follows.

Firstly, we estimated the computational complexity for our proposed approach for hexagonal-grid-layout microarray image registration. Considering a given $M \times N$-pixel image, the obtained results are compared with both the classical state-of-the-art approaches [1,33,52] and the machine learning approaches [67,68] for microarray spot segmentation. The results are summarized in Table 3. In our case, the computational cost for the image registration procedure is detailed as follows:

(i) The morphological opening procedure and the autocorrelation spot size estimation cost are given by the upper bound function $f(M,N) = (2S_e MN + 4MN)s$, with s representing one computational step, and S_e representing the size of the structural element used for morphological filtering.

(ii) The computational complexity of the shock-filter-based procedure for grid alignment is based on the number of microarray spots found on each line and in each column of spots, denoted by α and β, respectively. Let d be the average of the microarray spot diameter and $2d$ be the average width for a line or a column of spots. We computed for each spot line and spot column, the horizontal and vertical image profiles, respectively, with the total complexity of $2\alpha dM + 2\beta dN = 4MN$. Shock filters were applied to each of the determined profiles having a complexity of $p(\alpha M + \beta N)$, where $p\alpha M$ represents p iterations performed on the number of α profiles (i.e., one profile for each line of spots), and each profile was of size M. This led to the estimation of the computational cost given by $f(M,N) = 6MNs + pd(\alpha M + \beta N)s$, with $p > d$. Consequently, the order of growth for the total computational cost was approximated to $O(2S_e MN + p(\alpha M + \beta N))$, and it represents the total computational complexity of the proposed method.

In [1,33] the Voronoi diagrams are used for the grid alignment in for hexagonal-grid-layout microarray images. According to the analysis performed in [69], the computational complexity is given by the order of growth of the computational cost $O(f(S)) = O(S^2 log(S))$, where S represents the total number of spots (i.e., for our images $S = 44{,}000$). The main disadvantage is that a unique region is obtained if weekly expressed spots are grouped together in the same area. This is overcome by the approach proposed in [52], where a preliminary step is added to the Voronoi diagram algorithm. This step detects all the highly expressed spots, which represent starting points for growing similar hexagonal areas for weakly expressed spots. In terms of the computational cost, the following term dMN is added to the cost function, leading to the total computational cost of $f(M,N,S) = S^2 log(S) + dMN$.

Considering the machine learning approaches, the computational cost is given by both the training and prediction steps. Thus, according to Table 3, the order of growth for the machine learning-based grid alignment procedure has two terms, corresponding to the training and the prediction. For support vector machines, the total computational cost for grid alignment is given by $f(n,M,N) = n(MN)^2 + nNM$, where n is the number of grid lines [70].

Table 3. Computational complexity analysis for microarray grid alignment.

Reference	Method	Cost Arguments	Order of Growth
[1,33]	Voronoi diagrams	S	$O(S^2 \log S)$
[52]	Growing concentric hexagons	$S, (M, N)$	$O(S^2 \log S + dMN)$
[43,71]	Support vector machines	(M, N)	$O(n(MN)^2 + nMN)$
[72]	Evolutionary algorithms	$S, (M, N)$	$O(S^2 + dMN)$
[67,68]	Deep neural Networks	-	-
present	Shock filters driven by morphology	$S_e, (M, N)$	$O(2S_e MN + p(\alpha M + \beta N))$

Notes: S—represents total number of spots within the image under analysis (in round number 44,000); the pair $(M, N) = (1650, 4320)$ corresponds to the image size in pixels; $S_e = 144$ represents the size of the structuring element; the parameters denoted by lowercase letters are at least one order of magnitude smaller then the lowest one, S_e.

For the evolutionary algorithms, the gridding approach for microarrays [72] differs from the classical ones, since it does not involve any 1D projection of the image. The approach includes a measure of fitness for possible grids to achieve a robust grid alignment against high levels of image noise and a high percentage of weakly expressed spots. Considering the fitness function, the evolutionary algorithm locates the regular grid that best fits a set of spot center coordinates. According to [73], as referred to the algorithm's performance in terms of time complexity, the order of growth can be reduced to $O(m^2)$, with m being the total number of graph edges. By approximating m with S ($m > S$), the total number of spots, and considering the preliminary computational steps which consist of image dilation and an approximate spot spacing calculation [72], the order of growth for the computational complexity of $O(S^2 + dMN)$ is obtained.

Deep neural networks applied for microarray image analysis are discussed next in terms of computational complexity. To our knowledge, state-of-the-art research does not include deep neural networks applied for microarray grid alignment. Nevertheless, deep learning is used for bio-medical image segmentation [66], and, more precisely, it is also applied for microarray spot classification [68]. Since such approaches do not serve as grid alignment tools, the computational complexity levels of the deep learning approaches used for microarray spot segmentation were computed but not added to the Table 3 summary of grid alignment approaches [67,68]. Calling s the number of training samples, f the number of features and n_{l_i} the number of neurons in layer i, we have the approximation for the computational complexity given by $O(s^3 + fn_{l_1} + n_{l_1}n_{l_2} + \ldots)$, considering both the training procedure and the prediction. Taking into account the increased complexity, there is a great interest in reducing deep learning complexity, as shown in [66]. Herein, it is demonstrated that the computational complexity of the convolutional neural networks can be reduced by a factor of eight while achieving accurate bio-medical image segmentation. Even so, the computational complexity of our approach, which delivers segmentation information, as the results underline, is at least one order of magnitude lower than that of the deep learning approaches.

Let us consider the size of the image under analysis given by the (M, N) pair, with $M = 1650$ and $N = 4320$. As referred to in Table 3, we underline the cost arguments S, S_e and n having the values 44,000, 144 and 120, respectively. Consequently, as denoted in Table 3, reduced computational complexity is achieved by the proposed grid-alignment approach, despite the iterative nature, considering that shock filters are applied on 1D image profiles. Thus, if the training procedure is excluded, the computational complexity of the proposed approach is similar to that of the support vector machine [43,71], whereas compared with the other approach, the computational complexity is at least one order of magnitude lower. It is to be noticed that the grid alignment is accurately performed for weekly expressed spots, due to the autocorrelation refinement procedure, and the accuracy is comparable with the machine learning approaches for grid alignment.

4. Conclusions

In this paper, we presented a novel segmentation approach for estimation of gene expression levels based on shock filters, making it applicable to both hexagonal and rectangular grid layouts. For hexagonal grids, the original image is divided into two rectangular grid images, such that their overlap constitutes the initial image. The proposed method was validated using specific quality measures such as the coefficient of variation and mean absolute error, on a dataset which includes hexagonal-grid-layout microarray images. The spot segmentation results obtained were compared with the ones delivered by Agilent Feature Extraction platform. Correlation coefficients between spot features (e.g., foreground intensity) and the mean distance between spot location showed very good agreement. Moreover, the computational cost of the described method was analyzed and compared with state-of-the-art microarray spot segmentation methods, ranging from classical to deep learning ones. Significantly lower computational complexity was achieved compared with the discussed methods. The segmentation accuracy, however, was comparable with those of machine learning approaches.

Author Contributions: Conceptualization, A.B., C.C., R.G., O.B., F.T. and B.B.; Methodology, R.G., O.B., F.T. and B.B.; Software, A.B., C.C., R.G. and B.B.; Validation, A.B., C.C., R.G. and B.B.; Investigation, B.B.; Resources, O.B.; Writing—original draft, A.B. and C.C.; Writing—review & editing, R.G., O.B., F.T. and B.B.; Supervision, B.B.; Funding acquisition, F.T. All authors have read and agreed to the published version of the manuscript.

Funding: We would like to acknowledge the financial support from: the Ministry of Research, Innovation and Digitization, through Program 1 Development of the National Research and Development System—Funding Projects for Excellence in RDI—Contract No. 37PFE/30.12.2021, through "Nucleu" Programe within the National Plan for Research, Development and Innovation 2022-2027—project PN 23 24 02 01, through UEFISCDI project number PN-III-P2-2.1-SOL-2021-0084, and through European and International Cooperation Program—Horizon 2020 Subprogram-ERANET-TRANSCAN3-TANGERINE project—Contract No. 319/2022.

Conflicts of Interest: The authors declare no conflict of interest.

References

1. Galinsky, V.L. Automatic registration of microarray images. II. Hexagonal grid. *Bioinformatics* **2003**, *19*, 1832–1836. [CrossRef]
2. Chen, Y.; Hu, Z.; Zhao, D.; Zhou, K.; Huang, Z.; Zhao, W.; Yang, X.; Gao, C.; Cao, Y.; Hsu, Y.; et al. Self-assembled hexagonal superparamagnetic cone structures for fabrication of cell cluster arrays. *ACS Appl. Mater. Interfaces* **2021**, *13*, 10667–10673. [CrossRef] [PubMed]
3. Chen, H.; Gu, M.; Liu, X.; Zhang, J.; Liu, B.; Huang, S.; Ni, C. Simulated performances of pixelated CsI (Tl) scintillation screens with different micro-column shapes and array structures in X-ray imaging. *Sci. Rep.* **2018**, *8*, 16819. [CrossRef]
4. Ahmed, R.; Yetisen, A.K.; Butt, H. High numerical aperture hexagonal stacked ring-based bidirectional flexible polymer microlens array. *ACS Nano* **2017**, *11*, 3155–3165. [CrossRef]
5. Wu, H.; Pan, J.; Li, Z.; Wen, Z.; Qin, J. Automated Skin Lesion Segmentation Via an Adaptive Dual Attention Module. *IEEE Trans. Med. Imaging* **2021**, *40*, 357–370. [CrossRef] [PubMed]
6. Budai, A.; Suhai, F.I.; Csorba, K.; Toth, A.; Szabo, L.; Vago, H.; Merkely, B. Fully automatic segmentation of right and left ventricle on short-axis cardiac MRI images. *Comput. Med. Imaging Graph.* **2020**, *85*, 101786. [CrossRef] [PubMed]
7. Yu, F.; Liu, M.; Chen, W.; Wen, H.; Wang, Y.; Zeng, T. Automatic Repair of 3-D Neuron Reconstruction Based on Topological Feature Points and an MOST-Based Repairer. *IEEE Trans. Instrum. Meas.* **2021**, *70*, 5004913. [CrossRef]
8. Zhou, Y.; Chen, H.; Li, Y.; Liu, Q.; Xu, X.; Wang, S.; Yap, P.T.; Shen, D. Multi-task learning for segmentation and classification of tumors in 3D automated breast ultrasound images. *Med. Image Anal.* **2021**, *70*, 101918. [CrossRef]
9. Guo, Z.; Zhang, H.; Chen, Z.; van der Plas, E.; Gutmann, L.; Thedens, D.; Nopoulos, P.; Sonka, M. Fully automated 3D segmentation of MR-imaged calf muscle compartments: Neighborhood relationship enhanced fully convolutional network. *Comput. Med. Imaging Graph.* **2021**, *87*, 101835. [CrossRef]
10. Qin, F.; Lin, S.; Li, Y.; Bly, R.A.; Moe, K.S.; Hannaford, B. Towards Better Surgical Instrument Segmentation in Endoscopic Vision: Multi-Angle Feature Aggregation and Contour Supervision. *IEEE Robot. Autom. Lett.* **2020**, *5*, 6639–6646. [CrossRef]
11. Ren, H.; Hu, T. A Local Neighborhood Robust Fuzzy Clustering Image Segmentation Algorithm Based on an Adaptive Feature Selection Gaussian Mixture Model. *Sensors* **2020**, *20*, 2391. [CrossRef] [PubMed]
12. Zhang, X.; Sun, Y.; Liu, H.; Hou, Z.; Zhao, F.; Zhang, C. Improved clustering algorithms for image segmentation based on non-local information and back projection. *Inf. Sci.* **2021**, *550*, 129–144. [CrossRef]

13. Bai, X.; Zhang, Y.; Liu, H.; Wang, Y. Intuitionistic Center-Free FCM Clustering for MR Brain Image Segmentation. *IEEE J. Biomed. Health Inform.* **2019**, *23*, 2039–2051. [CrossRef] [PubMed]
14. Jiao, X.; Chen, Y.; Dong, R. An unsupervised image segmentation method combining graph clustering and high-level feature representation. *Neurocomputing* **2020**, *409*, 83–92. [CrossRef]
15. Daniels, C.J.; Ferdia, A.G. Unsupervised Segmentation of 5D Hyperpolarized Carbon-13 MRI Data Using a Fuzzy Markov Random Field Model. *IEEE Trans. Med. Imaging* **2018**, *37*, 840–850. [CrossRef]
16. Relan, D.; Relan, R. Unsupervised sorting of retinal vessels using locally consistent Gaussian mixtures. *Comput. Methods Programs Biomed.* **2021**, *199*, 105894. [CrossRef]
17. Xing, F.; Xie, Y.; Su, H.; Liu, F.; Yang, L. Deep Learning in Microscopy Image Analysis: A Survey. *IEEE Trans. Neural Netw. Learn. Syst.* **2018**, *29*, 4550–4568. [CrossRef]
18. Lazic, I.; Agullo, F.; Ausso, S.; Alves, B.; Barelle, C.; Berral, J.L.; Bizopoulos, P.; Bunduc, O.; Chouvarda, I.; Dominguez, D.; et al. The Holistic Perspective of the INCISIVE Project; Artificial Intelligence in Screening Mammography. *Appl. Sci.* **2022**, *12*, 8755. [CrossRef]
19. Hagerty, J.R.; Stanley, R.J.; Almubarak, H.A.; Lama, N.; Kasmi, R.; Guo, P.; Drugge, R.J.; Rabinovitz, H.S.; Oliviero, M.; Stoecker, W.V. Deep Learning and Handcrafted Method Fusion: Higher Diagnostic Accuracy for Melanoma Dermoscopy Images. *IEEE J. Biomed. Health Inform.* **2019**, *23*, 1385–1391. [CrossRef]
20. Sari, C.T.; Gunduz-Demir, C. Unsupervised Feature Extraction via Deep Learning for Histopathological Classification of Colon Tissue Images. *IEEE Trans. Med. Imaging* **2019**, *38*, 1139–1149. [CrossRef]
21. Singh, M.; Pujar, G.V.; Kumar, S.A.; Bhagyalalitha, M.; Akshatha, H.S.; Abuhaija, B.; Alsoud, A.R.; Abualigah, L.; Beeraka, N.M.; Gandomi, A.H. Evolution of Machine Learning in Tuberculosis Diagnosis: A Review of Deep Learning-Based Medical Applications. *Electronics* **2022**, *11*, 2634. [CrossRef]
22. Jeon, B.; Jang, Y.; Shim, H.; Chang, H.J. Identification of coronary arteries in CT images by Bayesian analysis of geometric relations among anatomical landmarks. *Pattern Recognit.* **2019**, *96*, 106958. [CrossRef]
23. Uslu, F.; Bharath, A.A. A recursive Bayesian approach to describe retinal vasculature geometry. *Pattern Recognit.* **2019**, *87*, 157–169. [CrossRef]
24. Obayya, M.; Haj Hassine, S.B.; Alazwari, S.; Nour, M.K.; Mohamed, A.; Motwakel, A.; Yaseen, I.; Sarwar Zamani, A.; Abdelmageed, A.A.; Mohammed, G.P. Aquila Optimizer with Bayesian Neural Network for Breast Cancer Detection on Ultrasound Images. *Appl. Sci.* **2022**, *12*, 8679. [CrossRef]
25. Pathan, S.; Kumar, P.; Pai, R.; Bhandary, S.V. Automated detection of optic disc contours in fundus images using decision tree classifier. *Biocybern. Biomed. Eng.* **2020**, *40*, 52–64. [CrossRef]
26. Liu, F.; Lin, G.; Qiao, R.; Shen, C. Structured learning of tree potentials in CRF for image segmentation. *IEEE Trans. Neural Netw. Learn. Syst.* **2017**, *29*, 2631–2637. [CrossRef] [PubMed]
27. Peptenatu, D.; Andronache, I.; Ahammer, H.; Taylor, R.; Liritzis, I.; Radulovic, M.; Ciobanu, B.; Burcea, M.; Perc, M.; Pham, T.D.; et al. Kolmogorov compression complexity may differentiate different schools of Orthodox iconography. *Sci. Rep.* **2022**, *12*, 10743. [CrossRef]
28. Dolz, J.; Gopinath, K.; Yuan, J.; Lombaert, H.; Desrosiers, C.; Ben Ayed, I. HyperDense-Net: A Hyper-Densely Connected CNN for Multi-Modal Image Segmentation. *IEEE Trans. Med. Imaging* **2019**, *38*, 1116–1126. [CrossRef]
29. Dai, Y.; Jin, T.; Song, Y.; Sun, S.; Wu, C. Convolutional Neural Network with Spatial-Variant Convolution Kernel. *Remote Sens.* **2020**, *12*, 2811. [CrossRef]
30. Ma, J.; Yang, X. Automatic dental root CBCT image segmentation based on CNN and level set method. In Proceedings of the Medical Imaging 2019: Image Processing, San Diego, CA, USA, 19–21 February 2019; Angelini, E.D., Landman, B.A., Eds.; International Society for Optics and Photonics, SPIE: Bellingham, WA, USA, 2019; Volume 10949, p. 109492N. [CrossRef]
31. Joseph, S.M.; Sathidevi, P.S. An Automated cDNA Microarray Image Analysis for the determination of Gene Expression Ratios. *IEEE ACM Trans. Comput. Biol. Bioinform.* **2021**, *20*, 136–150. [CrossRef]
32. Almugren, N.; Alshamlan, H. A Survey on Hybrid Feature Selection Methods in Microarray Gene Expression Data for Cancer Classification. *IEEE Access* **2019**, *7*, 78533–78548. [CrossRef]
33. Giannakeas, N.; Kalatzis, F.; Tsipouras, M.G.; Fotiadis, D.I. Spot addressing for microarray images structured in hexagonal grids. *Comput. Methods Programs Biomed.* **2012**, *106*, 1–13. [CrossRef]
34. Osher, S.; Rudin, L.I. Feature-Oriented Image Enhancement Using Shock Filters. *SIAM J. Numer. Anal.* **1990**, *27*, 919–940. [CrossRef]
35. van den Boomgaard, R.; Smeulders, A. The morphological structure of images: The differential equations of morphological scale-space. *IEEE Trans. Pattern Anal. Mach. Intell.* **1994**, *16*, 1101–1113. [CrossRef]
36. Kramer, H.P.; Bruckner, J.B. Iterations of a non-linear transformation for enhancement of digital images. *Pattern Recognit.* **1975**, *7*, 53–58. [CrossRef]
37. Alvarez, L.; Mazorra, L. Signal and Image Restoration Using Shock Filters and Anisotropic Diffusion. *SIAM J. Numer. Anal.* **1994**, *31*, 590–605. [CrossRef]
38. Vacavant, A. Smoothed Shock Filtering: Algorithm and Applications. *J. Imaging* **2021**, *7*, 56. [CrossRef] [PubMed]
39. Campbell, A.M.; Hatfield, W.T.; Heyer, L.J. Make microarray data with known ratios. *CBE Life Sci. Educ.* **2007**, *6*, 196–197. [CrossRef]

40. Xiao, J.; Lucas, A.; D'Andrade, P.; Visitacion, M.; Tangvoranuntakul, P.; Fulmer-Smentek, S. *Performance of the Agilent Microarray Platform for One-Color Analysis of Gene Expression*; Agilent Technologies, Inc.: Santa Clara, CA, USA, 2006; pp. 1–15.
41. Dobroiu, S.; van Delft, F.; Aveyard-Hanson, J.; Shetty, P.; Nicolau, D. Fluorescence Interference Contrast-enabled structures improve the microarrays performance. *Biosens. Bioelectron.* **2019**, *123*, 251–259. [CrossRef]
42. Bariamis, D.; Maroulis, D.; Iakovidis, D.K. Unsupervised SVM-based gridding for DNA microarray images. *Comput. Med. Imaging Graph.* **2010**, *34*, 418–425. [CrossRef]
43. Rueda, L.; Rezaeian, I. A fully automatic gridding method for cDNA microarray images. *BMC Bioinform.* **2011**, *12*, 113. [CrossRef] [PubMed]
44. Rueda, L.; Vidyadharan, V. A hill-climbing approach for automatic gridding of cDNA microarray images. *IEEE ACM Trans. Comput. Biol. Bioinform.* **2006**, *3*, 72–83. [CrossRef] [PubMed]
45. Bozinov, D.; Rahnenführer, J. Unsupervised technique for robust target separation and analysis of DNA microarray spots through adaptive pixel clustering. *Bioinformatics* **2002**, *18*, 747–756. [CrossRef] [PubMed]
46. Giannakeas, N.; Fotiadis, D.I. An automated method for gridding and clustering-based segmentation of cDNA microarray images. *Comput. Med. Imaging Graph.* **2009**, *33*, 40–49. [CrossRef]
47. Ho, J.; Hwang, W.L. Automatic Microarray Spot Segmentation Using a Snake-Fisher Model. *IEEE Trans. Med. Imaging* **2008**, *27*, 847–857. [CrossRef]
48. Ni, S.; Wang, P.; Paun, M.; Dai, W.; Paun, A. Spotted cDNA microarray image segmentation using ACWE. *Rom. J. Inf. Sci. Technol.* **2009**, *12*, 249–263.
49. Zacharia, E.; Maroulis, D. 3-D Spot Modeling for Automatic Segmentation of cDNA Microarray Images. *IEEE Trans. Nanobiosci.* **2010**, *9*, 181–192. [CrossRef]
50. Zahoor, J.; Zafar, K. Classification of Microarray Gene Expression Data Using an Infiltration Tactics Optimization (ITO) Algorithm. *Genes* **2020**, *11*, 819. [CrossRef]
51. Katzer, M.; Kummert, F.; Sagerer, G. Methods for automatic microarray image segmentation. *IEEE Trans. Nanobiosci.* **2003**, *2*, 202–214. [CrossRef]
52. Giannakeas, N.; Kalatzis, F.; Tsipouras, M.G.; Fotiadis, D.I. A generalized methodology for the gridding of microarray images with rectangular or hexagonal grid. *Signal Image Video Process.* **2016**, *10*, 719–728. [CrossRef]
53. Tcheslavski, G.V. *Morphological Image Processing: Gray-Scale Morphology*; Springer: London, UK, 2010.
54. Brändle, N.; Bischof, H.; Lapp, H. Robust DNA microarray image analysis. *Mach. Vis. Appl.* **2003**, *15*, 11–28. [CrossRef]
55. Belean, B.; Gutt, R.; Costea, C.; Balacescu, O. Microarray Image Analysis: From Image Processing Methods to Gene Expression Levels Estimation. *IEEE Access* **2020**, *8*, 159196–159205. [CrossRef]
56. Agilent Technologies. *Feature Extraction Reference Guide*; Agilent Technologies, Inc.: Santa Clara, CA, USA, 2012.
57. Handran, S.; Zhai, J.Y. *Biological Relevance of GenePix, Results*; Molecular Devices—Application Notes: San Jose, CA, USA, 2003.
58. Athanasiadis, E.; Cavouras, D.; Kostopoulos, S.; Glotsos, D.; Kalatzis, I.; Nikiforidis, G. A wavelet-based Markov random field segmentation model in segmenting microarray experiments. *Comput. Methods Programs Biomed.* **2011**, *104*, 307–315. [CrossRef] [PubMed]
59. Athanasiadis, E.I.; Cavouras, D.A.; Spyridonos, P.P.; Glotsos, D.T.; Kalatzis, I.K.; Nikiforidis, G.C. Complementary DNA Microarray Image Processing Based on the Fuzzy Gaussian Mixture Model. *IEEE Trans. Inf. Technol. Biomed.* **2009**, *13*, 419–425. [CrossRef] [PubMed]
60. Zacharia, E.; Maroulis, D. An Original Genetic Approach to the Fully Automatic Gridding of Microarray Images. *IEEE Trans. Med. Imaging* **2008**, *27*, 805–813. [CrossRef]
61. Jung, H.Y.; Cho, H.G. An automatic block and spot indexing with k-nearest neighbors graph for microarray image analysis. *Bioinformatics* **2002**, *18*, S141–S151. [CrossRef]
62. Zhao, J.; Yu, X.; Yang, X.; Xiang, Q.; Duan, H.; Yu, Y. Polarization independent subtractive color printing based on ultrathin hexagonal nanodisk-nanohole hybrid structure arrays. *Opt. Express* **2017**, *25*, 23137–23145. [CrossRef]
63. Singh, C.; Bhatia, N.; Kaur, A. Hough transform based fast skew detection and accurate skew correction methods. *Pattern Recognit.* **2008**, *41*, 3528–3546. [CrossRef]
64. Bao, W.; Yang, C.; Wen, S.; Zeng, M.; Guo, J.; Zhong, J.; Xu, X. A Novel Adaptive Deskewing Algorithm for Document Images. *Sensors* **2022**, *22*, 7944. [CrossRef]
65. Tai, Y.L.; Huang, S.J.; Chen, C.C.; Lu, H.H.S. Computational Complexity Reduction of Neural Networks of Brain Tumor Image Segmentation by Introducing Fermi–Dirac Correction Functions. *Entropy* **2021**, *23*, 223. [CrossRef]
66. Gadosey, P.K.; Li, Y.; Agyekum, E.A.; Zhang, T.; Liu, Z.; Yamak, P.T.; Essaf, F. SD-UNet: Stripping down U-Net for Segmentation of Biomedical Images on Platforms with Low Computational Budgets. *Diagnostics* **2020**, *10*, 110. [CrossRef] [PubMed]
67. Wang, Z.; Zineddin, B.; Liang, J.; Zeng, N.; Li, Y.; Du, M.; Cao, J.; Liu, X. A novel neural network approach to cDNA microarray image segmentation. *Comput. Methods Programs Biomed.* **2013**, *111*, 189–198. [CrossRef] [PubMed]
68. Rojas-Thomas, J.C.; Mora, M.; Santos, M. Neural networks ensemble for automatic DNA microarray spot classification. *Neural Comput. Appl.* **2019**, *31*, 2311–2327. [CrossRef]
69. Barequet, G.; Eppstein, D.; Goodrich, M.; Mamano, N. Stable-matching Voronoi diagrams: Combinatorial complexity and algorithms. *J. Comput. Geom.* **2020**, *11*, 26–59.

70. Zhang, K.; Lan, L.; Wang, Z.; Moerchen, F. Scaling up kernel SVM on limited resources: A low-rank linearization approach. In Proceedings of the Artificial Intelligence and Statistics, PMLR, La Palma, Spain, 21–23 April 2012; pp. 1425–1434.
71. Bariamis, D.; Iakovidis, D.K.; Maroulis, D. M3G: Maximum margin microarray gridding. *BMC Bioinform.* **2010**, *11*, 49. [CrossRef]
72. Morris, D. Blind Microarray Gridding: A New Framework. *IEEE Trans. Syst. Man Cybern. Part C Appl. Rev.* **2008**, *38*, 33–41. [CrossRef]
73. Oliveto, P.S.; He, J.; Yao, X. Time complexity of evolutionary algorithms for combinatorial optimization: A decade of results. *Int. J. Autom. Comput.* **2007**, *4*, 281–293. [CrossRef]

Disclaimer/Publisher's Note: The statements, opinions and data contained in all publications are solely those of the individual author(s) and contributor(s) and not of MDPI and/or the editor(s). MDPI and/or the editor(s) disclaim responsibility for any injury to people or property resulting from any ideas, methods, instructions or products referred to in the content.

Article

Hepatocellular Carcinoma Recognition from Ultrasound Images Using Combinations of Conventional and Deep Learning Techniques

Delia-Alexandrina Mitrea [1,*], Raluca Brehar [1], Sergiu Nedevschi [1], Monica Lupsor-Platon [2,3], Mihai Socaciu [2,3] and Radu Badea [2,3]

1 Department of Computer Science, Faculty of Automation and Computer Science, Technical University of Cluj-Napoca, 400114 Cluj-Napoca, Romania
2 Department of Medical Imaging, "Iuliu Hatieganu" University of Medicine and Pharmacy, 400347 Cluj-Napoca, Romania
3 "Prof. Dr. O. Fodor" Regional Institute of Gastroenterology and Hepatology, 400162 Cluj-Napoca, Romania
* Correspondence: delia.mitrea@cs.utcluj.ro

Abstract: Hepatocellular Carcinoma (HCC) is the most frequent malignant liver tumor and the third cause of cancer-related deaths worldwide. For many years, the golden standard for HCC diagnosis has been the needle biopsy, which is invasive and carries risks. Computerized methods are due to achieve a noninvasive, accurate HCC detection process based on medical images. We developed image analysis and recognition methods to perform automatic and computer-aided diagnosis of HCC. Conventional approaches that combined advanced texture analysis, mainly based on Generalized Co-occurrence Matrices (GCM) with traditional classifiers, as well as deep learning approaches based on Convolutional Neural Networks (CNN) and Stacked Denoising Autoencoders (SAE), were involved in our research. The best accuracy of 91% was achieved for B-mode ultrasound images through CNN by our research group. In this work, we combined the classical approaches with CNN techniques, within B-mode ultrasound images. The combination was performed at the classifier level. The CNN features obtained at the output of various convolution layers were combined with powerful textural features, then supervised classifiers were employed. The experiments were conducted on two datasets, acquired with different ultrasound machines. The best performance, above 98%, overpassed our previous results, as well as representative state-of-the-art results.

Keywords: convolutional neural networks (CNN); conventional machine learning (CML); advanced texture analysis methods; combination techniques; classification performance; hepatocellular carcinoma (HCC); ultrasound images

1. Introduction

Cancer is a severe affection which seriously threatens human health and sometimes leads to death. HCC is one of the biggest health problems in gastroenterology. It represents the most frequent primary cancer of the liver, the fourth most frequent cancer in men and the seventh most frequent cancer in women. It is also the third most frequent cancer-related cause of death, after lung cancer and colorectal cancer [1]. In the majority of cases, HCC evolves from cirrhosis, after a liver parenchyma restructuring phase at the end of which dysplastic nodules result, which can transform into HCC [2]. The presence of cirrhosis makes both the diagnosis and the treatment harder to perform: the presence of an underlying nodular pattern in cirrhosis makes the detection of the HCC nodular forms a daunting task. For many years, the golden standard for HCC diagnosis has been the needle biopsy, which is invasive and also raises risks, as it could generate infections and can lead to the spread of the tumor through the human body, respectively. However, the only viable way of detecting early HCC considered nowadays is medical imaging, because the clinical

and biological markers lack sensitivity in this case. One way of diagnosing HCC is by conducting vascular contrast-enhanced imaging studies, which rely on specific patterns of contrast enhancement in malignant tumors due to the effects of oncogenesis on local vascularity. The performance of these methods has gone so far that, in most cases, percutaneous biopsy is not even recommended for the definitive diagnosis [3]. However, of the three imaging methods, namely computed tomography (CT), magnetic resonance imaging (MRI) and ultrasonography (US), only the latter can be used as a screening option among cirrhotic patients, due to lack of availability for large populations (for both CT and MRI) and high radiation burden for CT, respectively. The B-mode (grayscale) ultrasound images are two-dimensional images that render the tissues and structures of interest as points of variable brightness. The ultrasound waves, generated by the transducer, are reflected by these tissues and structures, so the pixel values are related to the intensity of this reflection. These values contain important information concerning the nature of the corresponding tissues and structures. The resulting complex textures inside the tumors, different from those in the cirrhotic liver, might hold the answer for a better detection rate through computer analysis. Thus, noninvasive, computerized methods are due for detecting HCC as early and accurately as possible, revealing subtle aspects upon the tissue structure.

In ultrasound images, early HCC appears as a small, usually hypoechogenic nodule, without a visible capsule, having 2–3 cm in diameter. In more advanced stages, HCC increases in size, develops a hyperechogenic capsule, invades liver vessels and may present various visual US aspects, usually becoming overall hyperechogenic and heterogeneous due to the interleave of multiple tissue types, such as normal liver, fatty cells, active growth tissue or necrosis. Thus, in some situations, the increased echogenicity, heterogeneity and delimiting capsule of HCC are not that obvious. As both HCC and cirrhotic parenchyma represent forms of tissue restructuring, in many situations they can hardly be differentiated by the human eye [2]. An eloquent example is provided in Figure 1.

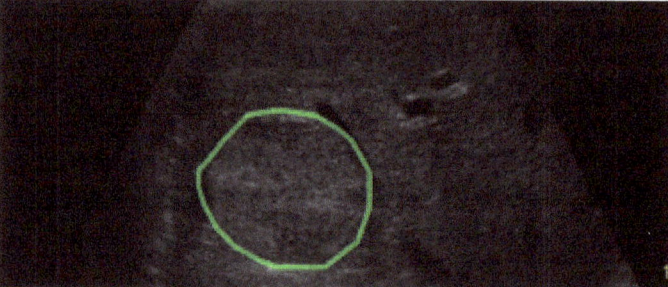

Figure 1. The visual aspect of HCC in ultrasound images: the HCC contour is marked with green.

In the current approach, we developed and assessed appropriate methods for combining conventional and deep learning techniques, the final purpose being to improve automatic HCC recognition based on medical images, with respect to the already existing results. These combinations were performed at classifier level. The values of the textural features were fused in various manners with those of the features obtained at the outputs of different layers of representative CNN architectures, then provided to a single supervised classifier. Appropriate dimensionality reduction methods, such as feature selection techniques and Kernel Principal Component Analysis (KPCA), were applied in this context, with the results being carefully analyzed. The relevance of the considered features and the correlations between the textural and deep learning features, as well as the corresponding medical significance, were discussed. The experiments were performed on two datasets, acquired with two different ultrasound machines. These datasets contained regions of interest corresponding to the HCC tumor and the cirrhotic parenchyma on which HCC had evolved, respectively.

In the context of our previous research, we developed and experimented with both conventional and deep learning methods for HCC recognition from ultrasound images. Regarding the conventional techniques, we defined the textural imagistic model of HCC, as described in [4]. Original and advanced texture analysis methods were developed and experimented with in this context, most of them based on Generalized Cooccurrence Matrices (GCM) of second and superior order [5]. The values of the most relevant textural features were provided at the entrances of powerful classifiers, such as Support Vector Machines (SVM), Multilayer Perceptron (MLP) and Random Forest (RF), along with AdaBoost combined with the C4.5 method for decision trees. The maximum accuracy was 84.09% when differentiated between HCC and the cirrhotic parenchyma, respectively, and it was 88.41% when differentiating between HCC and the hemangioma benign tumor [5]. As for the deep learning methods, we developed and assessed multiple techniques, mainly based on CNNs [6,7] but also on Stacked Denoising Autoencoders (SAE) [8]. A maximum classification accuracy of 91% resulted from the same dataset in the case when differentiating HCC from the cirrhotic parenchyma on which it had evolved.

1.1. The State of the Art

Regarding other representative methods belonging to the state of the art of the domain, conventional techniques combining texture analysis methods with traditional classifiers were widely applied previously for performing the automatic diagnosis of tumors and other affections within medical images [9–11]. Recently, the deep learning methods were extensively employed in the field of computer vision, particularly in medical imaging, leading to successful results. The CNNs demonstrated their value in both supervised and unsupervised approaches, such as those presented in [12–14]. Relevant approaches, referring to the automatic diagnosis of liver tumors, as well as of other affections (cirrhosis, which precedes liver cancer and other type of tumors), are described below.

1.1.1. Existing Approaches Targeting the Automatic Recognition of Liver Tumors from Medical Images

The conventional methods that combine texture analysis with supervised traditional classifiers were previously used in order to perform liver tumor recognition within medical images. The textural parameters were first employed by Raeth in [9] for differentiating between normal liver, diffuse liver diseases and malignant liver tumors from ultrasound images. Various types of textural features were employed for this purpose, such as those derived from the intensity histograms, those obtained from the run-length matrix, edge- and gradient-based features, the second-order Gray-Level Co-occurrence Matrix (GLCM) matrix and the associated Haralick features, co-occurrence matrices based on edge orientations and other gradient features and features derived through the Fourier transform, as well as parameters referring to speckle noise. All these features were provided to a decision-trees-based classifier that differentiated among pairs of classes, such as tumoral and nontumoral tissue, fatty and cirrhotic liver, normal liver tissue and hepatitis. In another similar approach, the run-length matrix and the corresponding parameters, in combination with the Haralick features resulted from GLCM, were experimented with in conjunction with ANN classifiers, SVM and Fisher Linear Discriminants (FLD), targeting the automatic diagnosis of the liver lesions within ultrasound images [15]. The ANN classifier, having a recognition rate close to 100%, overpassed the FLD technique, which yielded a classification accuracy of 79.6%. The Wavelet transform, applied in a recursive manner [10], as well as in combination with fractal features [16], was also involved in the ultrasound-images-based recognition of HCC. In the first case, a recognition rate of 90% resulted from employing an ANN classifier, while in the second case, an accuracy of 92% was achieved through the same type of classifier. More recent approaches that performed the automatic recognition of the liver tumors were based on CNNs. A relevant methodology that proposed a deep learning model for HCC automatic diagnosis was presented in [17]. The authors employed a ResNet18 CNN pretrained with the ImageNet dataset, the training being then refined using hematoxylinand eosin-stained pathological slides gathered from 592 HCC patients. At the end, a slide-level accuracy of 98.77% resulted. Another method, aiming for liver lesion segmentation from CT images, was presented in [18]. A specific CNN was trained by using image patches obtained from 67 tumors of 21 patients, in order to perform voxel classification as part of the segmentation process. These patches contained both tumor and healthy liver tissue. A success rate of 95.4% and an average overlap error of 16.3% resulted.

1.1.2. Existing Approaches that Employ Deep Learning for the Recognition of Other Affections Based on Medical Images

The estimation of the cirrhosis severity grades from 2D shear wave elastographic images for patients affected by chronic B-type hepatitis was performed in [12], where a CNN containing four convolution layers and a single fully connected layer was employed. The training set consisted of 1990 images corresponding to 398 cases. At the end, an AuC of 0.85 was achieved. A relevant approach involving CNN-based techniques was presented in [19]. The purpose was to detect breast tumor structures from ultrasound images using a CNN-based method called Single Shot MultiBox Detector (SSD). The corresponding dataset comprised 579 benign and 464 malignant breast lesion cases. The proposed method yielded better performance in terms of precision and recall as compared with the other existing state-of-the-art methods. A Deep Convolutional Neural Network (DCNN) was implemented in [20] for detecting incipient lung cancer from CT images. The experimental dataset consisted of 62,492 regions of interest extracted from 40,772 nodules, as well as of 21,720 non-nodules belonging to the Lung Image Database Consortium (LIDC) data store. A maximum classification accuracy of 86.4% resulted for this methodology. In the latest years, more complex approaches were developed for achieving an increased pathology recognition performance. The combination of multiple image modalities was analyzed in several studies, such as [21,22]. B-mode ultrasound images were combined with CEUS images through CNN-based techniques in order to automatically recognize breast tumors [21], and histological and immunohistochemical image data were fused in [22] through a CNN-based methodology for breast cancer diagnosis. Other approaches combined multiple types of deep learning features. In [23], aiming to detect breast cancer within histopathology images, the authors combined the deep learning features provided by the VGG16, InceptionV3 and ResNet50 architectures, the concatenated features being provided to a VirNet model, which performed the final feature fusion and classification. In the approach described in [24], the authors combined the deep learning features provided by the ResNet50 and DenseNet201 architectures for performing brain tumor classification. After a feature selection process, the relevant features were fused using a serial approach and provided to an SVM classifier that provided an 87.8% classification accuracy.

1.1.3. Existing Approaches that Combine Conventional and Deep Learning Techniques

Moreover, the combination between conventional and deep learning techniques was exploited in the domain for further improvement of the classification performance. As conventional features, radiomic features (intensity and texture), as well as other types of handcrafted features such as the Histogram of Oriented Gradients (HOG), were considered. These types of approaches are analyzed in the next paragraphs, mainly referring to the medical imaging domain. A relevant approach regarding classifier-level fusion is described in [25], where the authors studied the combination of deep learning and radiomic features for assessing PD-L1 expression level via preoperative MRI in HCC cases. An extended set of radiomic features were derived from the Volumes of Interest (VOI) using the *Pyradiomics* tool. These features included textural features, intensity features (first-order statistics) and geometric (shape) characteristics. The textural features comprised GLCM features, Gray-Level Size Zone Matrix (GLSZM) features, Neighboring Gray Tone Difference Matrix (NGTDM) features, Gray-Level Run-Length Matrix (GLRLM) features and Gray-Level Dependence Matrix (GLDM) features, respectively. In addition, the derived images were determined by applying eight types of image filters: gradient, wavelet, square, square root, logarithm, exponential, Laplacian of Gaussian (LoG) and 3D Local Binary Pattern (LBP-3D). The intensity and textural features were determined on the derived images as well. In order to obtain the deep learning features, an original 3D CNN architecture was developed, consisting of two 3D convolution layers and two fully connected layers. The deep learning features were extracted from the output of the first fully connected layer after applying the rectified linear unit (ReLu) activation function. The radiomic features were concatenated with the deep learning features, then a normalization procedure was applied, a redundant feature removal process was employed and the result was provided to a supervised classifier of the Support Vector Machine–Recursive Feature Elimination (SVM-RFE) type, which also performed feature selection. For eliminating the redundant features, the Pearson Correlation Coefficient (PCC) was implemented. The experiments were performed on an HCC dataset corresponding to 103 patients. The correlation between the relevant radiomic features and the PD-L1 expression level was also established, the considered classes being correlated with the PD-L1 expression level as well. An accuracy of 88.7% and a precision of 94.8% resulted, regarding the prediction of the PD-L1 expression level. Aiming to improve the state-of-the-art performance concerning the prediction of lymph node metastasis in head and neck cancer, in [26], the authors combined conventional radiomic and deep learning features resulted from CT and Positron Emission Tomography (PET) images. A new many-objective radiomics model (MO-radiomics) was designed for extracting valuable radiomic features, and an original 3D CNN architecture that fully utilized spatial and contextual information was employed for yielding the deep learning features. The MO-radiomics model consisted of textural features and intensity features (first-order statistics), respectively, as well as geometric features. The textural feature set comprised 3D GLCM features, a total of 257 features being finally extracted from the CT and PET images. Then, the SVM method was employed in order to build a predictive model, and an optimization problem was solved for selecting the final feature set and model parameters. Concerning the 3D CNN model, the corresponding architecture consisted of 12 convolutional layers, 2 max-pooling layers and 2 fully connected layers. The conventional and deep learning features were fused through an *evidential reasoning* method. The performance of this hybrid method was assessed for classifying normal, suspicious and lymph node metastasis. The proposed hybrid methodology finally led to a 92% accuracy, which overpassed the 79% accuracy of the conventional methods of the proposed 3D CNN. Another approach was illustrated in [27], where the authors fused deep learning features with radiomic features for predicting malignant lung nodules from CT images. The deep learning features provided by VGG-type CNNs, as well as by originally designed CNNs, were fused with classical radiomic features, including size, shape, GLCM, Laws and Wavelet features. A Symmetric Uncertainty technique was employed to select relevant attributes from both deep learning and conventional feature sets, then the fused set was given as input to an RF classifier. The best accuracy of 76.79% was obtained when employing VGG-type CNNs. The methodology presented in [28] also demonstrated that the combination of deep learning and radiomic features led to the highest performance in lung cancer survival prediction. Thus, several pretrained CNNs were adopted to obtain deep features from 40 contrast-enhanced CT images, representing non-small-cell adenocarcinoma lung cancer, these being combined with handcrafted features. The deep learning features were obtained before and after the last Rectified Linear Unit (ReLU) layer. Thereafter, multiple supervised classifiers, receiving relevant features at their inputs, were compared, achieving a maximum accuracy of 90% and a maximum AUC of 93.5%, respectively.

1.2. Contributions

According to the previous paragraph, there are many approaches that combine conventional features with deep learning features, demonstrating classification performance improvements. However, there are no relevant approaches that combine conventional and deep learning features for performing HCC automatic diagnosis within ultrasound images. We studied this possibility in our current work, aiming to further improve the accuracy of noninvasive HCC automatic diagnosis. Thus, the contributions of our current research are the following: (1) We combined conventional machine learning methods, involving advanced texture analysis methods with deep learning classifiers based on CNNs, to automatically recognize HCC tumors within B-mode ultrasound images. (2) We performed classifier-level fusion by experimenting with combination schemes involving various dimensionality reduction methods for obtaining the most valuable information from the whole data, such as relevant features or the main variation modes. An increase in the computational efficiency was an objective as well. Thus, we considered dimensionality reduction methods from both categories of feature selection and feature extraction techniques, the KPCA technique being taken into account for the second category. Regarding the feature selection techniques, we considered both classical techniques, as well bio-inspired approaches based on Particle Swarm Optimization (PSO). We mined for possible correlations among the textural and deep learning features. (3) As for the conventional techniques, we considered a large variety of textural features based on both classical and advanced original texture analysis methods, such as the superior-order Generalized Co-occurrence Matrices (GCM). We also considered multiresolution features, achieved after recursively applying the Wavelet transform. (4) Regarding the CNN-based techniques, we considered existing, representative deep learning architectures, as well as new architectures, improved by the authors in an original manner, starting from the standard architectures. (5) We updated the definition for the textural imagistic model of HCC [4], considering the combinations between the conventional and deep learning techniques, with appropriate experiments being performed. (6) We performed the experiments on two datasets of B-mode ultrasound images, constituted by the authors, acquired with two different ultrasound machines.

2. Materials and Methods

2.1. Description of the Experimental Datasets

For performing reliable experiments, two HCC B-mode ultrasound image datasets were exploited. The first one, denoted by *GE7*, contained B-mode ultrasound images corresponding to 200 HCC cases, acquired with a Logiq 7 (General Electric Healthcare, Chicago, IL, USA) ultrasound machine, under the same settings: frequency of 5.5 MHz, gain of 78, depth of 16 cm and a Dynamic Range (DR) of 111. The second dataset, denoted by *GE9*, consisted of B-mode ultrasound images belonging to 96 patients affected by HCC, acquired through a newer, Logiq 9 (General Electric Healthcare, Chicago, IL, USA) ultrasound machine, using the following set-up parameters: frequency of 6.0 MHz, gain of 58, depth of 16 cm and a DR of 69. These images were gathered by medical specialists at the 3rd Medical Clinic in Cluj-Napoca, respectively, at the "Octavian Fodor" Regional Institute of Gastroenterology and Hepatology in Cluj-Napoca. All the patients included in this study underwent biopsies for diagnostic confirmation. For each patient, multiple images were considered, corresponding to various orientations of the ultrasound transducer. Two classes were considered for differentiation in our study, these being HCC and the cirrhotic parenchyma on which HCC had evolved (denoted by PAR). The two classes were employed, as they are visually similar in many situations, it also being known that the HCC tumor usually evolves on cirrhotic parenchyma. This focus was suggested by the experienced radiologists, so no normal (healthy) cases were included in this study. The GE7 dataset included HCC tumors in various evolution phases. For this dataset, acquired previously, rectangular regions of interest having 50×50 pixels in size were manually selected by the specialized physicians inside the HCC tumor or on the cirrhotic parenchyma using a specific application implemented by the authors. The GE9 dataset comprised mostly advanced-stage HCC tumors. For this dataset, recently gathered, the HCC structures were manually delineated by the medical specialists, using the VGG Image Annotator (VIA) [29] application. Through the VIA interface, the specialists delimited the tumoral region through a polygon. According to these delimitations, rectangular regions of interest (patches) having 56×56 pixels in size were automatically extracted from the tumoral regions, respectively, from the cirrhotic parenchyma zone, using a sliding window algorithm, which assumed the traversal of the image with a window of 56 × 56 pixels in size. If the window was situated inside the delimiting polygon and its intersection with the non-HCC regions was smaller than 0.1%, the corresponding patch was assigned to the HCC class. If the window was situated outside the polygon and its intersection with the HCC zone was smaller than 0.1%, the current patch was integrated in the cirrhotic parenchyma class. Eloquent examples of patches from each dataset are illustrated in Table 1.

Table 1. Relevant examples from the two considered datasets GE7 and GE9.

Dataset	Class			
GE7	HCC			
	PAR			
GE9	HCC			
	PAR			

For performing a reliable computerized analysis, a patch size of around 50 × 50 pixels was chosen for being able to almost integrate in the tumor region entirely and to comprise a significant number of pixels. The initially generated patches were augmented through geometrical transforms (rotation, horizontal and vertical translation, scaling and horizontal flip). Finally, 6910 HCC patches, 7148 cirrhotic parenchyma patches resulting from the GE7 dataset and 10,000 patches/class from the GE9 dataset were obtained. The classes were almost equally distributed in both datasets.

2.2. The Proposed Solution

In our current study, the main objective was that of enhancing the HCC automatic recognition performance through the fusion between deep learning and CML methods at the classifier level, the newly obtained performance being compared with that achieved when employing only deep learning methods and CML methods, respectively, on the experimental datasets. This methodology is illustrated in Figure 2.

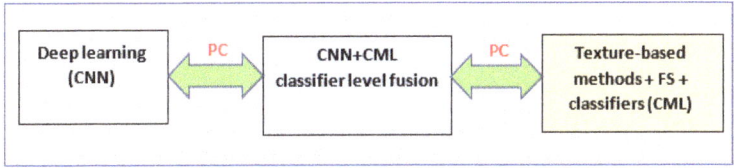

Figure 2. The graphical representation of the proposed methodology: the fusion between the CNN and CML methods (in the **middle**), as well as the performance comparison (PC) with the deep learning methods (**left**) and the CML methods (**right**), respectively.

2.2.1. Deep Learning Techniques Involved in Our Solution

CNNs constitute deep feed-forward ANNs adequate for image recognition. Their structure was inspired from biology, the organization of the connections between the neurons resembling that of the animal visual cortex [30]. With the appearance of powerful parallel computing devices such as graphics processing units (GPU), CNNs have started to be widely used, their value being emphasized in computer vision in 2012, in the context of the ImageNet competition. The main structural elements of CNNs are the convolutional layers that are employed for compressing the input data into recognized patterns to reduce the data size and to focus on the relevant patterns [31], respectively. As presented in [32], the main power of a CNN is achieved through its deep architecture that allows to extract discriminating features at multiple abstraction levels. As for the deep learning techniques involved in our solution, we assessed both relevant and newly developed CNN-based methods, considering both classical CNN architectures, as well as transformer-based methods. Thus, we experimented with several standard architectures of the ResNet [33], InceptionV3 [34], DenseNet [35], EfficientNet_b0 [36] and ResNext [37] type, the best performance being achieved for ResNet101 and InceptionV3, respectively, for the recently developed EfficientNet_b0. Thus, the residual connections of the ResNet architecture, the inception modules of InceptionV3 and also the scaling properties of EfficientNet_b0 led to the best results in the case of the current dataset. Regarding the transformer-based methods, the best performance was achieved for ConvNext_base, while other transformers such as the Vision Transformer (ViT) and ConvNext_small [38] were also assessed. Some of these architectures were enhanced for optimizing their performances. Thus, an improved version of EfficientNet_b0, denoted EfficientNet_ASPP, was designed by introducing, before the fully connected layer, an AtrousSpatial Pyramid Pooling (ASPP) module [30], in order to extract multiscale features, and a dropout layer was also added thereafter for avoiding the overfitting phenomenon. The ASPP module, which was inserted after the usual convolutional part of EfficientNet_b0, immediately before the fully connected layers, simultaneously performed a 1×1 convolution and two atrous convolutions of size 3×3 with the rates 2 and 3, respectively. At the end, a *depthcat* layer and a global average pooling layer were added, respectively. Regarding the dropout layer, an output probability of 0.5 was associated to it. A systematic description of all these CNN architectures is provided within Table 2. It also includes the size of the deep learning feature vector, as well the name of the layer at the end of which these features were extracted.

Table 2. The description of the CNN architectures.

CNN	Original Impr.	Last Layer	Vector Size
ResNet101	-	pool5	2048
InceptionV3	-	avg_pool	2048
EfficientNet_b0	dropout layer	GlobAvgPool	1283
EfficientNet_ASPP	ASPP module	gapool	1283
ConvNext_base	-	adaptiveAvgPool2d	1024

2.2.2. Conventional Techniques Involved in Our Solution

- Texture analysis methods

  Texture is an intuitive concept, inspired by the human perception, referring to the visual appearance of surfaces, particularly to the aspect of human body tissues represented within medical images. Texture can be characterized through statistical parameters, able to reveal subtle aspects upon the analyzed surface or tissue, overpassing human perception. Concerning the texture-based methods involved in our research as part of the CML approach, we analyzed both representative classical techniques, as well as more advanced techniques, developed by the authors. As classical textural features, we took into account first-order gray-level statistics, such as the corresponding arithmetic mean, maximum and minimum values, as well as second-order gray-level features, such as the Haralick parameters derived from GLCM [39], computed as described in [4]. In this group of features, we included homogeneity, energy, entropy, correlation, contrast and variance, which provided valuable information on the properties of the tissue referring to the echogenicity, heterogeneity, granularity and structural complexity. The autocorrelation index [39] was also considered, providing information on the granularity of the tissue. The Hurst fractal index was included as well in our feature set, providing information on the roughness and structural complexity of the tissue. Edge-based statistics, such as edge frequency and edge contrast [39], were also found useful in order to emphasize the structural complexity. The statistics of textural microstructures, which resulted after applying the laws convolution filters [40], were also involved in our research for the same reason. Features such as the frequency and density (arithmetic mean) of microstructures such as levels, edges, spots, waves and ripples were estimated as potentially relevant for the detection of the malignant tumors. Simultaneously, multiresolution features, in the form of the Shannon entropy computed after applying the Wavelet transform recursively twice [4], were considered able to derive subtle information on the malignant tissues, facilitating their differentiation from other tissue types. The Local Binary Pattern (LBP) also represents a powerful texture analysis method, invariant to illumination changes, particularly to those repetitive background changes due to wind or to water waves. It was firstly introduced in [41]. For obtaining these features, around each pixel, a circle of radius R can be considered. On this circle, N neighbors can be selected. For effectively achieving the LBP code, the difference between the central pixel and each of the N neighbors is computed. For each neighbor, if this difference is larger than 0, a code with the value of 1 is considered, otherwise, a 0 valued code is stored. The corresponding N codes constitute a number representing the LBP code. In our work, the LBP features were derived by varying the values of the R and N parameters. The following (R, N) value pairs were considered: (1,8), (2,16) and (3,24), respectively. Compressed LBP histograms with a smaller number of bins (100) were computed on each Region of Interest (ROI) of the dataset.

We also employed advanced, original textural features elaborated by the authors, such as the edge orientation variability [4] and GCM of superior order, respectively. The superior-order GCM were defined as described in (1). According to this mathematical formula, each element of this matrix was equal with the number of n-tuples of pixels, having the values $(a_1, a_2, \ldots, a_n)$ for the considered attribute A, which can stand for the intensity level, edge orientation, etc. These pixels are in a specific spatial report, defined by the displacement vectors.

$$C_D(a_1, a_2, \ldots, a_n) = \#\{((x_1, y_1), (x_2, y_2), \ldots, (x_n, y_n) : \\ A(x_1, y_1) = a_1, A(x_2, y_2) = a_2, \ldots, A(x_n, y_n) = a_n, \\ |x_2 - x_1| = |\overrightarrow{dx_1}|, |x_3 - x_1| = |\overrightarrow{dx_2}|, \ldots, |x_n - x_1| = |\overrightarrow{dx_{n-1}}|, \\ |y_2 - y_1| = |\overrightarrow{dy_1}|, |y_3 - y_1| = |\overrightarrow{dy_2}|, \ldots, |y_n - y_1| = |\overrightarrow{dy_{n-1}}|, \\ sgn((x_2 - x_1)(y_2 - y_1)) = sgn(\overrightarrow{dx_1} \cdot \overrightarrow{dy_1}), \ldots, \\ sgn((x_n - x_1)(y_n - y_1)) = sgn(\overrightarrow{dx_{n-1}} \cdot \overrightarrow{dy_{n-1}}))\} \quad (1)$$

The displacement vectors are defined by (2):

$$\overrightarrow{d} = ((\overrightarrow{dx_1}, \overrightarrow{dy_1}), (\overrightarrow{dx_2}, \overrightarrow{dy_2}), \ldots, (\overrightarrow{dx_{n-1}}, \overrightarrow{dy_{n-1}})) \quad (2)$$

In the current study, we included the Haralick features derived from the third-order GLCM. Regarding the spatial relation between the three considered pixels, they were either collinear, with the current pixel in the central position, or they formed a right-angle triangle, with the current pixel in the position of the 90° angle [4]. For each configuration, the third-order GLCM was computed, the Haralick feature values being provided separately for each direction combination. A newly defined form of Textural Microstructure Co-occurrence Matrix (TMCM) was employed for the first time in the current work, assuming to compute the co-occurrence matrix after applying the k-means algorithm [40] on the ROI. In this case, the considered attributes A were the cluster labels assigned to each pixel as the result of the grouping algorithm. The Haralick features yielded by the second- and third-order TMCM were derived thereafter, considering several values of k (250 and 500), which led to significant results. The Haralick features for both second-order and third-order TMCM were computed in the same manner as for the second- and third-order GLCM. A systematic description of the texture analysis methods involved in the current research, which highlights their classical or original character, is provided in Table 3.

Table 3. The texture analysis methods involved in our research.

Texture Analysis Method	Classical/Original
2nd-order GLCM and Haralick features	Classical
Autocorrelation index	Classical
Edge frequency, Edge contrast	Classical
Density and frequency of textural microstructures (Laws)	Classical
Shannon entropy computed after the application of the Wavelet transform	Classical
LBP features	Classical
Edge orientation variability	Original
GCM (3rd-order GLCM, 2nd- and 3rd-order TMCM) and Haralick features	Original

- Dimensionality reduction techniques

    After computing these potentially relevant textural features, the resulting vector was combined with the deep learning feature vector using specific fusion schemes described in the next subsection. These schemes involved dimensionality reduction methods from both classes of feature selection and feature extraction (KPCA). As feature selection methods, we employed both classical and bio-inspired techniques. Regarding the classical techniques, we employed Correlation-based Feature Subset (CFS) and Information Gain Attribute Selection (IGA) [42] that provided the best results in our previous research [5]. CFS represents a powerful method from the class of filters [42]. In the center of the corresponding algorithm, an appropriate heuristic is considered, which confers, to a certain attribute subset, a score that increases according to the strength of the correlation of this attribute with the class where the instance belongs and decreases when the same attribute is correlated with the other attributes, respectively. This method is employed together with an appropriate search algorithm (best first and genetic search) that provides all the potentially relevant attribute subsets [43]. Another representative method from the filters category is IGA. This technique assigns a score to each attribute reflecting the *Information Gain*, then it ranks the attributes in descending order based on this score. The method determines the entropy of the class C, before and after observing the attribute A [42]. The gain corresponding to the attribute A is given by the measure in which the attribute A conducts the decrease in the entropy of the class C. Thus, the score assigned to each attribute is computed as the difference between the entropy of the class C and the entropy of the class C obtained after observing the attribute A, respectively. The above-presented feature selection techniques were exploited in a combined manner by employing the intersection between the resulted feature subsets.
    From the class of the bio-inspired feature selection methods, the Particle Swarm Optimization (PSO) algorithm was considered. The elements of the particle swarm are associated to the items of the search space, these particles continuously changing their position in order to reach the optimal solution according to a well-defined criterion materialized through a fitness function [44]. In the context of the feature selection process, PSO is usually employed together with wrapper methods in order to search for

the best feature subset that maximizes the performance of a given classifier. The fitness function to be minimized usually refers to the classifier error rate (i.e., $1 - accuracy$). In the current work, our newly defined fitness function had the form provided by (3). The current fitness function represents the weighted mean between the classification error, the ratio between the number of the selected features and the total number of features, respectively, the classification error having a higher weight associated with it (0.8). The classification error was computed as the arithmetic mean between the errors of two basic classifiers, k-nn and the Bayesian classifier.

$$f = 0.8 \cdot error + 0.2 \cdot \frac{no\_selected\_features}{no\_features} \qquad (3)$$

The feature extraction methods project the original feature vectors onto a new feature space, having a lower dimensionality, simultaneously highlighting important characteristics of the data. A very popular feature extraction technique, the *Principal Component Analysis (PCA)*, performs the mapping of the initial data to a lower dimensional space, where the main variation modes are emphasized. *Kernel PCA (KPCA)*, the generalization of PCA, implies the transposition of PCA in a space of larger dimensionality, built by employing the kernel function K of the form $K = gram(X, X, kerneltype)$, where *kernetype* can be linear Gaussian of polynomial [45]. In our work, all the three versions of KPCA, linear, polynomial and Gaussian, respectively, were assessed. The fused vector was provided at the entrances of a powerful conventional classifier. The conventional supervised classifiers or metaclassifiers adopted in this situation were the Support Vector Machines (SVM), Random Forest (RF) and AdaBoost combined with the C4.5 algorithm for Decision Trees, respectively. These techniques, acknowledged in the domain for their increased performance, provided the best results in our former studies [5], as well as in our current study.

2.2.3. Combining the Traditional and Deep Learning Techniques at the Classifier Level

The combination (fusion) of the CNN-based methods and of the CML methods at the *classifier level* was assumed to provide the initial dataset consisting of HCC and PAR patches, at the input of a CNN classifier and to the texture analysis methods, respecitvely, as illustrated in Figure 3. Then, the deep learning features, extracted at the end of the convolutional part of the CNN, were fused with the textural feature vector through a simple concatenation or through a combination procedure that involved dimensionality reduction, such as Feature Selection (FS) or KPCA. At the end, a supervised traditional classifier was employed for completing the resulted hybrid classifier architecture to assess the classification performance. In this study, the above-described textural features formed the conventional feature vector. As for the deep learning features, they were gathered at the end of the last layer, which preceded the fully connected layers, as described within Section 2.2.1. In this context, appropriate fusion methods for yielding combined deep learning and textural feature vectors were elaborated by employing the following combination schemes: (1) the simple concatenation of the deep learning and textural feature vectors (*Concat*); (2) the concatenation of the deep learning and textural feature vectors, after the application of the classical feature selection procedure (*FS+Concat*); (3) the concatenation of the two feature vectors, followed by the application of the classical feature selection procedure (*Concat+FS*); (4) the concatenation of the two feature vectors, followed by the application of the PSO-based feature selection procedure (*Concat+PSO*); (5) the concatenation of the deep learning and textural feature vectors, after the application of the KPCA method, in order to yield the generalized principal components for each category, which were fused thereafter (*KPCA+Concat*); and (6) the concatenation of the two feature vectors, followed by the application of KPCA (*Concat+KPCA*).

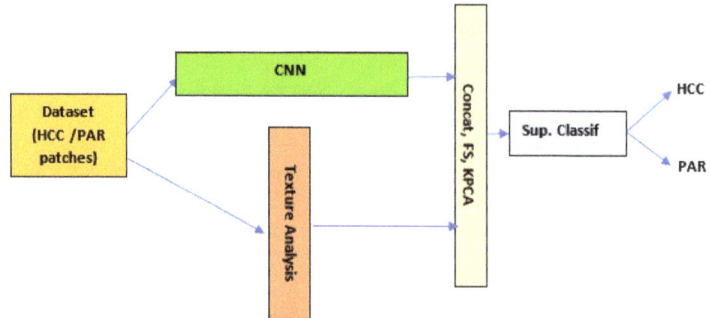

Figure 3. The graphical representation of the methodology for classifier-level fusion: (1) the images in the dataset are simultaneously provided to the CNN and to the texture analysis module; (2) then, the two resulting feature vectors are concatenated, feature selection methods or KPCA being eventually applied before or after the concatenation procedure; and (3) the fused feature vector is then provided at the entrance of a powerful conventional supervised classifier.

Concerning the classical feature selection methods, the CFS and IGA techniques were adopted, the PSO algorithm being implemented as described in Section 2.2.2. As for KPCA, in the case of the *Concat+KPCA* fusion scheme, 500 components were extracted, while in the case of the *KPCA+concat* fusion scheme, 300 components were derived from the deep learning, as well as from the textural feature vector, in order to balance the lengths of the final feature vectors that resulted in each case. Thereafter, the correlations between the deep learning features and the textural features were analyzed in order to explain the significance of the deep learning features with respect to the visual and physical properties of the malignant tissue. For this purpose, the Pearson correlation method was employed [40].

2.2.4. The Newly Defined Imagistic Model of HCC

In the context of our former research, we defined the imagistic textural model of HCC, consisting of (1) the complete set of relevant textural features which best differentiated among HCC and the visually similar classes: cirrhotic parenchyma on which HCC had evolved and benign liver tumors, respectively, as well as (2) the specific values associated with the relevant textural features: arithmetic mean, standard deviation and probability distribution [4]. In this study, this model was extended by adding the most relevant deep learning features extracted at the end from various levels of the CNNs, together with their specific values. Thus, the new set of best discriminative features (RelF) resulted by employing the most appropriate combination schemes (Comb) upon the textural (CML) and deep learning (DL) features, as illustrated in (4).

$$RelF = Comb(DL\_features, CML\_features) \qquad (4)$$

The newly resulting imagistic model of HCC consisted of the specific values associated to each feature of the *RelF* set and of the properties associated to the relevant feature map image, respectively, such as the arithmetic mean of the gray levels and the standard deviation, as depicted in (5). The feature map image resulted by transposing the final relevant feature vector into a gray-scale image. The discovered correlations between the textural and the deep learning features were also part of this model.

$$IM = \bigcup_{rf \in RF} (mean(rf), stdev(rf), prob\_distrib(rf)) \bigcup Prop(feature\_map\_img) \qquad (5)$$

2.2.5. Performance Assessment

For classification performance assessment, the following metrics, appropriate for automatic diagnosis in the medical domain, were approached: accuracy or recognition rate, sensitivity or True-Positive (TP) Rate, specificity or True-Negative (TN) Rate and Area under ROC (AuC), respectively [40]. In our experimental context, HCC was considered the positive class, while PAR was considered the negative class. As for the automatic cancer diagnosis, both sensitivity and specificity are important, referring to the probability of the presence and lack of disease, respectively. Thus, the presence of cancer should be detected as early as possible, but the situation of erroneous cancer detection should be avoided, as those patients who are not affected by malignancy should not be sent to specific, often harmful treatments.

2.3. Experimental Settings

The above-mentioned techniques were implemented as follows:

- Most of the CNNs were implemented in the Matlab R2021b environment, except ConvNext_base, which was available only in Python [37].
- The conventional classifiers and the classical feature selection methods were employed with the aid of the Weka 3.8. library [43].
- KPCA for feature extraction and PSO for feature selection were implemented in Matlab R2021b.
- Most of the texture analysis methods were implemented in Visual C++, except LBP, which was implemented in Python.

Thus, the majority of the CNNs, i.e., *ResNet101*, *InceptionV3*, *EfficientNet_b0* and the improved *EfficientNet_b0* were implemented in Matlab R2021b, with the aid of the Deep Learning Toolbox [46]. The improved *EfficientNet_b0* architecture, enhanced with an ASPP module and a dropout layer, was built in the *Deep Network Designer* environment, starting from the *EfficientNet_b0* architecture, as described in Section 2. All these networks were trained in the following conditions:

- The Stochastic Gradient Descent with Momentum (SGDM) strategy was employed;
- The learning rate was set to 0.0002;
- The momentum was set to 0.9;
- The minibatch size was set to 30;
- The duration of the training process was 100 epochs.

These hyperparameter values were set for achieving an accurate, efficient learning process and to simultaneously avoid overtraining, as well as considering the memory constraints of the computer (the minibatch size). All the above-mentioned networks were pretrained on the ImageNet dataset, the training being refined thereafter using the specific data from the B-mode ultrasound images of our datasets. The ConvNext-type CNN, as a recent, powerful architecture, was implemented in Python with the aid of the *Torchvision* library [37]. It was trained in a similar manner, using the same strategy and the same values of the hyperparameters as those adopted for the other CNN architectures. The last layer was reshaped for all the considered networks in order to provide only two outputs, which corresponded to the HCC and PAR classes. The feature maps were derived from the trained CNNs, as mentioned within Section 2.2.1, using specific Matlab and Python functions (*activations* and *get_activations*, respectively). Regarding the dimensionality reduction techniques, the method of KPCA was employed in Matlab 2021, with the aid of the Matlab-Kernel-PCA toolbox [47], the linear, third-degree polynomial and Gaussian kernels being experimented on. The PSO-based feature selection method was implemented in Matlab as well, using a specific framework [48]. The classical feature selection methods were implemented by using the Weka 3.8. library [43]. Thus, the CfsSubsetEval(CFS) technique was implemented with BestFirst search, while the InfoGainAttributeEval method was employed in conjunction with Ranker search. The conventional classifiers were employed, as well, using the Weka 3.8. library [43], as follows:

- The John Platt's Sequential Minimal Optimization (SMO) algorithm [43], the Weka equivalent of SVM, was assessed, the best performance resulting for the polynomial kernel of 3rd degree.
- The AdaBoost metaclassifier was assessed for 100 iterations in conjunction with the J48 method, the equivalent of the C4.5 algorithm in Weka.
- The RandomForest (RF) technique of Weka was adopted as well.

Some of the textural features were computed using our own Visual C++ software modules, as described in Section 2.3, independently on orientation, illumination and scale, after applying a median filter for speckle noise reduction. The LBP features were computed in Python using the *Numpy* library.

All these experiments were conducted on a computer having an i7 processor of 2.60 GHz, 8 GB of internal (RAM) memory and an Nvidia Geforce GTX 1650 Ti GPU. Regarding *the performance evaluation strategy* for the CNN-based methods, 75% of the data constituted the training set, 8% of the data stood for the validation set and 17% of the data were integrated in the test set. For the conventional classifiers, 75% of the data constituted the training set, while 25% of the data were integrated in the test set.

3. Results

3.1. CNN Performance Assessment

In Table 4, the values of the classification performance parameters for the individual CNNs, obtained through transfer learning, on both considered datasets were provided. The maximum values resulted for each classification performance parameter, for each dataset, were highlighted with *bold*. Thus, for the first dataset (GE7), the highest classification accuracy, the highest sensitivity, the most increased specificity and the best AUC resulted for the ResNet101 architecture. EfficientNet_ASPP, the improved version of the EfficientNet_b0 architecture, led to an increase in the classification performance in comparison with EfficientNet_b0 regarding all the assessed metrics.

For the second dataset, GE9, InceptionV3 provided the best classification accuracy, followed by ResNet101. The best sensitivity resulted for ResNet101, while the highest specificity was achieved for ConvNext_base. The most increased AuC was obtained for InceptionV3. For the GE9 dataset, EfficientNet_ASPP, the enhanced version of EfficientNet, led, once again, to an increased classification performance in terms of accuracy, sensitivity and AUC, in comparison with the original, EfficientNet_b0 architecture. As we can notice, the values of the classification performance parameters achieved for the first dataset, GE7, were higher than the values resulted for the same parameters in the case of the GE9 dataset. The reason could be the fact that the GE7 dataset included a smaller number of HCC patches that were manually selected, emphasizing a specific HCC region that in many cases was visually different from the cirrhotic parenchyma, while in the case of the GE9 dataset, the patches were automatically selected from the entire tumor surface.

Table 4. Results obtained using transfer learning.

Dataset	Method	Accuracy	Sensitivity	Specificity	AUC
GE7	ResNet101	**95.9%**	**95.6%**	**91.2%**	**93.4%**
	InceptionV3	88.7%	88.8%	88.6%	89%
	EfficientNet_b0	74.93%	72.9%	77.5%	75.2%
	EfficientNet_ASPP	76.9%	77.4%	76.1%	76.75%
	ConvNext_base	83%	78%	88%	83%
GE9	ResNet101	78.4%	**82.0%**	75.5%	78.75%
	InceptionV3	80.39%	81.63%	79%	**86%**
	EfficientNet_b0	74.32%	75.22%	73.22%	82%
	EfficientNet_ASPP	76.2%	79.8%	73.22%	76.51%
	ConvNext_base	**81%**	75%	**86%**	80.50%

3.2. Assessing the Performance of the Textural Features through Conventional Classifiers

In Table 5, the values of the classification performance parameters obtained on each dataset by providing the relevant textural features at the entrances of conventional classifiers are depicted. The maximum values are highlighted in *bold* for each parameter for each dataset. For the first dataset, GE7, the highest classification accuracy and the best sensitivity, as well as the best AUC, resulted for AdaBoost, while the highest specificity was obtained for SVM. As for the GE9 dataset, the most increased accuracy, the best sensitivity and the best specificity, as well as the highest AUC, resulted for AdaBoost. As we can infer by comparing Tables 4 and 5, the values of the classification performance parameters achieved through the CNN techniques were comparable to those obtained when employing conventional CML for both datasets. However, the maximum values were achieved for the CNNs in most of the situations.

Table 5. Results obtained on the set of relevant textural features through conventional classifiers.

Dataset	Method	Acc.	Sens.	Spec.	AUC
GE7	SMO (poly grd.3)	92.85%	92.6%	**93.1%**	92.85%
	AdaBoost+J48	**92.92%**	**94.1%**	92.8%	**93.45%**
	RF	89.9%	93.3%	88.5%	90.9%
GE9	SMO (poly grd. 3)	78.136%	77.9%	78.4%	78.1%
	AdaBoost+J48	**82.5%**	**81.1%**	**83.2%**	**89.7%**
	RF	75.85%	69.4%	82.1%	84.5%

3.3. Assessing the Performance of the Combination between the Textural and CNN Features

3.3.1. Performance Assessment on the GE7 Dataset

Within Table 6, the arithmetic mean of the values of the performance parameters obtained on the GE7 dataset trough the three considered conventional classifiers for each combination of the textural features with deep learning features derived from a certain type of CNN are depicted. For each parameter, the highest values are emphasized in bold in the case of each CNN. As we can notice, the absolute maximum of the mean accuracy, 97.47%, as well as the absolute maximum of the mean sensitivity, 97.53%, resulted when combining ResNet101 with the textural features through the *Concat+FS* fusion scheme; the absolute maximum of the average specificity, 98.63%, resulted when combining InceptionV3 with the textural features for the *KPCA+concat* fusion scheme, while the absolute maximum of mean AUC, 97.86%, resulted when combining ResNet101 with the textural features through the PSO scheme. The best overall accuracy of 98.23% resulted when the InceptionV3 CNN architecture was involved for the *KPCA+concat* combination scheme. In the case of AdaBoost, the best overall sensitivity of 98.2% resulted when ResNet101 was involved; in the case of the *KPCA+concat*, for the AdaBoost metaclassifier, the highest overall specificity of 98.9% was achieved for the *KPCA+concat* fusion scheme in the case of the RF classifier when the InceptionV3 CNN was involved, while the highest AUC of 99.3% resulted for *KPCA+concat*, in the case of the RF classification technique, when the EfficientNet_ASPP CNN architecture was employed.

In Figure 4, the comparisons between the average accuracy values corresponding to the considered combination schemes, in the case of each CNN, are illustrated. These values are also compared with the accuracy values obtained when using only the CNN by itself. Above each group, which corresponds to a certain combination scheme, the arithmetic mean of the accuracy values per group was depicted. As it can be noticed, the performance of the considered combination schemes overpassed that of the individual CNNs in most of the situations. Moreover, all the combination schemes involving feature selection and KPCA provided a better performance than that achieved when employing a simple concatenation between the CNN and the textural feature vectors. Thus, a maximum average accuracy of 93.46% was achieved in the case of *KPCA+Concat*, followed by an average accuracy of 91.13% achieved for the *Concat+KPCA* combination. Regarding the CNN architectures,

ResNet101, followed by InceptionV3, provided the highest accuracy values for most of the considered fusion schemes. In Appendix A, within Figure A1, the standard deviations of the classification accuracy values for each combination scheme are provided. As it can be noticed, in the case of the GE7 dataset, the smallest standard deviation of 1.3 was achieved for the *Concat+FS* combination scheme and then by the PSO combination scheme (1.43), followed by the *FS+Concat* fusion scheme (1.51). On the last position, the *KPCA+concat* fusion scheme was situated, the standard deviation being 5.83.

Table 6. Results obtained on GE7 through various combination methods.

Combination	Fusion Method	Acc.	Sens.	Spec.	AUC
ResNet101+TF	Concat	95.25%	95.4%	95.13%	93.03%
	FS+Concat	96.71%	95.8%	97.76%	96.1%
	Concat+FS	**97.48%**	**97.53%**	**97.56%**	92.1%
	PSO	96.65%	95.26%	98%	**97.86%**
	KPCA+Concat	97.01%	97.26%	96.8%	93.16%
	Concat+KPCA	96.92%	95%	96.76%	91.53%
InceptionV3+TF	Concat	91.74%	92.43%	91.1%	95.7%
	FS+Concat	91.69%	95%	93.8%	**97.06%**
	Concat+FS	94.39%	95.2%	90.2%	94.53%
	PSO	93.87%	**96.4%**	91.33%	95.86%
	KPCA+Concat	**95.49%**	92.16%	**98.63%**	96.6%
	Concat+KPCA	86.87%	88.36%	85.4%	90.96%
EfficientNet_b0+TF	Concat	77.42%	74.06%	81.1%	81%
	FS+Concat	78.48%	77.76%	79.26%	82.78%
	Concat+FS	77.03%	71.43%	80.9%	80.66%
	PSO	78.1%	78%	78.2%	82.6%
	KPCA+Concat	**93.22%**	**90.63%**	95.26%	**94.33%**
	Concat+KPCA	92.8%	90.33%	**95.63%**	94.26%
EfficientNet_ASPP+TF	Concat	72.75%	71.13%	74.4%	78.2%
	FS+Concat	79.67%	81.06%	78.3%	84%
	Concat+FS	78.7%	79.6%	77.66%	83.23%
	PSO	78.1%	78%	78.2%	82.6%
	KPCA+Concat	**94.99%**	**92.53%**	**96.9%**	**95.96%**
	Concat+KPCA	90.01%	90.43%	91.53%	93.53%
Convnext_base+TF	Concat	69.33%	69.2%	69.43%	74.1%
	FS+Concat	79.31%	83.4%	80.7%	88.4%
	Concat+FS	73.97%	73.16%	74.8%	78.3%
	PSO	72.77%	72.23%	73.5%	76.66%
	KPCA+Concat	86.57%	86.23%	82.53%	87.16%
	Concat+KPCA	**89.03%**	**88.5%**	**90.23%**	93.4%

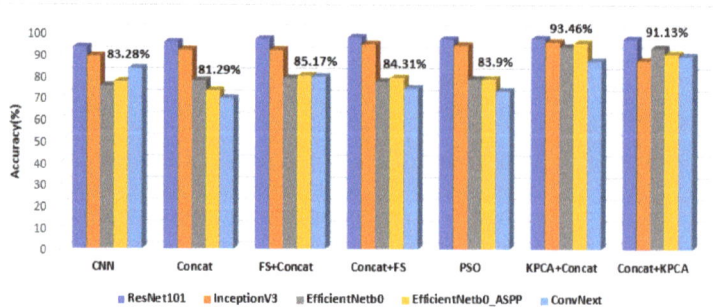

Figure 4. The comparisons between the average accuracy values obtained for each combination scheme for the considered CNN architectures in the case of the GE7 dataset.

3.3.2. Performance Assessment on the GE9 Dataset

Table 7 illustrates the arithmetic mean of the performance parameters resulted from the GE9 dataset for each combination of the textural features, with deep learning features extracted from a certain CNN, for each fusion scheme. For each parameter, for each type of CNN, the highest values were emphasized in bold. We can infer that the maximum overall value of the mean accuracy of 98.01%, the maximum mean sensitivity of 98.26%, the maximum mean specificity of 97.9% and the maximum mean AUC of 94.16% were obtained for the *KPCA+concat* fusion scheme when ResNet101 was involved. As for the individual values, obtained through each conventional classifier, the best overall accuracy of 98.9% and the best overall specificity of 98.6% resulted for the combination between InceptionV3 and the textural features for the *KPCA+concat* combination scheme when employing the AdaBoost metaclassifier. The best sensitivity of 99.2% was achieved in the case of *KPCA+concat* for AdaBoost for the combination between ResNet101 and the textural features, while the most increased AUC of 99.7% resulted for *KPCA+concat* for the RF classification technique in the case when InceptionV3 was combined with the textural features.

Table 7. Results obtained on GE9 through various combination methods.

Combination	Fusion Method	Acc.	Sens.	Spec.	AUC
ResNet101+TF	Concat	86.3%	84.6%	86.73%	90.26%
	FS+Concat	83.9%	88%	79.8%	88.96%
	Concat+FS	84.22%	87.4%	81.03%	89.1%
	PSO	75.42%	76.73%	74.16%	79.56%
	KPCA+Concat	**98.01%**	**98.26%**	**97.9%**	**94.16%**
	Concat+KPCA	96.92%	96%	97.76%	92.53%
InceptionV3+TF	Concat	82.85%	86.93%	78.66%	87.8%
	FS+Concat	84.21%	87.4%	81.03%	89.1%
	Concat+FS	82.04%	86.9%	76.73%	85%
	PSO	84.2%	87.46%	80.93%	**89.36%**
	KPCA+Concat	**87.23%**	**89.9%**	82.86%	86.36%
	Concat+KPCA	83.33%	82.9%	**84.16%**	88.66%
EfficientNet_ASPP+TF	Concat	82.04%	83.36%	77.56%	87%
	FS+Concat	83.83%	88.16%	79.3%	88.73%
	Concat+FS	85.72%	88.66%	80.66%	89.43%
	PSO	85.2%	88.46%	81.93%	90.36%
	KPCA+Concat	**88.86%**	**89.8%**	**87.76%**	**92.53%**
	Concat+KPCA	81.29%	87.06%	71.93%	84.33%
ConvNext_base+TF	Concat	85.63%	84.6%	86.73%	90.26%
	FS+Concat	86.33%	87.06%	**87.06%**	91.36%
	Concat+FS	85.79%	86.7%	84.93%	89.2%
	PSO	**86.94%**	87.06%	79.8%	**91.93%**
	KPCA+Concat	76.22%	76.63%	75.53%	78.16%
	Concat+KPCA	85.48%	**88.76%**	81.86%	90.16%

Within Figure 5, the comparison among the arithmetic mean of the accuracy values for each combination scheme for the considered CNN architectures is depicted, the arithmetic mean of the accuracy values per fusion scheme being illustrated above each corresponding group. The best average accuracy of 87.58% was achieved in the case of KPCA, followed by concatenation, while the second best mean value of 86.71% was obtained in the case of concatenation followed by KPCA. The information inferred by Figure 5 confirms that provided by Figure 4, the ranking of the fusion schemes being almost similar according to these figures. It must also be noticed that simply performing concatenation led to worse results than all the other combination schemes. Concerning the best performing CNN architectures, ResNet101, as well as ConvNext_base, provided very good performances in most of the situations. Regarding the standard deviations of the accuracy values achieved for each fusion scheme in the case of the current dataset, according to Figure A1, the smallest standard deviation resulted from the *FS+Concat* fusion scheme (1.27), followed by PSO (1.66) and then by *Concat+FS* (1.89), *Concat+KPCA* (11.83) being situated on the last position.

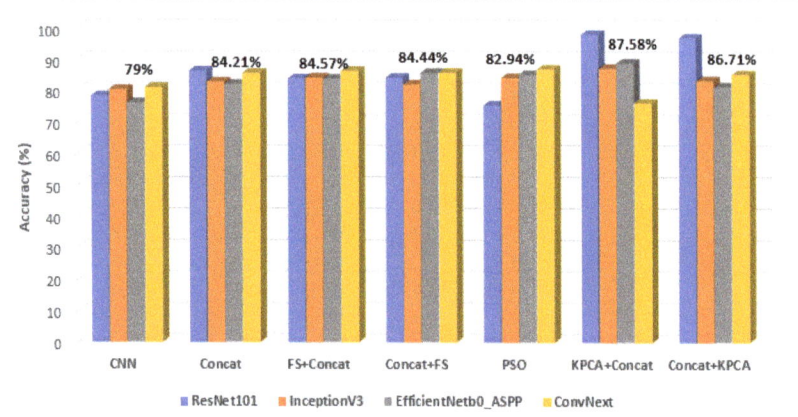

Figure 5. The comparisons between the average accuracy values obtained for each combination scheme for the considered CNN architectures in the case of the GE9 dataset.

3.4. The Newly Defined Imagistic Model of HCC

The relevance of the considered textural features in the classification process was assessed in the context of the entire feature vector when considering their combination with the CNN features. In Figures 6 and 7, the ranking of the most relevant textural features from each dataset is provided, this ordering being derived after the application of the IGA technique upon the combined feature vector, containing both textural and CNN features. The length of each line of the graphic represents the arithmetic mean of the particular scores resulted from the application of IGA upon the combination between the textural features and the CNN features provided by each CNN technique. In both these figures, we notice, on the first positions, the presence of the features derived from the generalized co-occurrence matrices, including the second- and third-order TMCM and GLCM features.

For the GE7 dataset, the most relevant feature is the contrast obtained from the TMCM matrix, computed for k = 500, having an average score of 0.066, the maximum average score among the entire feature set being 0.357. The Haralick features derived from the GLCM matrices of order 2 and 3, computed for various directions of the displacement vectors, followed thereafter, emphasizing the heterogeneous, chaotic structure of the tumor tissue through the GLCM_Energy, GLCM_Entropy and GLCM_Variance. They also revealed differences in granularity between the HCC and PAR tissue classes, through the GLCM_Correlation. Towards the end of the ranking, we notice the presence of the entropy computed after the application of the Wavelet transform at the first level on the third component (high–low) and of the LBP features, respectively, emphasizing again the chaotic structure, as well as the complexity of the malignant tumor.

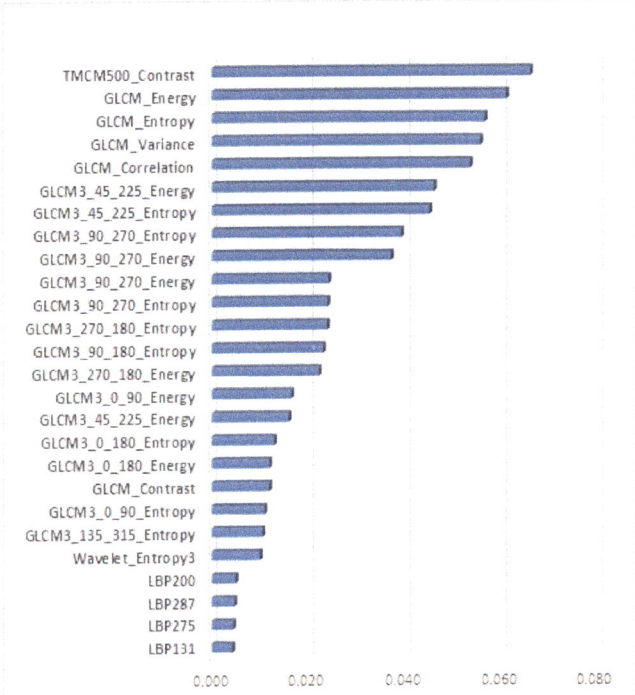

Figure 6. The ranking of the most relevant textural features among the combined feature vector for the GE7 dataset.

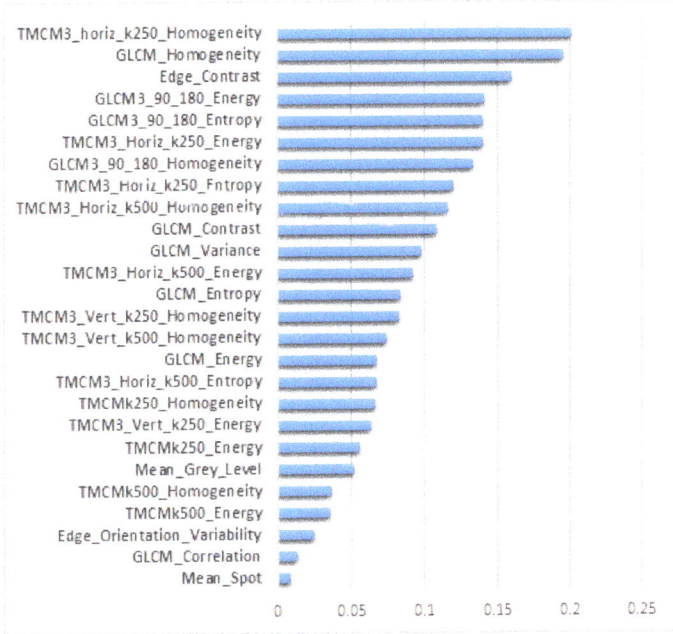

Figure 7. The ranking of the most relevant textural features through the combined feature vector for the GE9 dataset.

As for the GE9 dataset, as depicted in Figure 7, the first position among the whole feature vector was occupied by the homogeneity derived from the third-order TMCM matrix when the displacement vectors were collinear on the horizontal direction and the value of k was 250, being associated with the highest relevance score of 0.2, immediately followed by the same parameter derived from the second-order GLCM matrix. These attributes emphasized the differences in homogeneity between HCC and the cirrhotic parenchyma on which it had evolved. Towards the end of the ranking, we notice the presence of the edge orientation variability, of the GLCM correlation, of the density of the spot microstructures computed after employing the Laws' convolution filters, denoting both the complexity of the HCC tissue as well as the difference in granularity between the HCC and PAR tissues, respectively. The correlations between the textural features and the CNN features were evaluated as well for each CNN architecture on both datasets. The plots of the pairwise correlations between the considered textural features and the CNN features, assessed with the aid of the Pearson correlation method, are depicted in Appendix B, Fiugre A2. The fact that there exist increased correlations among the CNN features themselves can be noticed; there are some medium correlations among the textural features, as well as smaller correlations between the textural features and the CNN features, respectively. As for the correlations between the textural features and the CNN features, the highest correlations were those met for the GLCM_variance with three ResNet101 features for the GE9 dataset, the maximum correlation coefficients being 0.429, 0.26 and 0.17, respectively, followed by the correlations met between the TMCM500_contrast and the InceptionV3 features on the GE7 dataset of 0.197, 0.194, 0.184, 0.176 and 0.171, then by the correlations obtained on the GE7 dataset between the GLCM3_45_225_energy, the GLCM_90_270_energy with the EfficientNet_ASPP features of 0.179, by those between the TMCM500_contrast and the ResNet101 features on the GE9 dataset of 0.176, respectively, and by those between the GLCM_homogeneity and the InceptionV3 features on the GE7 dataset of 0.124. As part of the newly approached textural model, the comparisons between the activation maps corresponding to the CNN features derived from EfficientNet_ASPP and those obtained from the fusion of these types of CNN features with the textural features when employing the *Concat+KPCA* combination scheme, for both datasets, GE7 and GE9, are depicted in Figure 8. The *Concat+KPCA* combination scheme was taken into account as being one of the best performing fusion schemes that also transformed the elements of the original concatenated vector, yielding more refined fused features, which emphasized the main variation modes. These activation maps were achieved by adequately reshaping the feature vectors for obtaining a maximal square image. The first and second lines correspond to the results achieved on the first dataset, GE7, while the third and fourth lines correspond to the results obtained on the GE9 dataset. The left-hand-side column corresponds to the HCC class, while the right-hand-side column stands for the PAR class. The first and third lines correspond to the activation maps achieved in the case of the fusion between the EfficientNet_ASPP CNN features and the textural features, while the second and fourth lines to the activation maps obtained in the cases when only the EfficientNet_ASPP CNN features were taken into account. It can be noticed that, in the case of HCC, the patterns are more heterogeneous and the frequency of the increased pixel values is larger than in the case of PAR, these differences being more emphasized for the activation maps corresponding to the fusion between the CNN and textural features. This remark is confirmed by the numerical differences obtained between the corresponding values of the mean gray levels and standard deviations of these maps, respectively. Thus, in the case of the EfficientNet_ASPP activation maps obtained for the GE7 dataset, the difference between the standard deviations for the HCC and PAR classes was 0.0034, while the difference between the HCC and PAR gray-level means was 0.0091. For the same dataset, when considering the activation maps for the fusion between the CNN and textural features, the difference between the standard deviations corresponding to the HCC and PAR classes was 0.0169, while the difference between the gray-level means was 0.2545. In the case of the EfficientNet_ASPP activation maps achieved for the GE9 dataset, the difference between the standard deviations for the HCC and PAR classes was 0.0042, while the difference between the HCC and PAR gray-level means was 0.0079. When taking into account the activation maps for the combination between the CNN and textural features, the difference between the HCC and PAR standard deviations was 0.0129, the difference between the gray level means being 0.0759.

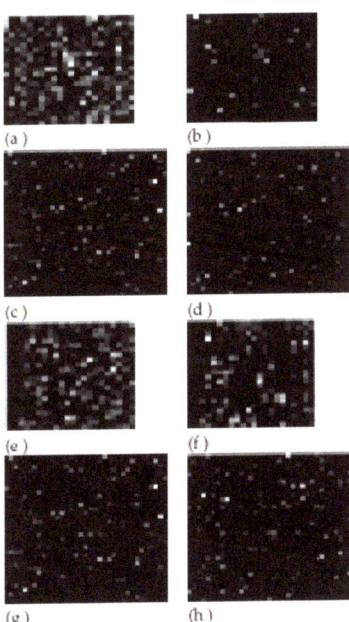

Figure 8. The comparisons between the activation maps for EfficientNet_ASPP for the combination between EfficientNet_ASPP and the textural features through the *Concat+KPCA* fusion scheme, respectively: (**a**,**b**,**e**,**f**) the maps for EfficientNet_ASPP combined with the textural features for HCC (**a**,**e**), respectively, PAR (**b**,**f**); (**c**,**d**,**g**,**h**) the maps achieved in the case of EfficientNet_ASPP for HCC (**c**,**g**), respectively, for PAR (**d**,**h**); (**a**–**d**) stand for GE7; (**e**–**h**) stand for GE9.

4. Discussion

As it results from the previous sections, as well as from Figure 9, the combinations between the CNN-based techniques and the CML techniques at the classifier level achieved better classification performances in terms of accuracy, sensitivity, specificity and AUC, in comparison with the individual application of each class of methods.

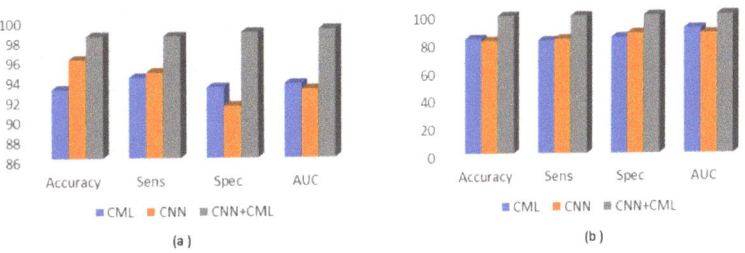

Figure 9. Comparison between the maximum value of the classification performance metrics for the three considered classes of methods: CML, CNN and the combination between CML and CNN (CML+CNN): (**a**) on the GE7 dataset; (**b**) on the GE9 dataset.

The best classification performances were achieved for *KPCA+concat*, followed by *concat+KPCA*, *FS+concat* being situated on the third position. However, the standard deviations of the accuracy values were best when employing feature selection before or after concatenation, as it results from Figure A1 form Appendix A. Regarding the KPCA technique, the best results, in all the considered situations, were achieved when employing the Gaussian kernel. It can also be noticed that firstly applying FS and KPCA, followed by concatenation, provided better classification performances than when applying concatenation and then FS and KPCA, respectively. The conventional dimensionality reduction methods led to higher classification performances than the bio-inspired feature selection method, based on PSO, when being applied in the same conditions. Concerning the CNN architectures that were combined with the conventional techniques, the best results were provided when involving ResNet101, followed by those obtained when InceptionV3 was involved. Thus, the residual connections that contributed to overpassing the gradient vanishing problem, as well as the inception modules, significantly contributed to the enhancement of the classification performance. Convnext_base, the transformer-based architecture involved in our experiments, yielded a very good classification performance as well, especially in the case of the GE9 dataset. The newly designed EfficientNet_ASPP architecture also provided satisfying results, overpassing EfficientNet_b0 in many situations. As for the conventional classifiers, as it results from the previously presented experiments, the best classification performance was achieved by AdaBoost combined with decision trees, followed by RF and SVM, respectively. For the last-mentioned classifier, the best performance resulted when considering the 3rd-degree polynomial kernel. The computational efficiency of our solution is also satisfying, as most of the CNN architectures and most of the conventional classifiers that led to the best solution were based on less complex algorithms, dimensionality reduction also being performed upon the involved feature vectors. Regarding the textural features involved in the current work, they demonstrated their importance when assessed together with the deep learning features, achieving relevance values situated in most cases in the interval 0.05–0.2, slightly below the maximum relevance value, around 0.3, of the entire feature vector. The contrast and homogeneity derived from the TMCM matrix confirmed the heterogeneity, as well as the complex structure of the HCC tissue, as compared with that of the cirrhotic parenchyma on which it had evolved. These properties are due to the coexistence of multiple tissue types, as well as to the rich vascularization of HCC. Other relevant textural features, such as the correlation derived from the GCM, revealed differences in granularity between HCC and the cirrhotic parenchyma on which it had evolved. Some correlations between the textural features and the deep learning features, assessed through the Pearson Correlation Coefficient, resulted as well, especially between the second- and superior-order textural features, derived from the newly defined GCM, and the ResNet101, InceptionV3 and EfficientNet_ASPP CNN features, respectively. These correlations revealed the capacity of the deep learning features to emphasize the properties of the HCC and cirrhotic parenchyma tissues.

The comparisons with the already existing state-of-the-art results, in terms of classification performance metrics, are depicted in Table 8. For assessing the significance of the improvements for each state-of-the-art method, a specific metric was computed, expressed as an average difference according to the formula (6). In (6), Acc, $Sens$, $Spec$ and AUC were the performance metrics corresponding to the current work, while Acc_{sa}, $Sens_{sa}$, $Spec_{sa}$ and AUC_{sa} were the metric values corresponding to the state-of-the-art techniques.

$$Avg\_dif = \frac{(Acc - Acc_{sa}) + (Sens - Sens_{sa}) + (Spec - Spec_{sa}) + (AUC - AUC_{sa})}{4} \quad (6)$$

Thus, the maximum obtained classification performances in the current work overpassed the maximum performance achieved in the research paper [7]. In [7], the authors assessed the combinations, at the classifier level, between representative CNN architectures, such as ResNet101, InceptionV3 and EfficientNet_b0, on the GE9 dataset. In this case, the average difference in the performances was −1.002. The maximum classification performance achieved in the current work also overpassed that reported in [24], the average difference between the corresponding metrics being −14.38, which is larger than that which previously resulted. Thus, the method described in [24] was reproduced in the current work in the following manner: the DenseNet201 and ResNet50 CNN architectures were trained in Matlab2021 using the Deep-Learning toolbox, then our conventional FS methodology based on the combination between the CFS and IGA techniques was applied on each CNN feature vector. The experiments were performed on the most recently acquired dataset, GE9. Thereafter, the results were concatenated and the SMO classifier was applied with the aid of the Weka 3.8 library. The method described in [27] was also considered for comparison. In order to reproduce this method, we trained a VGG-16 CNN on the GE9 dataset, using the same training parameters as specified in this paper. Concerning the conventional feature vector, for characterizing the local shape, we also derived HOG features on the GE9 dataset, using Matlab-specific functions, which were added to the previously extracted textural features. Thereafter, the Symmetrical Uncertainty feature selection technique in conjunction with Ranker was applied in Weka 3.8 [43] in order to retain the most important features from each feature vectors, the resulting conventional and deep learning features being concatenated thereafter. As it can be noticed from Table 8, an average difference of −7.43 resulted between the corresponding classification performance parameters of this state-of-the-art method and the maximum performance our current research, respectively. Regarding the computational complexity of the methods analyzed in this paragraph, we can infer that our methodology is more efficient. Thus, the approaches presented in [7] and [24] required training two types of CNNs, as well as a conventional classifier, while our technique required training only one CNN and a traditional classifier for the best solution. Moreover, in the case of the technique described in [27], a VGG-type network corresponded to the best solution, this being a complex network having many parameters, so that the training time was much more increased than that required for our CNNs.

Table 8. Comparisons with other relevant state-of-the-art approaches.

Method	Accuracy	Sensitivity	Specificity	AUC	Avg_dif
D. Mitrea et al., 2022 [7]	97.79%	97.9%	98.9%	97.8%	−1.002
Aziz et al , 2021 [24]	84.77%	90%	79.4%	84.7%	−14.38
Paul et al., 2018 [27]	90.28%	94.1%	86.4%	95.9%	−7.43
Current approach	**98.9%**	**98.6%**	**99.2%**	**99.7%**	-

Moreover, the current work approaches a similar subject and methodology as the research described in [25]. However, in our approach, the HCC automatic recognition was performed in a noninvasive and efficient manner based on ultrasound images, while in [25], the authors conducted their analysis on MRI images, which might involve additional costs and risks. One of the similarities between these two approaches is the application of feature selection on the concatenated vector, followed by the employment of a traditional classifier. While in [25], a single, complex feature selection procedure was applied upon the combined feature vector, followed by the employment of the SVM classifier, in the current approach, multiple fusion methods, involving various dimensionality reduction methods were assessed, followed by the application of various conventional classification techniques, including SVM.

5. Conclusions

The fusion between CNN-based techniques and conventional ML methods based on advanced texture analysis proved to be very efficient, leading to increased classification accuracies higher than 95% in many situations. The combination schemes that provided the best results were *KPCA+Concat*, *Concat+KPCA* and *FS+Concat*, respectively, highlighting the role of the KPCA technique in this context. The computational efficiency of our solution was also satisfying, as discussed in Section 4. A new approach of the imagistic textural model of HCC was also elaborated, emphasizing the relevant textural features and the capacity of the newly resulted hybrid feature maps to differentiate between the HCC and PAR classes, as well the correlations between the deep learning features and the textural features. It resulted that many deep learning features were correlated with the textural features, the deep learning features also being able to reveal the HCC and PAR tissue properties. All the resulted relevant features confirmed the heterogeneous and complex structure of the HCC tissue, also revealing differences in granularity between the HCC and PAR tissue classes. Thus, the newly elaborated methodology can be appropriate for the computer-aided and automatic diagnosis of HCC. The corresponding classification performances overpassed those obtained when considering the CNN methods and conventional ML methods by themselves, as well as those resulted from the case of some representative state-of-the-art methods. As future developments, we aim to involve multiple medical image types, such as CT and MRI in this analysis, as well as to refer to multiple classes of tumors, such as pancreatic and renal tumors, including benign tumors as well. Concerning the fusion methods, the Canonical Correlation Analysis [49] is also targeted for classifier-level fusion, while decision-level fusion will also be considered.

Author Contributions: Conceptualization, D.-A.M., S.N., M.L.-P. and R.B. (Radu Badea); methodology, D.-A.M., R.B. (Raluca Brehar) and S.N.; software, D.-A.M. and R.B. (Raluca Brehar); validation, D.-A.M., S.N. and M.S. and R.B. (Radu Badea); formal analysis, D.-A.M., R.B. (Raluca Brehar) and M.L.-P.; investigation, D.-A.M., S.N. and R.B. (Radu Badea); resources, M.S., M.L.-P. and R.B. (Radu Badea); data curation, M.S., M.L.-P. and R.B. (Radu Badea); writing—original draft preparation, D.-A.M. and R.B. (Raluca Brehar); writing—review and editing, D.-A.M., R.B. (Raluca Brehar), S.N. and M.L.-P.; visualization, M.S. and R.B. (Radu Badea); supervision, S.N. and R.B. (Radu Badea); project administration, D.-A.M.; funding acquisition, D.-A.M., R.B. (Raluca Brehar), S.N. and R.B. (Radu Badea). All authors have read and agreed to the published version of the manuscript.

Funding: This research was funded by the Ministry of Research, Innovation and Digitization, CNCS—UEFISCDI, project number PN-III-P-1.1-TE-2021-1293, within PNCDI III.

Informed Consent Statement: Informed consent was obtained from all subjects involved in the study.

Acknowledgments: This work was supported by a grant of the Ministry of Research, Innovation and Digitization, CNCS—UEFISCDI, project number PN-III-P-1.1-TE-2021-1293, TE 156/2022, within PNCDI III. This research was also supported by the CLOUDUT Project, cofunded by the European Fund of Regional Development through the Competitiveness Operational Programme 2014–2020, contract no. 235/2020.

Conflicts of Interest: The authors declare no conflict of interest.

Abbreviations

The following abbreviations are used in this manuscript:

HCC	Hepatocellular Carcinoma
PAR	Cirrhotic Parenchyma on which HCC had evolved
CNN	Convolutional Neural Networks
ML	Machine Learning
US	Ultrasonography
CT	Computer Tomography
MRI	Magnetic Resonance Imaging
AUC	Area under the ROC
FS	Feature Selection
RF	Random Forest
SVM	Support Vector Machines
SMO	The John Platt's Sequential Minimal Optimization algorithm
PSO	Particle Swarm Optimization

Appendix A

In Figure A1, the average standard deviations of the accuracy values for each dataset, GE7 and GE9, are depicted. The values for the average standard deviations resulted from computing the arithmetic mean of the standard deviations of the classification accuracies for each considered CNN for each considered fusion scheme.

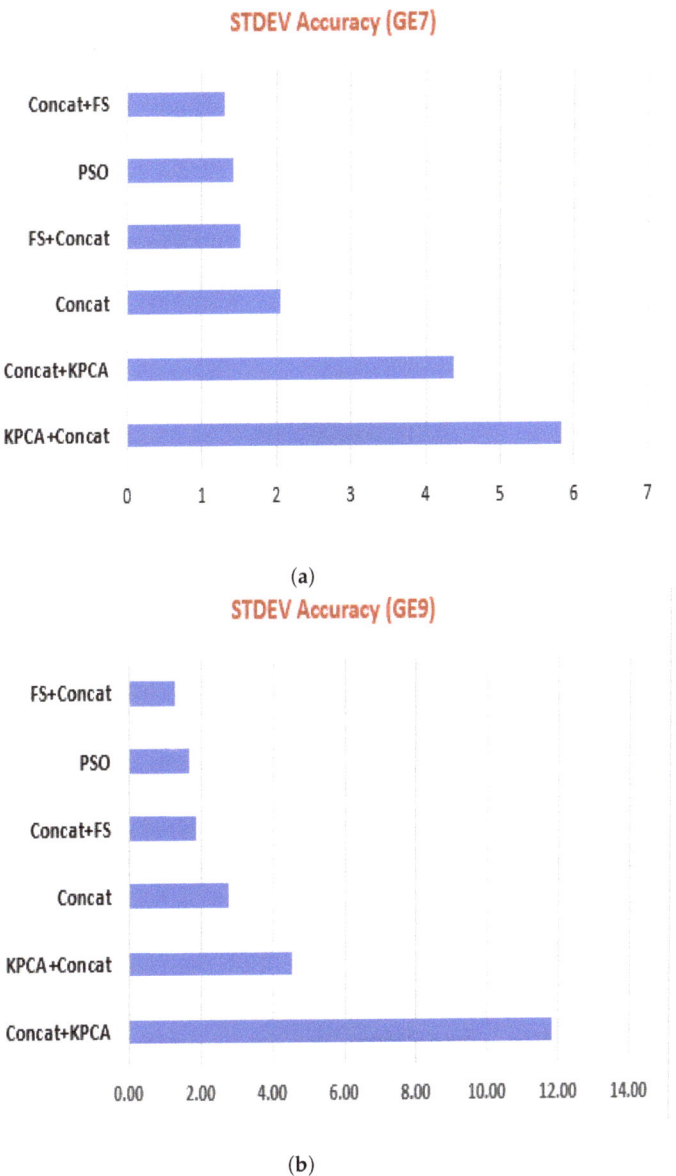

(a)

(b)

Figure A1. The standard deviations of the accuracy values for each combination scheme: (**a**) for the GE7 dataset; (**b**) for the GE9 dataset.

Appendix B

In Figure A2, the plots of the pairwise correlations between the textural features and the CNN features are depicted. In each plot, each feature is denoted by an index. In the case of the first dataset, GE7, the textural features have associated indexes from 1 to 550, while the CNN features have indexes from 551 to 2598 in the case of InceptionV3 and ResNet101, from 551 to 1788 in the case of EfficientNet_ASPP and from 551 to 1575 in the case of Convnext_base, respectively. In the case of the GE9 dataset, the textural features have associated indexes from 1 to 98, while the CNN features have indexes from 99 to 2047 in the case of InceptionV3 and ResNet101, from 99 to 1237 in the case of EfficientNet_ASPP and from 99 to 1122 in the case of Convnext_base, respectively. The fact that there exist high correlations (≥ 0.5) between the CNN features themselves can be noticed. There also exist correlations between the textural features and the CNN features but of smaller values (≤ 0.429). In Figure A2, the first column corresponds to the correlations derived from the first dataset, while the second column to the correlations computed on the second dataset; the first row corresponds to the correlations between the InceptionV3 features and the textural features, the second row to the correlations between the ResNet101 features and the textural features and the third row corresponds to the correlations between the EfficeintNetb0_ASPP features and the textural features, while the last row stands for the correlations between the the Convnext_base CNN features and the textural features.

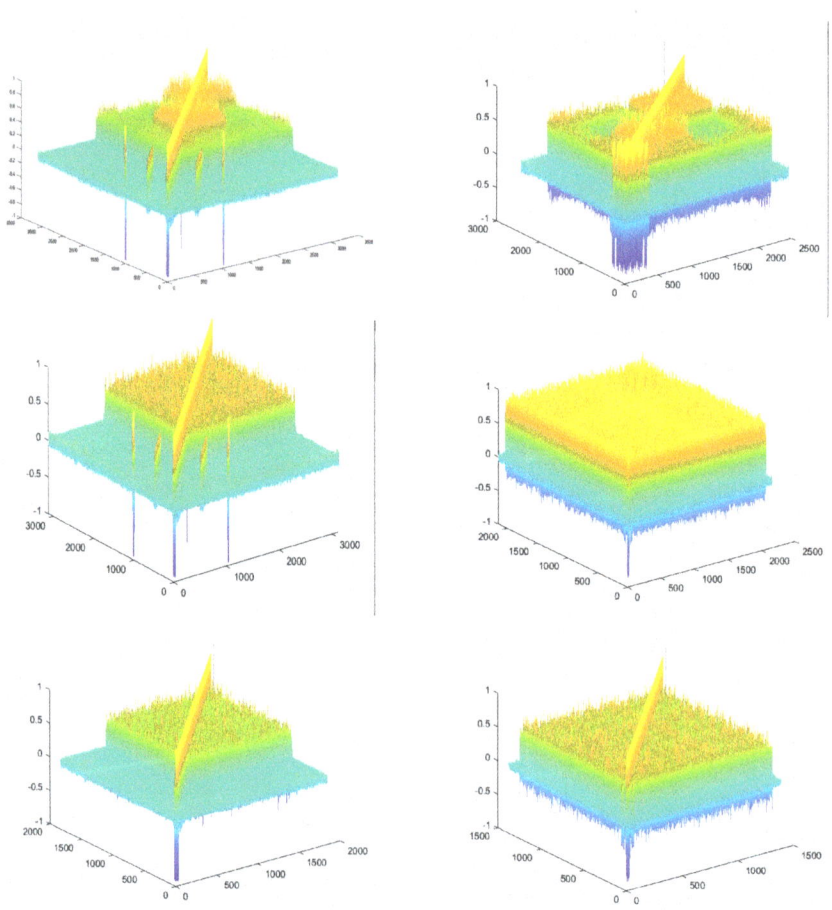

Figure A2. *Cont.*

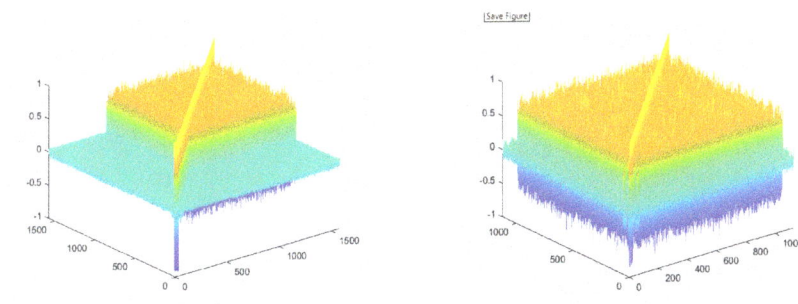

Figure A2. The correlations between the textural and CNN features.

References

1. European Association for the Study of the Liver; European Organisation for Research and Treatment of Cancer. EASL-EORTC clinical practice guidelines: Management of hepatocellular carcinoma. *J. Hepatol.* **2012**, *56*, 908–943. [CrossRef]
2. Sherman, M. Approaches to the diagnosis of hepatocellular carcinoma. *Curr. Gastroenterol. Rep.* **2005**, *7*, 11–18. [CrossRef]
3. Elmohr, M.; Elsayes, K.M.; Chernyak, V. LI-RADS: Review and updates. *Clin. Liver Dis.* **2021**, *17*, 108–112. [CrossRef]
4. Mitrea, D.; Mitrea, P.; Nedevschi, S.; Badea, R.; Lupsor, M.; Socaciu, M.; Golea, A.; Hagiu, C.; Ciobanu, L. Abdominal Tumor Characterization and Recognition Using Superior-Order Cooccurrence Matrices, Based on Ultrasound Images. *Comput. Math. Methods Med.* **2012**, *2012*, 348135. [CrossRef]
5. Mitrea, D.; Nedevschi, S.; Badea, R. Automatic Recognition of the Hepatocellular Carcinoma from Ultrasound Images using Complex Textural Microstructure Co-Occurrence Matrices (CTMCM). In Proceedings of the 7th International Conference on Pattern Recognition Applications and Methods—Volume 1: ICPRAM, INSTICC, Funchal, Portugal, 16–18 January 2018; SciTePress: Setubal, Portugal, 2018; pp. 178–189. [CrossRef]
6. Brehar, R.; Mitrea, D.A.; Vancea, F.; Marita, T.; Nedevschi, S.; Lupsor-Platon, M.; Rotaru, M.; Badea, R. Comparison of Deep-Learning and Conventional Machine-Learning Methods for the Automatic Recognition of the Hepatocellular Carcinoma Areas from Ultrasound Images. *Sensors* **2020**, *20*, 3085. [CrossRef]
7. Mitrea, D.; Brehar, R.; Nedevschi, S.; Socaciu, M.; Badea, R. Hepatocellular Carcinoma recognition from ultrasound images through Convolutional Neural Networks and their combinations. In Proceedings of the International Conference on Advancements of Medicine and Health Care through Technology, Cluj-Napoca, Romania, 20–22 October 2022; IFMBE Proceedings Series; Springer: Berlin/Heidelberg, Germany, 2022; pp. 1–6.
8. Mitrea, D.; Mendoiu, C.; Mitrea, P.; Nedevschi, S.; Lupsor-Platon, M.; Rotaru, M.; Badea, R. HCC Recognition within B-mode and CEUS Images using Traditional and Deep Learning Techniques. In Proceedings of the 7th International Conference on Advancements of Medicine and Health Care through Technology, Cluj-Napoca, Romania, 12–15 October 2020; IFMBE Proceedings Series; Springer: Berlin/Heidelberg, Germany, 2020; pp. 1–6.
9. Raeth, U.; Schlaps, D. Diagnostic accuracy of computerized B-scan texture analysis and conventional ultrasonography in diffuse parenchymal and malignant liver disease. *J. Clin. Ultrasound* **1985**, *13*, 87–89. [CrossRef]
10. Yoshida, H.; Casalino, D.; Keserci, B.; Coskun, A.; Ozturk, O.; Savranlar, A. Wavelet-packet-based texture analysis for differentiation between benign and malignant liver tumours in ultrasound images. *Phys. Med. Biol.* **2003**, *48*, 3735–3753. [CrossRef]
11. Duda, D.; Kretowski, M.; Bezy-Vendling, J. Computer aided diagnosis of liver tumors based on multi-image texture analysis of contrast-enhanced CT. Selection of the most appropriate texture features. *Stud. Log. Gramm. Rhetor.* **2013**, *35*, 49–70. [CrossRef]
12. Hui, Z.; Liu, C.; Fankun, M. Deep learning Radiomics of shear wave elastography significantly improved diagnostic performance for assessing liver fibrosis in chronic hepatitis B: A prospective multicentre study. *Gut* **2019**, *68*, 729–741. [CrossRef]
13. Liu, X.; Song, J.; Wang, S.; Zhao, J.; Chen, Y. Learning to Diagnose Cirrhosis with Liver Capsule Guided Ultrasound Image Classification. *Sensors* **2017**, *17*, 149. [CrossRef]
14. Koutrintzes, D.; Mathe, E.; Spyrou, E. Boosting the Performance of Deep Approaches through Fusion with Handcrafted Features. In Proceedings of the 11th International Conference on Pattern Recognition Applications and Methods, Online, 3–5 February 2022; Scitepress Digital Library: Setubal, Portugal, 2022; pp. 370–377.
15. Sujana, H.; Swarnamani, S. Application of Artificial Neural Networks for the classification of liver lesions by texture parameters. *Ultrasound Med. Biol.* **1996**, *22*, 1177–1181. [CrossRef]
16. Lee, W.; Hsieh, K.; Chen, Y. A study of ultrasonic liver images classification with artificial neural networks based on fractal geometry and multiresolution analysis. *Biomed. Eng. Appl. Basis Commun.* **2004**, *16*, 59–67. [CrossRef]
17. Feng, S.; Yu, X.; Liang, W.; Li, X.; Zhong, W.; Hu, W.; Zhang, H.; Feng, Z.; Song, M.; Zhang, J.; et al. Development of a Deep Learning Model to Assist With Diagnosis of Hepatocellular Carcinoma. *Front. Oncol.* **2021**, *11*, 4990. [CrossRef]

18. Vivanti, R.; Epbrat, A. Automatic liver tumor segmentation in follow-up CT studies using convolutional neural networks. In Proceedings of the Patch-Based Methods in Medical Image Processing Workshop, Munich, Germany, 9 October 2015; pp. 45–54.
19. Zhantao, C.; Lixin, D.; Guowu, Y. Breast Tumor Detection in Ultrasound Images Using Deep Learning. In *Patch-Based Techniques in Medical Imaging*; Lecture Notes in Computer Science; Springer: Cham, Switzerland, 2017.
20. Li, W.; Cao, P. Pulmonary Nodule Classification with Deep Convolutional Neural Networks on Computed Tomography Images. *Comput. Math. Methods Med.* **2016**, *2016*, 6215085. [CrossRef]
21. Yang, Z.; Gong, X.; Guo, Y.; Liu, W. A Temporal Sequence Dual-Branch Network for Classifying Hybrid Ultrasound Data of Breast Cancer. *IEEE Access* **2020**, *8*, 82688–82699. [CrossRef]
22. Pradhan, P.; Kohler, K.; Guo, S.; Rosin, O.; Popp, J.; Niendorf, A.; Bocklitz, T. Data Fusion of Histological and Immunohistochemical Image Data for Breast Cancer Diagnostics using Transfer Learning. In Proceedings of the 10th International Conference on Pattern Recognition Applications and Methods, Online, 4–6 February 2021; Scitepress Digital Library: Setubal, Portugal, 2021; pp. 495–506.
23. Cheng, X.; Tan, L. Feature Fusion Based on Convolutional Neural Network for Breast Cancer Auxiliary Diagnosis. *Math. Probl. Eng.* **2021**, *2021*, 7010438. [CrossRef]
24. Aziz, A.; Attique, M.; Tariq, U.; Nam, Y.; Nazir, M.; Jeong, C.W.; Mostafa, R.R.; Sakr, R.H. An Ensemble of Optimal Deep Learning Features for Brain Tumor Classification. *Comput. Mater. Contin.* **2021**, *69*, 2653–2670. [CrossRef]
25. Tian, Y.; Komolafe, T.E.; Zheng, J.; Zhou, G.; Chen, T.; Zhou, B.; Yang, X. Assessing PD-L1 Expression Level via Preoperative MRI in HCC Based on Integrating Deep Learning and Radiomics Features. *Diagnostics* **2021**, *11*, 1875. [CrossRef]
26. Chen, L.; Zhou, Z. Combining Many-objective Radiomics and 3-dimensional Convolutional Neural Network through Evidential Reasoning to Predict Lymph Node Metastasis in Head and Neck Cancer. *J. Med. Imaging* **2021**, *5*, 011021.
27. Paul, R. Predicting malignant nodules by fusing deep features with classical radiomics features. *J. Med. Imaging* **2018**, *5*, 011021. [CrossRef]
28. Paul, R.; Hawkins, S.H.; Balagurunathan, Y.; Schabath, M.; Gillies, R.J.; Hall, L.O.; Goldgof, D.B. Deep Feature Transfer Learning in Combination with Traditional Features Predicts Survival Among Patients with Lung Adenocarcinoma. *Tomography* **2016**, *2*, 388–395. [CrossRef]
29. Dutta, A.; Gupta, A.; Zissermann, A. VGG Image Annotator (VIA). Version 2.0.9. 2016. Available online: http://www.robots.ox.ac.uk/vgg/software/via/ (accessed on 10 March 2020).
30. Chatterjee, H.S. Various Types of Convolutional Neural Network. 2019. Available online: https://towardsdatascience.com/various-types-of-convolutional-neural-network-8b00c9a08a1b (accessed on 15 July 2022).
31. *Tutorial of Deep Learning*; Release 0.1; University of Montreal: Montreal, QC, Canada, 2015.
32. Litjens, G.; Kooi, T.; Bejnordi, B.E.; Setio, A.A.A.; Ciompi, F.; Ghafoorian, M.; van der Laak, J.A.; van Ginneken, B.; Sánchez, C.I. A survey on deep learning in medical image analysis. *Med. Image Anal.* **2017**, *42*, 60–88. [CrossRef]
33. He, K.; Zhang, X.; Ren, S.; Sun, J. Deep Residual Learning for Image Recognition. In Proceedings of the 2016 IEEE Conference on Computer Vision and Pattern Recognition (CVPR), Las Vegas, NV, USA, 27–30 June 2016; pp. 770–778.
34. Szegedy, C.; Vanhoucke, V.; Ioffe, S.; Shlens, J.; Wojna, Z. Rethinking the Inception Architecture for Computer Vision. In Proceedings of the IEEE Conference on Computer Vision and Pattern Recognition (CVPR), Las Vegas, NV, USA, 27–30 June 2016; pp. 2818–2826.
35. Huang, G.; Liu, Z.; Weinberger, K.Q. Densely Connected Convolutional Networks. *arXiv* **2016**, arXiv:1608.06993.
36. Tan, M.; Le, Q. EfficientNet: Rethinking Model Scaling for Convolutional Neural Networks. *arXiv* **2020**, arXiv:1905.11946v5.
37. Torchvision Library for Python. 2022. Available online: https://pytorch.org/vision/stable/index.html (accessed on 19 April 2022).
38. Li, Z.; Gu, T.; Li, B.; Xu, W.; He, X.; Hui, X. ConvNeXt-Based Fine-Grained Image Classification and Bilinear Attention Mechanism Model. *Appl. Sci.* **2022**, *12*, 9016. [CrossRef]
39. Materka, A.; Strzelecki, M. *Texture Analysis Methods—A Review*; Technical Report; Institute of Electronics, Technical University of Lodz: Lodz, Poland, 1998.
40. Meyer-Base, A. *Pattern Recognition for Medical Imaging*; Elsevier: Amsterdam, The Netherlands, 2009.
41. Ojala, T.; Pietikäinen, M.; Harwood, D. A comparative study of texture measures with classification based on featured distributions. *Pattern Recognit.* **1996**, *29*, 51–59. [CrossRef]
42. Hall, M. Benchmarking attribute selection techniques for discrete class data mining. *IEEE Trans. Knowl. Data Eng.* **2003**, *15*, 1–16. [CrossRef]
43. Waikato Environment for Knowledge Analysis (Weka 3). 2022. Available online: http://www.cs.waikato.ac.nz/ml/weka/ (accessed on 10 May 2022).
44. Gaber, T.; Hassanien, T. *Particle Swarm Optimization: A Tutorial*; IGI Global: Manchester, UK, 2017.
45. Van Der Maaten, L.; Postma, E.; Van den Herik, J. Dimensionality reduction: A comparative review. *J. Mach. Learn. Res.* **2009**, *10*, 66–71.
46. Deep Learning Toolbox for Matlab. 2022. Available online: https://it.mathworks.com/help/deeplearning/index.html (accessed on 14 April 2022).
47. Kitayama, M. Matlab-Kernel-PCA Toolbox. 2017. Available online: https://it.mathworks.com/matlabcentral/fileexchange/71647-matlab-kernel-pca (accessed on 20 August 2020).

48. Too, J. Particle Swarm Optimization for Feature Selection. 2022. Available online: https://github.com/JingweiToo/-Particle-Swarm-Optimization-for-Feature-Selection (accessed on 3 April 2022).
49. Gao, L.; Qi, L.; Chen, E.; Guan, L. Discriminative Multiple Canonical Correlation Analysis for Information Fusion. *IEEE Trans. Image Process.* **2018**, *27*, 1951–1965. [CrossRef]

Disclaimer/Publisher's Note: The statements, opinions and data contained in all publications are solely those of the individual author(s) and contributor(s) and not of MDPI and/or the editor(s). MDPI and/or the editor(s) disclaim responsibility for any injury to people or property resulting from any ideas, methods, instructions or products referred to in the content.

Article

Using Sparse Patch Annotation for Tumor Segmentation in Histopathological Images

Yiqing Liu [1,†], Qiming He [1,†], Hufei Duan [1], Huijuan Shi [2], Anjia Han [2,*] and Yonghong He [1,*]

1. Institute of Biopharmaceutical and Health Engineering, Tsinghua Shenzhen International Graduate School, Shenzhen 518055, China
2. Department of Pathology, The First Affiliated Hospital, Sun Yat-sen University, Guangzhou 510080, China
* Correspondence: hananjia@mail.sysu.edu.cn (A.H.); heyh@sz.tsinghua.edu.cn (Y.H.)
† These authors contributed equally to this work.

Abstract: Tumor segmentation is a fundamental task in histopathological image analysis. Creating accurate pixel-wise annotations for such segmentation tasks in a fully-supervised training framework requires significant effort. To reduce the burden of manual annotation, we propose a novel weakly supervised segmentation framework based on sparse patch annotation, i.e., only small portions of patches in an image are labeled as 'tumor' or 'normal'. The framework consists of a patch-wise segmentation model called PSeger, and an innovative semi-supervised algorithm. PSeger has two branches for patch classification and image classification, respectively. This two-branch structure enables the model to learn more general features and thus reduce the risk of overfitting when learning sparsely annotated data. We incorporate the idea of consistency learning and self-training into the semi-supervised training strategy to take advantage of the unlabeled images. Trained on the BCSS dataset with only 25% of the images labeled (five patches for each labeled image), our proposed method achieved competitive performance compared to the fully supervised pixel-wise segmentation models. Experiments demonstrate that the proposed solution has the potential to reduce the burden of labeling histopathological images.

Keywords: histology images; tumor segmentation; sparse annotation; weakly-supervised learning; semi-supervised learning

Citation: Liu, Y.; He, Q.; Duan, H.; Shi, H.; Han, A.; He, Y. Using Sparse Patch Annotation for Tumor Segmentation in Histopathological Images. Sensors 2022, 22, 6053. https://doi.org/10.3390/s22166053

Academic Editors: Mitrea Delia-Alexandrina and Sergiu Nedevschi

Received: 19 July 2022
Accepted: 10 August 2022
Published: 13 August 2022

Publisher's Note: MDPI stays neutral with regard to jurisdictional claims in published maps and institutional affiliations.

Copyright: © 2022 by the authors. Licensee MDPI, Basel, Switzerland. This article is an open access article distributed under the terms and conditions of the Creative Commons Attribution (CC BY) license (https://creativecommons.org/licenses/by/4.0/).

1. Introduction

Deep learning has made rapid development and remarkable progress in pathological image analysis in recent years [1–7]. The application of deep learning in pathological diagnosis and prognosis cannot be imagined without high-quality annotations. However, acquiring precise annotations is difficult since it requires knowledge of pathology and is time-consuming and labor-intensive, particularly for segmentation tasks that involve manually outlining the specific structures.

Unfortunately, experts with a wealth of pathological knowledge, the source of high quality and clean clinical tagging of key data, are often scarce and have limited energy to spend on data labeling. Therefore, deep-learning methods based on sparsely annotated labels are critical to reducing their workload of labeling and pushing the application of deep learning in the field of pathology. Tumor segmentation has been one of the most fundamental tasks in digital pathology for accurate diagnosis.

Since a whole slide image (WSI) usually has an extremely high resolution, e.g., 50,000 × 50,000 pixels, common practice is to crop it into smaller images and assign each of them a label for model training. There are two typical models, including an image-wise segmentation model [8–13] and pixel-wise segmentation model [14–18]. An image-wise segmentation model predicts whether the given image contains tumorous regions.

A binary label ('tumor' or 'normal') is assigned to each image in the training set to train these models. However, the performance of an image-wise segmentation model is limited by the insufficiency of the labeling information. Since a mere binary label 'tumor' cannot reflect the location and proportion of the tumor, assigning the same label 'tumor' to different images as long as they contain the tumor may confuse the network training and lead to inaccurate segmentation results, which is unacceptable—in particular for small tumors.

In contrast, a pixel-wise segmentation model can produce more accurate segmentation results. However, pathologists must annotate the tumor regions as masks to train the model, which takes much more time and energy. More importantly, unlike other medical images, such as MRI and CT images, pathology images usually lack a clear distinction between the normal and tumor areas [19], which imposes additional difficulties for labeling.

To compensate for the shortcomings of the above two methods, we propose the concept of the patch-level label. Note that, in our proposed method, a patch refers to a grid cell of an image, which is different from the definition in other articles [11,18]. Suppose we divide the image with the size of 224 × 224 pixels into a 14 × 14 grid, then the patch size is 16 × 16 pixels. For each image in the training set, pathologists only need to annotate several (usually 5–10) patches as the label, significantly saving the annotation cost. The left of Figure 1 shows different types of labels.

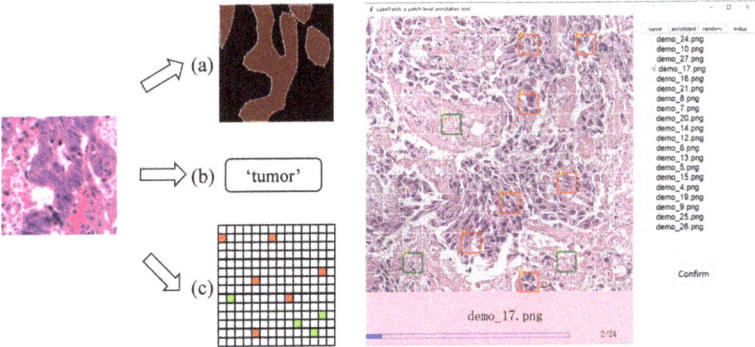

Figure 1. Left: The illustration of different types of labels. (**a**) Pixel-level label, where the red area denotes tumor region and the black area denotes non-tumor region. (**b**) Image-level label, suggesting that the image contains tumor region. (**c**) Patch-level label (proposed), where the red patches and green patches are manual annotations indicating tumor and non-tumor regions, respectively. **Right:** A software we developed for sparse patch annotation.

We designed a patch-wise segmentation model called Pseger to accommodate this new label. It has two branches for image classification and patch classification, respectively. The image classification is an auxiliary task that helps improve the performance of the patch classification branches. Due to the superior performance the Trasformer-based networks [20] have achieved in recent years, we select Swin Transformer [21], a representation of them as the backbone of the model. Moreover, this method can be easily extended to other backbones.

To take advantage of the unlabeled data, we trained our Pseger with an innovative semi-supervised algorithm. The algorithm is developed based on the characteristics of the patch-level label, integrating the ideas of consistent learning [22] and self-training [23]. The contributions of this paper are summarized as follows:

- We proposed the concept of sparse patch annotation for tumor segmentation, which can significantly reduce the annotation burden. To achieve this new way of labeling, we developed an annotation tool (Figure 1, right).
- In order to handle this new label, we created a patch-wise segmentation model called Pseger, which was equipped with an innovative semi-supervised algorithm to make full use of the unlabeled data.
- We comprehensively evaluated our proposed method on two datasets. The experimental results showed that when trained with only 25% labeled data (five patches

for each labeled image), our approach can yield a competitive result compared to the pixel-wise segmentation models trained using 100% labeled data. The ablation study showed the effectiveness of the semi-supervised algorithm.

2. Related Works

2.1. Weakly-Supervised Learning

Pixel-level labels require a considerable amount of time and effort, and the frequently occurring manual errors may give the network the wrong guidance. Weakly-supervised learning (WSL) has recently emerged as a paradigm to relieve the burden of dense pixel-wise annotations [24]. Many WSL techniques have been proposed, including global image-level labels [25,26], scribbles [19,27], points [28,29], bounding boxes [30,31], and global image statistics, such as the target-region size [32,33].

Although these weakly supervised methods have achieved good performance in natural and medical image segmentation, most weak annotations may not necessarily be best or most suited for tumor segmentation. As mentioned above, the image-level label cannot reflect the location and proportion of the tumor, which may result in inaccurate segmentation results. Other label types are more suitable for segmentation tasks where the instances have clear boundaries, such as glands and nuclei. Nevertheless, the boundary between the normal and the tumor area in pathology images is usually fuzzy and ambiguous. Unlike existing weak annotations, we propose patch-level annotation for patch-wise tumor segmentation.

2.2. Multi-Task Learning

Multi-task learning is an emerging field in machine learning that seeks to improve the performance of multiple related tasks by leveraging useful information among them [34]. A deep-learning model for multi-task learning usually consists of a feature extractor shared by all the tasks and multiple branches for each task. In recent years, multi-task learning has been widely exploited in the field of pathological image analysis [18,35,36]. For example, Wang et al. [18] proposed a hybrid model for pixel-wise HCC segmentation of H&E-stained WSIs.

The model had three subnetworks sharing the same encoder, corresponding to three associated tasks. Guo et al. [37] employed a classification model to filter images containing tumorous regions and subsequently refined the segmentation results by a pixel-wise segmentation model. Inspired by these seminal works, we adopted a two-branch model, one branch for image classification and another for patch segmentation, to learn more general features and thus reduce the risk of overfitting.

2.3. Semi-Supervised Learning

Semi-supervised learning (SSL) is a combination of both supervised and unsupervised learning methods, in which the network is trained with a small amount of labeled data and a large amount of unlabeled data. SSL methods can make full use of the information provided by unlabeled data, thereby improving the model performance. In recent years, SSL methods have been widely used in the computer vision field [38–43].

There are two common SSL strategies, including consistent learning [22] and self-training [23]. The general idea of consistent learning is that model prediction should keep constant under different perturbations to the input. This method allows for various perturbations to be designed depending on the characteristics of the data and the network. For instance, Xu et al. [40] proposed two novel data augmentation mechanisms and incorporated them into the consistency learning framework for prostate ultrasound segmentation.

Another strategy, self-training, can be broadly divided into four steps. First, train a teacher model using labeled data. Second, use a trained teacher model to generate pseudo labels for unlabeled images. Third, learn an equal-or-larger student model on labeled and unlabeled images. Finally, use the student as a teacher and repeat the above procedures several times. Wang et al. [41] proposed a few-shot learning framework by combining ideas of semi-supervised learning and self-training. They first adopted a teacher-student model in the initial semi-supervised learning stage and obtained pseudo labels for unlabeled data. Then, they designed a self-training method to update pseudo labels and the segmentation model by alternating downsampling and cropping strategies.

3. Materials and Methods

Here, we propose a novel patch-wise segmentation model called PSeger. Equipped with an innovative semi-supervised algorithm, it can learn from the patch-level label and take advantage of the unlabeled data. Figure 2 gives an overview of the training procedure. It involves three steps: (1) basic training; (2) pseudo label generation; and (3) consistency learning. They are described in detail in the following. The information about the two datasets we used is also described later.

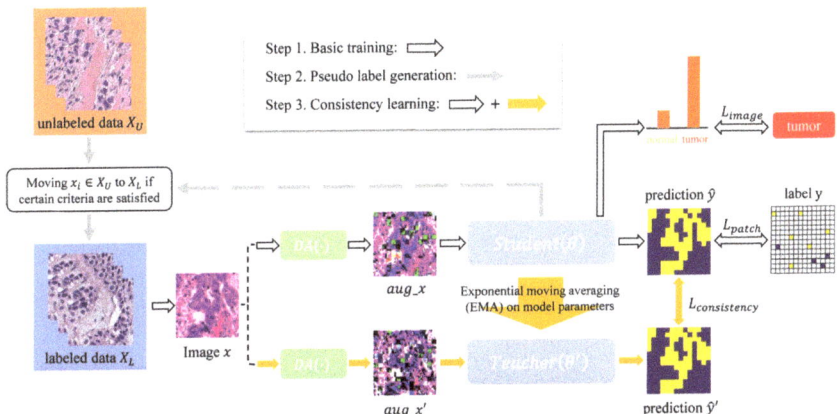

Figure 2. Overview of the framework for training PSeger. $DA(\cdot)$ indicates data augmentation module.

3.1. Basic Training

Since the idea of patch-level label is inspired by Vision Transformer (ViT) [20], we take it as the backbone of PSeger to illustrate the process of basic training. An overview of the model is depicted in Figure 3, which consists of an embedding projection module, a sequence of transformer encoder blocks, and two classifiers for image classification and patch classification, respectively. In the process of forward propagation, an input image $x \in \mathbb{R}^{H \times W \times N_C}$ (H, W, and N_C represent the height, width, and number of channels of x, respectively) is first flattened into $M = HW/P^2$ non-overlapped patches with the size of $P \times P$ pixels. Then, a 2-D convolution operation is employed to obtain patch embeddings, supplemented with position encoding:

$$z_0 = \left[x^1 P_E; x^2 P_E; \ldots; x^M P_E\right] + P_E^{pos}, \tag{1}$$

where $z_0 \in \mathbb{R}^{M \times L}$ (L represents the embedding length) is the input of the first transformer encoder block, $x^k \in \mathbb{R}^{P \times P \times C}$ is the kth patch, P_E is the embedding projection, and P_E^{pos} is the position encoding. Then, the embeddings are processed by the transformer encoder blocks. Each block includes a multi-head self-attention (MSA) [44] module and a multi-layer perceptron (MLP) module, both of which are operating as residual operators, and with a layer normalization (LN) [45]. The output of the lth transformer encoder block can be described as follows,

$$z'_l = MSA(LN(z_{l-1})) + z_{l-1}, \quad l = 1 \ldots L, \tag{2}$$

$$z_l = MLP(LN(z'_l)) + z'_l, \quad l = 1 \ldots L, \tag{3}$$

where z_L is the final output of the transformer encoder. Each element of the output $z_L^k \in z_L$ contains contextual features due to the attention mechanism, which makes it possible to classify a patch based on the information of the related patches. We adopt an *MLP* head H_{patch} for patch classification. By these means, z_l processed by an *LN* is sent to H_{patch} before applying a softmax function to obtain predictions of each patch:

$$\hat{y} = Softmax(H_{patch}(LN(z_L))), \tag{4}$$

where $\hat{y} \in \mathbb{R}^{M \times C}$ are the patch predictions, and C is the number of categories.

In addition to the patch classifier, we introduce an auxiliary image classifier H_{image} to the network, which determines whether an input image has a tumor or not. The main motivation for use of image classifier is to help the patch classifier achieve better performance, since in multi-task learning the network tends to find more representative features shared by different tasks [18]. Similar to the patch classifier, the image classifier receives the average of the Lth transformer encoder output $z_L \in \mathbb{R}^{M \times L}$ with an *LN*, and produces the classification result $\hat{y}_{img} \in \mathbb{R}^C$ through a softmax function:

$$\hat{y}_{img} = Softmax\left(H_{image}\left(LN\left(\sum_{i=1}^{M} z_L^k / M\right)\right)\right). \tag{5}$$

The loss function for the basic training is defined as:

$$L_{sup} = L_{patch} + \alpha L_{img}, \tag{6}$$

where L_{img} and L_{patch} are the losses for image classification task and patch classification task, respectively. α is a weighting factor for the two losses. Both L_{img} and L_{patch} are cross-entropy loss functions; however, L_{patch} only considers the annotated patches. Specifically, L_{patch} is defined as:

$$L_{patch} = -\frac{1}{K} \sum_{k}^{K} \sum_{c}^{C} y^k \log \hat{y}^{(k,c)}, \tag{7}$$

where K is the number of the labeled patches in the sample x, C is the number of classes, y^k is the binary indicator (0 or 1) if class label c is the correct classification for the kth patch. $\hat{y}^{(k,c)}$ is the prediction of the kth patch at the cth class.

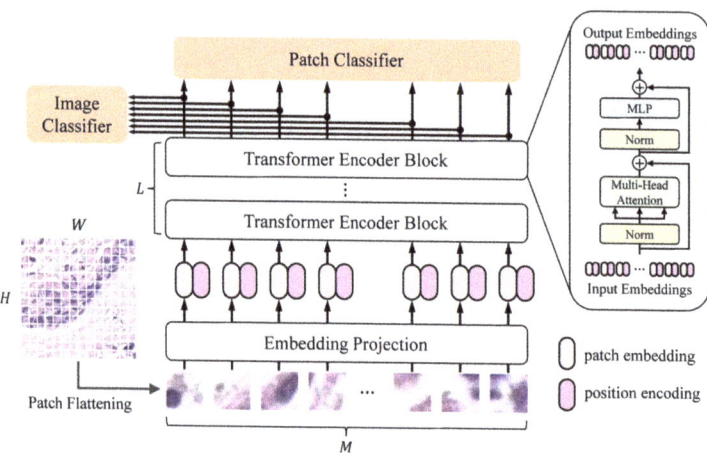

Figure 3. Illustration of PSeger (using Vistion Transformer as backbone).

3.2. Pseudo Label Generation

After the basic training process, the model with the best patch classification accuracy on the validation set is used to generate the pseudo labels for samples in the unlabeled data X_U, as is depicted in Figure 4. The trained model receives as input an image $x_i \in X_U$ and infers the image prediction $\hat{y}_{i,img}$ and patch predictions $\hat{y}_i$, which are subsequently transformed into the image probability $p_{i,img}$ and patch probabilities p_i by the softmax function. The latter are then ranked by their dominant values. We move x_i from X_U to X_L along with its pseudo label if $p_{i,img}$ and ranked p_i (denoted as $r(p_i)$) meet the following criteria:

1. $max(p_{i,img}) > \tau_1$, where τ_1 is the confidence threshold for the image prediction.
2. $max(r(p_i[K])) > \tau_2$, where τ_2 is the confidence threshold for the patch prediction.
3. $\forall k \in [1, K]$, $\texttt{argmax}(r(p_i)[k]) = \texttt{argmax}(p_{i,img})$, which means the patch predictions should remain consistent with the image prediction.

We made some attempts with small-scale data in the early stage and found that the image prediction confidence scores were high (usually above 0.9); however, the patch prediction confidence scores were relatively low (usually below 0.7). Therefore, we empirically set τ_1 to 0.8 and τ_2 to 0.6.

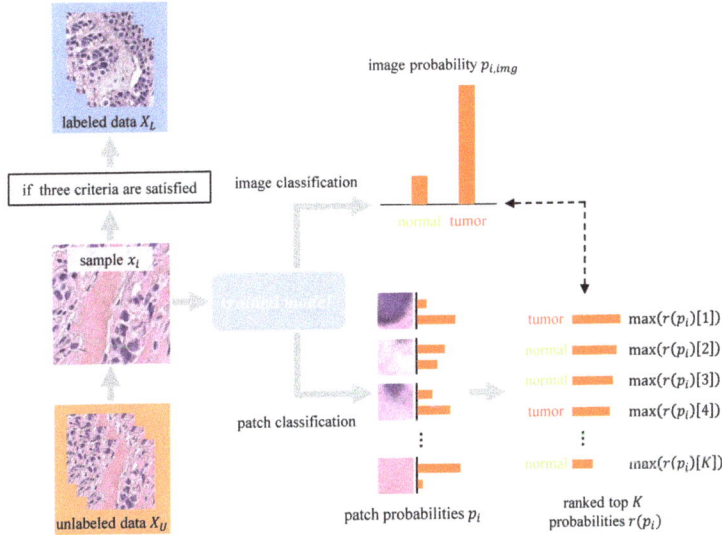

Figure 4. Illustration of the pseudo label generation process. Note that the ranked top K probabilities $r(p_i)$ only displays the dominant values for each $r(p_i)[k], k \in [1, K]$. For example, if $r(p_i)[k]$ is 'tumor': 0.6, 'normal': 0.4, then the dominant value of $r(p_i)[k]$ is 'tumor': 0.6. Thus, $max(r(p_i)[k]) = 0.6$ and $argmax(r(p_i)[k]) = $ 'tumor'.

3.3. Consistency Learning

When the step of pseudo label generation is finished, the model begins to retrain on the updated training set X_L. The details are as follows. First, for an input image $x \in X_L$, it is transformed into aug_x and aug_x' by twice independent data augmentation operation. Then, the student model and the teacher model take them as input and output two sets of patch predictions $\hat{y}$ and $\hat{y}'$, respectively. These two sets should remain consistent based on the smoothness assumption in semi-supervised learning [46]. Therefore, we apply the KL divergence consistency loss between $\hat{y}$ and $\hat{y}'$:

$$L_{cons} = -\frac{1}{M} \sum_{m}^{M} \sum_{c}^{C} \hat{y}^{(m,c)} \log \frac{\hat{y}^{(m,c)}}{\hat{y}'^{(m,c)}}. \qquad (8)$$

where M is the number of patches in the sample x; C is the number of categories; $\hat{y}^{(m,c)}$ and $\hat{y}'^{(m,c)}$ are the predictions of the mth patches at the cth category. Thus, the total loss function can be written as,

$$L_{total} = L_{sup} + \lambda(E) L_{cons}, \quad (9)$$

where L_{sup} is previously defined in Equation (6). $\lambda(E)$ is a function of training epoch index E, which helps control the balance between the supervised loss and the consistency loss. As is the case with other consistency learning methods [40,47], we use a Gaussian ramp-up function as $\lambda(E)$:

$$\lambda(E) = \begin{cases} \lambda_{max} \cdot exp[-5(1 - \frac{E}{E_{max}})^2], & E < E_{max} \\ \lambda_{max}, & \text{otherwise} \end{cases}, \quad (10)$$

where E is the epoch index. When $E = E_{max}$, λ reaches the maximum weight λ_{max} for the consistency loss. We empirically set λ_{max} to 1 and E_{max} to 20 epochs. For the student model, the parameters θ are updated through back-propagation algorithm by minimizing L_{total}. For the teacher model, the parameter θ' are initially set to θ_0 and updated by computing the exponential moving average of θ:

$$\theta'_t = \alpha \theta'_{t-1} + (1 - \alpha) \theta_t. \quad (11)$$

where t represents the index of the global training steps. α helps control the speed at which the teacher model parameters θ' are updated, and we empirically set it to 0.99.

3.4. Datasets

We evaluated our proposed method on a public dataset BCSS [48] and an in-house dataset. BCSS dataset includes 151 hematoxylin and eosin-stained images corresponding to 151 histologically-confirmed breast cancer cases. The mean image size is 1.18 mm^2 (SD = 0.80 mm^2). We followed the train-test splitting rule (https://bcsegmentation.grand-challenge.org/Baseline/ (accessed on 1 June 2022)) that the images from these institutes were used as an unseen testing set to report accuracy: OL, LL, E2, EW, GM, and S3. (The abbreviations stand for tissue source sites (For more details, see https://docs.gdc.cancer.gov/Encyclopedia/pages/TCGA_Barcode/) (accessed on 1 June 2022)). Then, the remained 108 images were cropped into 27,207 smaller images (with the size of 224 × 224). We used 1018 of these smaller images for validation and the remained were for training.

The in-house dataset came from Department of Pathology, the First Affiliated Hospital of Sun Yat-sen University, China. This study was approved by the Ethics Committee of First Affiliated Hospital of Sun Yat-sen University, and data collection were performed in accordance with relevant guidelines and regulations. The dataset contains 28,187 images from 111 cases (WSIs). We used the images of 84 cases for training and validation, and the images from the remaining cases for test. For the training set, 292 images were from the non-tumor regions, labeled as 'normal'.

A total of 24,971 images were from tumor regions but many of them did not contain any tumor cells. We selected 407 out of these images and labeled 10 patches for each images using our self-developed annotation tool. Among these labeled images, if one contains any tumor cells, then at least one patch will be labeled as 'tumor', and the image label will be 'tumor', as well. Details about the BCSS dataset and the in-house dataset are shown in Tables 1 and 2, respectively.

Table 1. Summary of the BCSS dataset.

Cases/WSIs/ROIs	151
ROIs for training and validation	106
Images (224 × 224) for training	26,189
Images (224 × 224) for validation	1018
ROIs for test	45
Images (224 × 224) for test	9444

Table 2. Summary of the in-house dataset.

Cases/WSIs	111
Cases for training and validation	84
Patch-level-labeled Images (224 × 224) for training	407
Patch-level-labeled Images (224 × 224) for validation	292
Images (224 × 224) from non-tumor regions for training	222
Unlabeled images (224 × 224) for training	24,564
Cases for test	27
Patch-level-labeled Images (224 × 224) for test	2702

4. Results

4.1. Experimental Setup

4.1.1. Training Settings

In the training step, we employed the AdamW optimizer [49] with a base learning rate of 5×10^{-4}. For the learning rate schedule, we adopted a linear warmup for five epochs (the warmup learning rate was 5×10^{-7}), followed by cosine annealing for 20 epochs. The batch size was 16, and the backbones used for Pseger were pre-trained on ImageNet. All experiments were done with a RTX 3090. There are five training strategies for PSeger:

- *Baseline:* train the model only on the labeled data.
- *Baseline+CL:* train the model only on the labeled data with consistency learning.
- *Baseline+CL with X_u:* train the model on both the labeled data and unlabeled data with consistency learning.
- *Baseline+ST with X_u:* first train the model on the labeled data, then use the trained model to infer the pseudo labels of the unlabeled data, and finally retrain the model on both the labeled data and pseudo-labeled data.
- *Baseline+ST+CL with X_u:* first train the model on the labeled data, then use the trained model to infer the pseudo labels of the unlabeled data, and finally retrain the model on both the labeled data and pseudo-labeled data with consistency learning.

4.1.2. Evaluation Metrics

In the experiment of comparison with segmentation models, we choose Intersection over Union (IoU) as the evaluation indicator, which is calculated as follows,

$$IoU = \frac{A \cap B}{A \cup B}, \quad (12)$$

where A and B are the predicted tumor area and ground truth, respectively. The final IoU score is obtained by averaging the IoU for each RoI in the BCSS test set.

In the ablation study, since our in-house dataset has no pixel-wise annotations, we select patch-level and image-level Acc, AUC, and F1 as evaluation indicators. AUC (Area Under the Curve) score is simply the area under the Receiver Operating Characteristic (ROC) curve. Acc and F1 are calculated as follows,

$$Acc = \frac{TP + TN}{TP + TN + FP + FN}, \quad (13)$$

$$F1 = \frac{2TP}{2TP + FP + FN}, \quad (14)$$

where TP, TN, FP, and FN are true positive, true negative, false positive, and false negative, respectively. The final scores of each evaluation indicator are calculated by averaging the score for each image in BCSS or the in-house test set.

4.2. Comparison with Segmentation Models

We compared our proposed method to a variety of segmentation models on the BCSS dataset (Figure 5). We trained PSeger with two strategies: *Baseline* and *Baseline+ST+CL with X_u*. Five patches were labeled for each images in the labeled training set, and ratios of labeled training data were from 1% to 25%. In comparison, we chose two architectures of segmentation models, DeepLabv3+ [50] and Unet++ [51], and equipped them with six backbones: ResNet18, ResNet34, ResNet50 [52], EfficientNet-B1, EfficientNet-B3 [53], and RegNetX-1.6GF [54], respectively.

Therefore, 12 segmentation models were trained and tested on the BCSS dataset. These segmentation models and the training and test steps were implemented base on SegmentationModels [55]. By comparing the two graphs in Figure 5, we can see that when the proportion of labeled training data reaches 25%, our proposed method can achieve 80.31 ± 0.23% IoU on the test set, comparable with the third-best model (DeepLabv3plus+EfficientNet-b1: IoU = 80.31 ± 0.95%) out of 12 segmentation models.

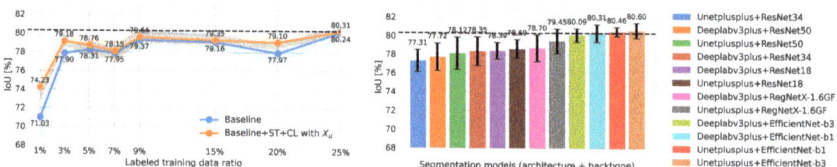

Figure 5. Comparison between our proposed method and pixel-wise segmentation models on the BCSS dataset. **Left:** IoU values of PSeger trained on different ratios of labeled training data by two training strategies (*Baseline, Baseline+ST+CL with X_u*). **Right:** IoU values of different segmentation models trained on the full training set. The values of the black dotted lines in the left and right are both 80.31, representing the IoU that PSeger (trained by *Baseline+ST+CL with X_u* on the training set with 25% labeled data) and the third-best segmentation model (DeepLabv3plus+EfficientNet-b1) have achieved.

4.3. Visualization of Segmentation Results

To further compare our proposed method with the pixel-wise segmentation method, we selected one of the best performing PSegers (trained by *Baseline+ST+CL with X_u* with 25% images in the training set labeled, IoU = 80.65%) and compared it with the best performing model in segmentation models (Unetplusplus+EfficientNet-b3, IoU = 81.74%), as is shown in Figure 6.

In general, the performance of PSeger is comparable to that of Unetplusplus+EfficientNet-b3. The largest prediction differences aroused in case 1 and case 4. In case 1, PSeger performed worse because of more false detection on non-tumorous area; in case 4, Unetplusplus+EfficientNet-b3 performed poorly because of more false positive regions and much more missed detection on tumorous area.

In addition, Figures 7 and 8 display some segmentation results on our in-house dataset. Red and green overlays are tumor regions and non-tumor regions judged by PSeger, respectively, while regions not covered by any overlay are background areas. It can be seen from Figure 8 that our method can accurately segment the invasive tumor and distinguish some non-tumor structures easily confused with tumors.

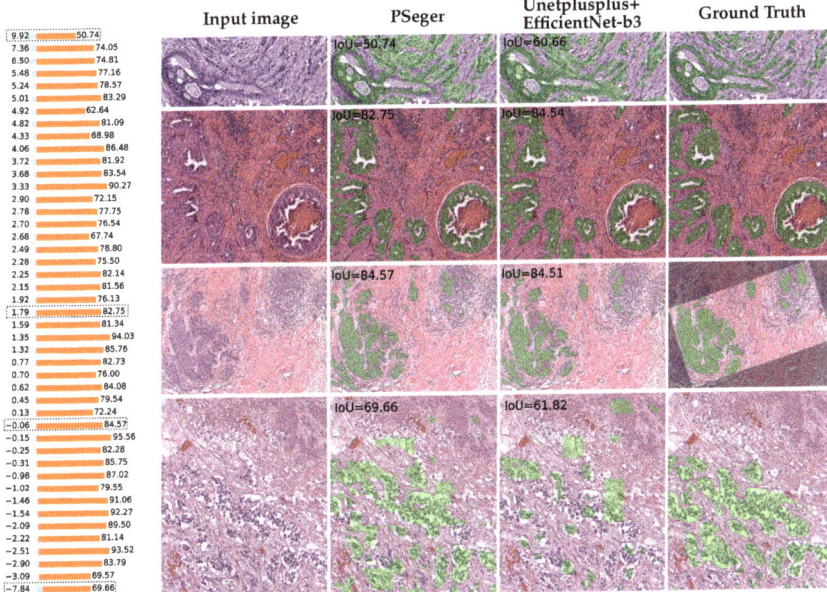

Figure 6. Comparison between PSeger and Unetplusplus+EfficientNet-b3. **Left:** IoU values (orange bars) of PSeger on 45 tested ROIs and their differences (sky-blue bars) with those of Unetplusplus+EfficientNet-b3. The bar pairs are sorted in descending order of the values of the blue bars. **Right:** Images of four representative cases. From top to bottom, rows are case 1–4, also framed by black dotted rectangles in the bar graph on the left. From left to right, columns are input images, segmentation results by PSeger, segmentation results by Unetplusplus+EfficientNet-b3, and ground truths. Green overlays are annotated or predicted tumor regions, black overlays are ignored regions, and others are non-tumor regions.

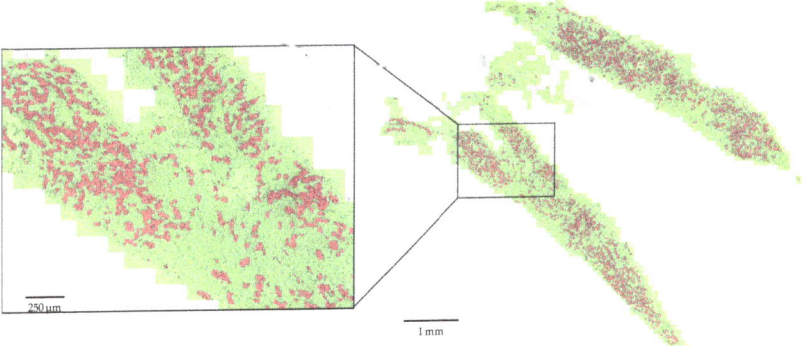

Figure 7. Segmentation results on a whole slide image.

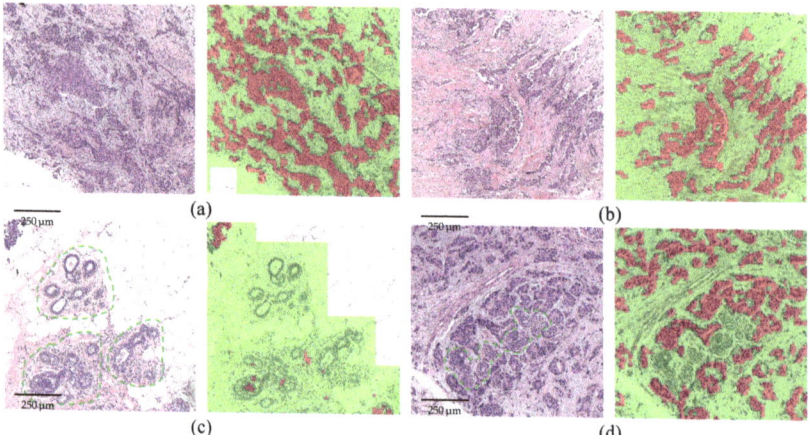

Figure 8. Segmentation results of some ROIs. (**a**,**b**) Examples of invasive tumor. (**c**) An example of lobules (a normal structure in breast tissue). (**d**) An example of lobules surrounded by the invasive tumor. Lobules in (**c**,**d**) are outlined by green dashed polygons.

4.4. Ablation Study

4.4.1. The Effect of the Amount of Labeling

As an important factor affecting model performance, the amount of labeling is reflected in two aspects: the ratio of annotated training samples to all training samples (denoted as $X_l\%$), and the number of the labeled patches in each sample (denoted as K). We conducted experiments on the BCSS dataset to examine the effect of $X_l\%$ and K on the model performance. Figure 9 shows the patch-level AUC values and the image-level AUC values of Baseline and Baseline+ST+CL with X_u under different $X_l\%$ and K, respectively, and the results are given as the mean of three experiments performed in duplicate.

Overall, the two AUC values have increased with increased $X_l\%$ and K. However, the increase has slowed down with higher $X_l\%$ and K. More importantly, Baseline+ST+CL with X_u always outperforms Baseline on image-level AUC, while the former has better patch-level AUC than the latter only when $X_l = 1\%$ or $K = 3$.

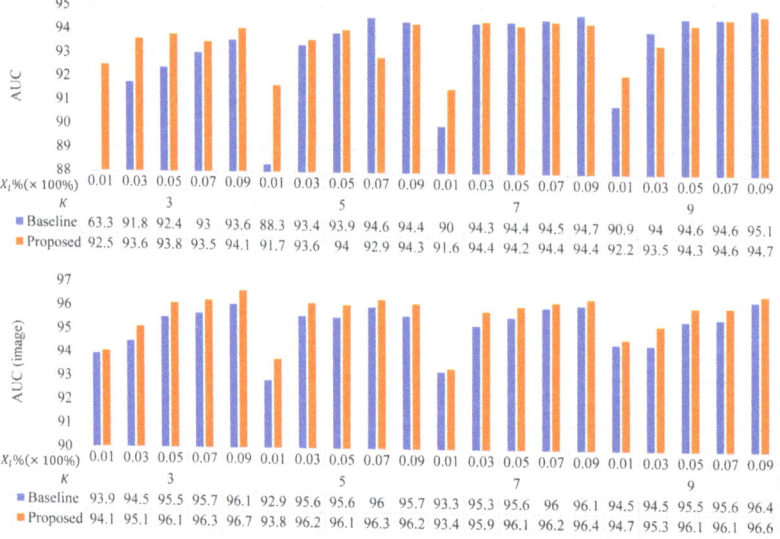

Figure 9. The effect of the amount of labeling.

4.4.2. Training with Different Strategies

To assess the contributions of self-training and consistency learning separately, we performed experiments on the BCSS dataset and the in-house dataset with five different training strategies mentioned before. Each experiment was repeated five times independently and the results are summarized in Tables 3 and 4, where bold and underlined values represent the best and second-best results on a metric, respectively.

Table 3. Model performance on the BCSS dataset with different training strategies.

Training Strategy	AUC	Acc	F1	AUC_{img}	Acc_{img}	$F1_{img}$
Baseline	88.62 ± 0.99	84.28 ± 0.68	78.63 ± 1.44	93.25 ± 0.71	85.91 ± 0.62	87.21 ± 0.67
Baseline+CL	88.41 ± 1.20	83.71 ± 1.43	77.62 ± 2.45	93.23 ± 0.84	86.06 ± 0.63	87.41 ± 0.70
Baseline+CL with X_u	88.67 ± 0.82	84.02 ± 0.74	78.29 ± 1.68	93.09 ± 0.81	**86.17 ± 0.59**	**87.55 ± 0.61**
Baseline+ST with X_u	<u>91.98 ± 0.49</u>	<u>85.58 ± 0.57</u>	<u>80.05 ± 1.28</u>	<u>94.05 ± 0.74</u>	85.48 ± 1.57	86.39 ± 1.83
Baseline+ST+CL with X_u	**92.04 ± 0.36**	**85.72 ± 0.65**	**80.40 ± 1.53**	**94.31 ± 0.32**	<u>85.89 ± 1.49</u>	<u>86.85 ± 1.75</u>

Table 4. Model performance on the in-house dataset with different training strategies.

Training Strategy	AUC	Acc	F1	AUC_{img}	Acc_{img}	$F1_{img}$
Baseline	89.73 ± 0.60	81.79 ± 0.61	82.98 ± 0.47	**97.07 ± 0.72**	92.28 ± 0.76	92.36 ± 0.74
Baseline+CL	89.92 ± 0.52	81.96 ± 0.41	83.17 ± 0.27	96.57 ± 0.67	92.46 ± 0.62	92.53 ± 0.61
Baseline+CL with X_u	89.64 ± 0.24	82.12 ± 0.58	<u>83.28 ± 0.4</u>	96.70 ± 1.07	<u>92.84 ± 0.27</u>	<u>92.90 ± 0.26</u>
Baseline+ST with X_u	**90.26 ± 0.45**	**82.97 ± 0.52**	**83.9 ± 0.37**	<u>97.11 ± 1.03</u>	92.65 ± 0.42	92.72 ± 0.41
Baseline+ST+CL with X_u	89.26 ± 0.74	<u>82.14 ± 0.4</u>	83.22 ± 0.4	96.78 ± 0.42	**92.86 ± 0.26**	**92.92 ± 0.25**

From Table 3, the strategy of Baseline+ST+CL with X_u helps PSeger achieve the best performance on four of the six indicators (AUC = 92.04%, Acc = 85.72%, F1 = 80.4%, AUC_{img} = 94.31%), significantly higher than the value that the strategy of Baseline has achieved (AUC = 88.62%, Acc = 84.28%, F1 = 78.63%, AUC_{img} = 93.25%). The strategy of Baseline+ST with X_u achieves the second-best performance (AUC = 91.98%, Acc = 85.58%, F1 = 80.05%, AUC_{img} = 94.05%), which is roughly similar to that of Baseline+ST+CL with X_u. Additionally, the performance of Baseline+CL is inferior to that of Baseline. Furthermore, when X_u is involved in the training procedure, the model (Baseline+CL with X_u) performs better than Baseline and has reached the highest in the two indicators of Acc_{img} (86.17%) and $F1_{img}$ (87.55%).

From Table 4, while the performance of PSeger trained by Baseline+ST+CL with X_u on the in-house dataset is still better than that trained by Baseline, combining the two semi-supervised strategies (consistency learning and self-training) does not achieve better performance than either.

4.4.3. Backbone Selections

In this experiment, we used all labeled data in the BCSS training set to train the models with different backbones, including DenseNet121 [56], EfficientNet-B0, EfficientNet-B1 [53], HRNet-w18 [57], ResNet18, ResNet34, ResNet50 [52], ResNeXt-101 (32 × 8d) [58], ViT-base [20], and Swin-base [21] and tested their performance on the BCSS test set (Table 5). The experiment was repeated five times. From the results, the model using Swin-base as backbone achieves the best performance, significantly better than other models.

Nevertheless, the CNN-based models still achieve decent outcomes. It is somewhat surprising that the model using ViT-base as the backbone is not as good as the models using the CNN architecture in the patch-level evaluation indexes; however, it can surpass most CNN architecture models in the image-level evaluation indexes (second only to ResNeXt-101 (32 × 8d)).

Table 5. Model performance on the BCSS dataset using different backbones.

Backbone	AUC	ACC	F1	AUC (Image)	ACC (Image)	F1 (Image)
DenseNet121	94.33 ± 0.06	87.47 ± 0.09	83.04 ± 0.13	95.68 ± 0.12	89.67 ± 0.22	91.22 ± 0.20
EfficientNet-B0	94.76 ± 0.12	87.57 ± 0.19	83.00 ± 0.37	95.66 ± 0.08	89.57 ± 0.24	91.14 ± 0.18
EfficientNet-B1	94.57 ± 0.04	87.30 ± 0.04	82.71 ± 0.17	95.80 ± 0.10	89.60 ± 0.07	91.15 ± 0.02
HRNet-w18	94.31 ± 0.09	87.21 ± 0.14	82.47 ± 0.27	95.99 ± 0.10	89.99 ± 0.18	91.39 ± 0.14
ResNet18	94.03 ± 0.09	87.04 ± 0.12	82.26 ± 0.21	95.35 ± 0.19	88.96 ± 0.26	90.56 ± 0.22
ResNet34	94.35 ± 0.05	87.37 ± 0.09	82.85 ± 0.16	95.72 ± 0.16	89.62 ± 0.28	91.17 ± 0.21
ResNet50	94.11 ± 0.12	87.33 ± 0.13	82.76 ± 0.29	95.94 ± 0.18	90.01 ± 0.38	91.48 ± 0.30
ResNeXt-101 (32 × 8d)	94.64 ± 0.09	87.58 ± 0.09	83.11 ± 0.13	96.25 ± 0.07	90.34 ± 0.14	91.64 ± 0.09
ViT-base	94.47 ± 0.07	87.39 ± 0.06	82.94 ± 0.08	96.16 ± 0.09	90.20 ± 0.21	91.66 ± 0.17
Swin-base	**95.41 ± 0.05**	**88.40 ± 0.08**	**84.29 ± 0.12**	**96.64 ± 0.10**	**91.47 ± 0.04**	**92.70 ± 0.05**

5. Discussion

In the ablation study, we first investigated the effect of the amount of labeling on model performance (Figure 9). On the image-level AUC, the model trained by *Baseline+ST+CL* with X_u was always better than that trained by *Baseline* under otherwise equal conditions. However, on the patch-level AUC, that was not always true, particularly when $K > 3$ and $X_l\% > 1\%$. This meant that the proposed semi-supervised method can effectively improve the image classification performance; however, it enhanced the patch classification performance only when the amount of annotation was small. When the annotation amount increased, the semi-supervised learning method was not as good as the fully-supervised learning method. Further study is therefore needed to optimize semi-supervised training.

Next, we performed experiments on different training strategies (Tables 3 and 4). Both consistency learning and self-training benefited the model, and self-training improved the model performance more significantly. Additionally, combining the consistency learning strategy with the self-training strategy has the potential to fully utilize the pseudo-annotated data and further improve model performance. However, it depends on the dataset and requires appropriate parameter settings to achieve the expected result.

Finally, the experiment of training with different backbones (Table 5) proves that our proposed method is suitable for transformer-based models and models with CNN architecture. By comparing the performance of different models, we found that Swin Transformer was better than CNN models on both image-level metrics and patch-level metrics.

In comparison, Vision Transformer was only better than most CNNs on image-level metrics and inferior to many CNNs on patch-level metrics. This may because the patch classification accuracy depends on the ability to capture localized features and the sensitivity to context-driven features. Although Vision Transformer is more sensitive to contextual features than CNN models, its local feature extraction ability is poorer, which affects the final patch classification accuracy.

Our proposed method can be improved in several ways:

- **Hierarchical patch-level label.** Here, we only considered the annotation form at a single scale, which did not take advantage of the information at different magnifications of the pathological images. Therefore, the annotation can be extended to multiple scales, allowing the model to learn from hierarchical information.
- **Automatic patch selection for labeling.** Choosing which patches to label is subjective and will affect the learning effect of the model. Hence, an active learning mechanism [59] can be introduced to automatically find the most informative patches to label, improving learning efficiency.
- **Hybrid CNN-transformer architecture.** In terms of local feature extraction and global feature capture, CNN and transformer have respective advantages, as analyzed before. Therefore, a hybrid CNN-transformer architecture, like in [60,61], might combine the benefits of the two better to achieve greater performance.
- **More advanced semi-supervised algorithm.** Our semi-supervised algorithm still has problems, such as being sensitive to hyperparameters. In the future, ideas from some

advanced semi-supervised algorithms in recent years, such as Mixmatch [62], can be introduced into the training algorithm. At the same time, some constraints can be added to prevent the model from overfitting, such as the consistency of prediction results between the patch classification branch and the image classification branch.

6. Conclusions

In this work, we proposed a novel form of annotation, sparse patch annotation, and developed an annotation tool to achieve this new way of labeling. We created a patch-wise segmentation model called Pseger to handle this new label, which was equipped with an innovative semi-supervised algorithm to fully utilize the unlabeled data. We compared the proposed method to various pixel-wise segmentation models (Figure 5). It was shown that, when trained with only 25% labeled data (five patches for each labeled image), our model achieved comparable segmentation results with the semantic segmentation models trained on fully pixel-level labeled data.

Our proposed method enables pathologists to focus their time and energy on labeling the representative parts of the image rather than carefully delineating complex boundaries, significantly reducing the annotation burden.

Author Contributions: Conceptualization, Y.L.; methodology, Y.L. and Q.H.; software, H.D.; validation, Q.H., H.D. and H.S.; formal analysis, Q.H.; investigation, Y.L.; resources, A.H.; data curation, H.S.; writing—original draft preparation, Y.L.; writing—review and editing, Q.H. and H.D.; visualization, Q.H.; supervision, A.H. and Y.H.; funding acquisition, Y.H. All authors have read and agreed to the published version of the manuscript.

Funding: This work was supported by National Science Foundation of China (61875102), Science and Technology Research Program of Shenzhen City (JCYJ20180508152528735), Oversea cooperation foundation, Graduate School at Shenzhen, Tsinghua University (HW2018007), and Tsinghua University Spring Breeze Fund (2020Z99CFZ023).

Institutional Review Board Statement: Not applicable.

Informed Consent Statement: Not applicable.

Data Availability Statement: Our annotation tool is available at: https://github.com/FHDD/PSeger-LabelMe (accessed on 1 June 2022). The public dataset used in this study can be accessed at the following link: https://bcsegmentation.grand-challenge.org/ (accessed on 1 June 2022). The private dataset is available upon reasonable request to the corresponding authors.

Conflicts of Interest: The authors declare no conflict of interest.

Abbreviations

The following abbreviations are used in this manuscript:

AUC	Area Under the Curve
BCSS	Breast Cancer Semantic Segmentation
CT	Computed Tomography
CL	Consistency Loss
HCC	Hepatocellular Carcinoma
H&E	Hematoxylin and Eosin
IoU	Intersection over Union
LN	Layer Normalization
MLP	Multi-layer Perceptron
MRI	Magnetic Resonance Imaging
MSA	Multi-head Self-attention
ROC	Receiver Operating Characteristic
ROI	Region of Interests

SD	Standard Deviation
SSL	Semi-supervised Learning
ST	Self-training
ViT	Vision Transformer
WSI	Whole Slide Image
WSL	Weakly-supervised Learning

References

1. Campanella, G.; Hanna, M.G.; Geneslaw, L.; Miraflor, A.; Silva, V.W.K.; Busam, K.J.; Brogi, E.; Reuter, V.E.; Klimstra, D.S.; Fuchs, T.J. Clinical-grade computational pathology using weakly supervised deep learning on whole slide images. *Nat. Med.* **2019**, *25*, 1301–1309. [CrossRef] [PubMed]
2. Lu, M.Y.; Williamson, D.F.; Chen, T.Y.; Chen, R.J.; Barbieri, M.; Mahmood, F. Data-efficient and weakly supervised computational pathology on whole-slide images. *Nat. Biomed. Eng.* **2021**, *5*, 555–570. [CrossRef] [PubMed]
3. Coudray, N.; Ocampo, P.S.; Sakellaropoulos, T.; Narula, N.; Snuderl, M.; Fenyö, D.; Moreira, A.L.; Razavian, N.; Tsirigos, A. Classification and mutation prediction from non–small cell lung cancer histopathology images using deep learning. *Nat. Med.* **2018**, *24*, 1559–1567. [CrossRef] [PubMed]
4. Courtiol, P.; Maussion, C.; Moarii, M.; Pronier, E.; Pilcer, S.; Sefta, M.; Manceron, P.; Toldo, S.; Zaslavskiy, M.; Le Stang, N.; et al. Deep learning-based classification of mesothelioma improves prediction of patient outcome. *Nat. Med.* **2019**, *25*, 1519–1525. [CrossRef] [PubMed]
5. Kather, J.N.; Pearson, A.T.; Halama, N.; Jäger, D.; Krause, J.; Loosen, S.H.; Marx, A.; Boor, P.; Tacke, F.; Neumann, U.P.; et al. Deep learning can predict microsatellite instability directly from histology in gastrointestinal cancer. *Nat. Med.* **2019**, *25*, 1054–1056. [CrossRef] [PubMed]
6. Lu, M.Y.; Chen, T.Y.; Williamson, D.F.; Zhao, M.; Shady, M.; Lipkova, J.; Mahmood, F. AI-based pathology predicts origins for cancers of unknown primary. *Nature* **2021**, *594*, 106–110. [CrossRef] [PubMed]
7. Naik, N.; Madani, A.; Esteva, A.; Keskar, N.S.; Press, M.F.; Ruderman, D.; Agus, D.B.; Socher, R. Deep learning-enabled breast cancer hormonal receptor status determination from base-level H&E stains. *Nat. Commun.* **2020**, *11*, 5727. [PubMed]
8. Wang, D.; Khosla, A.; Gargeya, R.; Irshad, H.; Beck, A.H. Deep learning for identifying metastatic breast cancer. *arXiv* **2016**, arXiv:1606.05718.
9. Qaiser, T.; Tsang, Y.W.; Taniyama, D.; Sakamoto, N.; Nakane, K.; Epstein, D.; Rajpoot, N. Fast and accurate tumor segmentation of histology images using persistent homology and deep convolutional features. *Med. Image Anal.* **2019**, *55*, 1–14. [CrossRef] [PubMed]
10. Ni, H.; Liu, H.; Wang, K.; Wang, X.; Zhou, X.; Qian, Y. WSI-Net: Branch-based and hierarchy-aware network for segmentation and classification of breast histopathological whole-slide images. In *International Workshop on Machine Learning in Medical Imaging*; Springer: Berlin/Heidelberg, Germany, 2019; pp. 36–44.
11. Hou, L.; Samaras, D.; Kurc, T.M.; Gao, Y.; Davis, J.E.; Saltz, J.H. Patch-based convolutional neural network for whole slide tissue image classification. In Proceedings of the IEEE Conference on Computer Vision and Pattern Recognition, Las Vegas, NV, USA, 27–30 June 2016; pp. 2424–2433.
12. Liu, Y.; Gadepalli, K.; Norouzi, M.; Dahl, G.E.; Kohlberger, T.; Boyko, A.; Venugopalan, S.; Timofeev, A.; Nelson, P.Q.; Corrado, G.S.; et al. Detecting cancer metastases on gigapixel pathology images. *arXiv* **2017**, arXiv:1703.02442.
13. Mi, W.; Li, J.; Guo, Y.; Ren, X.; Liang, Z.; Zhang, T.; Zou, H. Deep learning-based multi-class classification of breast digital pathology images. *Cancer Manag. Res.* **2021**, *13*, 4605. [CrossRef] [PubMed]
14. Li, J.; Tao, R.; Wu, Q.; Li, B. Da-refinenet: A dual input whole slide image segmentation algorithm based on attention. *arXiv* **2019**, arXiv:1907.06358.
15. Dong, N.; Kampffmeyer, M.; Liang, X.; Wang, Z.; Dai, W.; Xing, E. Reinforced auto-zoom net: Towards accurate and fast breast cancer segmentation in whole-slide images. In *Deep Learning in Medical Image Analysis and Multimodal Learning for Clinical Decision Support*; Springer: Berlin/Heidelberg, Germany, 2018; pp. 317–325.
16. Van Rijthoven, M.; Balkenhol, M.; Siliņa, K.; Van Der Laak, J.; Ciompi, F. HookNet: Multi-resolution convolutional neural networks for semantic segmentation in histopathology whole-slide images. *Med. Image Anal.* **2021**, *68*, 101890. [CrossRef] [PubMed]
17. Chan, L.; Hosseini, M.S.; Rowsell, C.; Plataniotis, K.N.; Damaskinos, S. Histosegnet: Semantic segmentation of histological tissue type in whole slide images. In Proceedings of the IEEE/CVF International Conference on Computer Vision, Seoul, Korea, 27 October–2 November 2019; pp. 10662–10671.
18. Wang, X.; Fang, Y.; Yang, S.; Zhu, D.; Wang, M.; Zhang, J.; Tong, K.y.; Han, X. A hybrid network for automatic hepatocellular carcinoma segmentation in H&E-stained whole slide images. *Med. Image Anal.* **2021**, *68*, 101914. [PubMed]
19. Cho, S.; Jang, H.; Tan, J.W.; Jeong, W.K. DeepScribble: Interactive Pathology Image Segmentation Using Deep Neural Networks with Scribbles. In Proceedings of the 2021 IEEE 18th International Symposium on Biomedical Imaging (ISBI), Nice, France, 13–16 April 2021; pp. 761–765.
20. Dosovitskiy, A.; Beyer, L.; Kolesnikov, A.; Weissenborn, D.; Zhai, X.; Unterthiner, T.; Dehghani, M.; Minderer, M.; Heigold, G.; Gelly, S.; et al. An image is worth 16x16 words: Transformers for image recognition at scale. *arXiv* **2020**, arXiv:2010.11929.

21. Liu, Z.; Lin, Y.; Cao, Y.; Hu, H.; Wei, Y.; Zhang, Z.; Lin, S.; Guo, B. Swin transformer: Hierarchical vision transformer using shifted windows. In Proceedings of the IEEE/CVF International Conference on Computer Vision, Montreal, BC, Canada, 11–17 October 2021; pp. 10012–10022.
22. Tarvainen, A.; Valpola, H. Mean teachers are better role models: Weight-averaged consistency targets improve semi-supervised deep learning results. In Proceedings of the Advances in Neural Information Processing Systems, Long Beach, CA, USA, 4–9 December 2017; pp. 1195–1204.
23. Yalniz, I.Z.; Jégou, H.; Chen, K.; Paluri, M.; Mahajan, D. Billion-scale semi-supervised learning for image classification. *arXiv* **2019**, arXiv:1905.00546.
24. Belharbi, S.; Ben Ayed, I.; McCaffrey, L.; Granger, E. Deep active learning for joint classification & segmentation with weak annotator. In Proceedings of the IEEE/CVF Winter Conference on Applications of Computer Vision, Waikoloa, HI, USA, 3–8 January 2021; pp. 3338–3347.
25. Pinckaers, H.; Bulten, W.; van der Laak, J.; Litjens, G. Detection of prostate cancer in whole-slide images through end-to-end training with image-level labels. *IEEE Trans. Med. Imaging* **2021**, *40*, 1817–1826. [CrossRef]
26. Zhou, C.; Jin, Y.; Chen, Y.; Huang, S.; Huang, R.; Wang, Y.; Zhao, Y.; Chen, Y.; Guo, L.; Liao, J. Histopathology classification and localization of colorectal cancer using global labels by weakly supervised deep learning. *Comput. Med. Imaging Graph.* **2021**, *88*, 101861. [CrossRef]
27. Lin, D.; Dai, J.; Jia, J.; He, K.; Sun, J. Scribblesup: Scribble-supervised convolutional networks for semantic segmentation. In Proceedings of the IEEE Conference on Computer Vision and Pattern Recognition, Las Vegas, NV, USA, 26 June–1 July 2016; pp. 3159–3167.
28. Bearman, A.; Russakovsky, O.; Ferrari, V.; Fei-Fei, L. What is the point: Semantic segmentation with point supervision. In Proceedings of the European Conference on Computer Vision, Amsterdam, The Netherlands, 11–14 October 2016; pp. 549–565.
29. Qu, H.; Wu, P.; Huang, Q.; Yi, J.; Yan, Z.; Li, K.; Riedlinger, G.M.; De, S.; Zhang, S.; Metaxas, D.N. Weakly supervised deep nuclei segmentation using partial points annotation in histopathology images. *IEEE Trans. Med. Imaging* **2020**, *39*, 3655–3666. [CrossRef]
30. Mahani, G.K.; Li, R.; Evangelou, N.; Sotiropolous, S.; Morgan, P.S.; French, A.P.; Chen, X. Bounding Box Based Weakly Supervised Deep Convolutional Neural Network for Medical Image Segmentation Using an Uncertainty Guided and Spatially Constrained Loss. In Proceedings of the 2022 IEEE 19th International Symposium on Biomedical Imaging (ISBI), Kolkata, India, 28–31 March 2022; pp. 1–5.
31. Liang, Y.; Yin, Z.; Liu, H.; Zeng, H.; Wang, J.; Liu, J.; Che, N. Weakly Supervised Deep Nuclei Segmentation with Sparsely Annotated Bounding Boxes for DNA Image Cytometry. *IEEE ACM Trans. Comput. Biol. Bioinform.* **2022**, early access. [CrossRef]
32. Jia, Z.; Huang, X.; Eric, I.; Chang, C.; Xu, Y. Constrained deep weak supervision for histopathology image segmentation. *IEEE Trans. Med. Imaging* **2017**, *36*, 2376–2388. [CrossRef] [PubMed]
33. Kervadec, H.; Dolz, J.; Tang, M.; Granger, E.; Boykov, Y.; Ayed, I.B. Constrained-CNN losses for weakly supervised segmentation. *Med. Image Anal.* **2019**, *54*, 88–99. [CrossRef] [PubMed]
34. Zhang, Y.; Yang, Q. An overview of multi-task learning. *Natl. Sci. Rev.* **2018**, *5*, 30–43. [CrossRef]
35. Graham, S.; Vu, Q.D.; Jahanifar, M.; Minhas, F.; Snead, D.; Rajpoot, N. One Model is All You Need: Multi-Task Learning Enables Simultaneous Histology Image Segmentation and Classification. *arXiv* **2022**, arXiv:2203.00077.
36. Cheng, J.; Liu, J.; Kuang, H.; Wang, J. A Fully Automated Multimodal MRI-based Multi-task Learning for Glioma Segmentation and IDH Genotyping. *IEEE Trans. Med. Imaging* **2022**, *41*, 1520–1532. [CrossRef]
37. Guo, Z.; Liu, H.; Ni, H.; Wang, X.; Su, M.; Guo, W.; Wang, K.; Jiang, T.; Qian, Y. A fast and refined cancer regions segmentation framework in whole-slide breast pathological images. *Sci. Rep.* **2019**, *9*, 882. [CrossRef]
38. Shi, F.; Chen, B.; Cao, Q.; Wei, Y.; Zhou, Q.; Zhang, R.; Zhou, Y.; Yang, W.; Wang, X.; Fan, R.; et al. Semi-Supervised Deep Transfer Learning for Benign-Malignant Diagnosis of Pulmonary Nodules in Chest CT Images. *IEEE Trans. Med. Imaging* **2021**, *41*, 771–781. [CrossRef]
39. Nguyen, H.H.; Saarakkala, S.; Blaschko, M.B.; Tiulpin, A. Semixup: In-and out-of-manifold regularization for deep semi-supervised knee osteoarthritis severity grading from plain radiographs. *IEEE Trans. Med. Imaging* **2020**, *39*, 4346–4356. [CrossRef]
40. Xu, X.; Sanford, T.; Turkbey, B.; Xu, S.; Wood, B.J.; Yan, P. Shadow-consistent Semi-supervised Learning for Prostate Ultrasound Segmentation. *IEEE Trans. Med. Imaging* **2021**, *41*, 1331–1345. [CrossRef]
41. Wang, W.; Xia, Q.; Hu, Z.; Yan, Z.; Li, Z.; Wu, Y.; Huang, N.; Gao, Y.; Metaxas, D.; Zhang, S. Few-shot learning by a Cascaded framework with shape-constrained Pseudo label assessment for whole Heart segmentation. *IEEE Trans. Med. Imaging* **2021**, *40*, 2629–2641. [CrossRef]
42. Zhang, Y.; Li, M.; Ji, Z.; Fan, W.; Yuan, S.; Liu, Q.; Chen, Q. Twin self-supervision based semi-supervised learning (TS-SSL): Retinal anomaly classification in SD-OCT images. *Neurocomputing* **2021**, *462*, 491–505. [CrossRef]
43. Li, D.; Yang, J.; Kreis, K.; Torralba, A.; Fidler, S. Semantic segmentation with generative models: Semi-supervised learning and strong out-of-domain generalization. In Proceedings of the IEEE/CVF Conference on Computer Vision and Pattern Recognition, Virtual, 19–25 June 2021; pp. 8300–8311.
44. Vaswani, A.; Shazeer, N.; Parmar, N.; Uszkoreit, J.; Jones, L.; Gomez, A.N.; Kaiser, Ł.; Polosukhin, I. Attention is all you need. In Proceedings of the Advances in Neural Information Processing Systems, Long Beach, CA, USA, 4–9 December 2017; pp. 6000–6010.

45. Touvron, H.; Cord, M.; Douze, M.; Massa, F.; Sablayrolles, A.; Jégou, H. Training data-efficient image transformers & distillation through attention. In Proceedings of the International Conference on Machine Learning, Vienna, Austria, 18–24 July 2021; pp. 10347–10357.
46. Laine, S.; Aila, T. Temporal ensembling for semi-supervised learning. *arXiv* **2016**, arXiv:1610.02242.
47. Li, X.; Yu, L.; Chen, H.; Fu, C.W.; Xing, L.; Heng, P.A. Transformation-consistent self-ensembling model for semisupervised medical image segmentation. *IEEE Trans. Neural Netw. Learn. Syst.* **2020**, *32*, 523–534. [CrossRef] [PubMed]
48. Amgad, M.; Elfandy, H.; Hussein, H.; Atteya, L.A.; Elsebaie, M.A.; Abo Elnasr, L.S.; Sakr, R.A.; Salem, H.S.; Ismail, A.F.; Saad, A.M.; et al. Structured crowdsourcing enables convolutional segmentation of histology images. *Bioinformatics* **2019**, *35*, 3461–3467. [CrossRef] [PubMed]
49. Loshchilov, I.; Hutter, F. Decoupled weight decay regularization. *arXiv* **2017**, arXiv:1711.05101.
50. Chen, L.C.; Zhu, Y.; Papandreou, G.; Schroff, F.; Adam, H. Encoder-decoder with atrous separable convolution for semantic image segmentation. In Proceedings of the European Conference on Computer Vision (ECCV), Munich, Germany, 8–14 September 2018; pp. 801–818.
51. Zhou, Z.; Siddiquee, M.M.R.; Tajbakhsh, N.; Liang, J. Unet++: A nested u-net architecture for medical image segmentation. In *Deep Learning in Medical Image Analysis and Multimodal Learning for Clinical Decision Support*; Springer: Berlin/Heidelberg, Germany, 2018; pp. 3–11.
52. He, K.; Zhang, X.; Ren, S.; Sun, J. Deep residual learning for image recognition. In Proceedings of the IEEE Conference on Computer Vision and Pattern Recognition, Las Vegas, NV, USA, 26 June–1 July 2016; pp. 770–778.
53. Tan, M.; Le, Q. Efficientnet: Rethinking model scaling for convolutional neural networks. In Proceedings of the International Conference on Machine Learning, Long Beach, CA, USA, 10–15 June 2019; pp. 6105–6114.
54. Radosavovic, I.; Kosaraju, R.P.; Girshick, R.; He, K.; Dollár, P. Designing network design spaces. In Proceedings of the IEEE/CVF Conference on Computer Vision and Pattern Recognition, Seattle, WA, USA, 13–19 June 2020; pp. 10428–10436.
55. Yakubovskiy, P. Segmentation Models Pytorch. 2020. Available online: https://github.com/qubvel/segmentation_models.pytorch (accessed on 1 June 2022).
56. Huang, G.; Liu, Z.; Van Der Maaten, L.; Weinberger, K.Q. Densely connected convolutional networks. In Proceedings of the IEEE Conference on Computer Vision and Pattern Recognition, Honolulu, HI, USA, 21–26 July 2017; pp. 4700–4708.
57. Wang, J.; Sun, K.; Cheng, T.; Jiang, B.; Deng, C.; Zhao, Y.; Liu, D.; Mu, Y.; Tan, M.; Wang, X.; et al. Deep high-resolution representation learning for visual recognition. *IEEE Trans. Pattern Anal. Mach. Intell.* **2020**, *43*, 3349–3364. [CrossRef] [PubMed]
58. Xie, S.; Girshick, R.; Dollár, P.; Tu, Z.; He, K. Aggregated residual transformations for deep neural networks. In Proceedings of the IEEE Conference on Computer Vision and Pattern Recognition, Honolulu, HI, USA, 21–26 July 2017; pp. 1492–1500.
59. Yang, L.; Zhang, Y.; Chen, J.; Zhang, S.; Chen, D.Z. Suggestive annotation: A deep active learning framework for biomedical image segmentation. In *International Conference on Medical Image Computing And Computer-Assisted Intervention*; Springer: Berlin/Heidelberg, Germany, 2017; pp. 399–407.
60. Xie, Y.; Zhang, J.; Shen, C.; Xia, Y. Cotr: Efficiently bridging cnn and transformer for 3d medical image segmentation. In Proceedings of the International Conference on Medical Image Computing and Computer-Assisted Intervention, Strastbourg, France, 27 September–1 October 2021; Springer: Berlin/Heidelberg, Germany, 2021; pp. 171–180.
61. Dalmaz, O.; Yurt, M.; Çukur, T. ResViT: Residual vision transformers for multi-modal medical image synthesis. *arXiv* **2021**, arXiv:2106.16031.
62. Berthelot, D.; Carlini, N.; Goodfellow, I.; Papernot, N.; Oliver, A.; Raffel, C.A. Mixmatch: A holistic approach to semi-supervised learning. In Proceedings of the 33rd Conference on Neural Information Processing Systems (NeurIPS), Vancouver, BC, Canada, 8–14 December 2019; pp. 1–11.

Article

Explainable Transformer-Based Deep Learning Model for the Detection of Malaria Parasites from Blood Cell Images

Md. Robiul Islam [1,*], Md. Nahiduzzaman [1], Md. Omaer Faruq Goni [1], Abu Sayeed [2], Md. Shamim Anower [3], Mominul Ahsan [4,*] and Julfikar Haider [5]

1. Department of Electrical & Computer Engineering, Rajshahi University of Engineering & Technology, Rajshahi 6204, Bangladesh; mdnahiduzzaman320@gmail.com (M.N.); omaerfaruq0@gmail.com (M.O.F.G.)
2. Department of Computer Science & Engineering, Rajshahi University of Engineering & Technology, Rajshahi 6204, Bangladesh; abusayeed.cse@gmail.com
3. Department of Electrical & Electronic Engineering, Rajshahi University of Engineering & Technology, Rajshahi 6204, Bangladesh; md.shamimanower@yahoo.com
4. Department of Computer Science, University of York, Deramore Lane, York YO10 5GH, UK
5. Department of Engineering, Manchester Metropolitan University, John Dalton Building, Chester Street, Manchester M1 5GD, UK; j.haider@mmu.ac.uk
* Correspondence: robiulruet00@gmail.com (M.R.I.); md.ahsan2@mail.dcu.ie (M.A.)

Abstract: Malaria is a life-threatening disease caused by female anopheles mosquito bites. Various plasmodium parasites spread in the victim's blood cells and keep their life in a critical situation. If not treated at the early stage, malaria can cause even death. Microscopy is a familiar process for diagnosing malaria, collecting the victim's blood samples, and counting the parasite and red blood cells. However, the microscopy process is time-consuming and can produce an erroneous result in some cases. With the recent success of machine learning and deep learning in medical diagnosis, it is quite possible to minimize diagnosis costs and improve overall detection accuracy compared with the traditional microscopy method. This paper proposes a multiheaded attention-based transformer model to diagnose the malaria parasite from blood cell images. To demonstrate the effectiveness of the proposed model, the gradient-weighted class activation map (Grad-CAM) technique was implemented to identify which parts of an image the proposed model paid much more attention to compared with the remaining parts by generating a heatmap image. The proposed model achieved a testing accuracy, precision, recall, f1-score, and AUC score of 96.41%, 96.99%, 95.88%, 96.44%, and 99.11%, respectively, for the original malaria parasite dataset and 99.25%, 99.08%, 99.42%, 99.25%, and 99.99%, respectively, for the modified dataset. Various hyperparameters were also finetuned to obtain optimum results, which were also compared with state-of-the-art (SOTA) methods for malaria parasite detection, and the proposed method outperformed the existing methods.

Keywords: malaria parasite; image analysis; deep learning; transformer-based model; grad-cam visualization

1. Introduction

The World Health Organization states that about 438,000 and 620,000 people died from malaria in 2015 and 2017, respectively, whereas 300 to 500 million people are infected by malaria [1]. Malaria virus transmission is influenced by weather conditions that are suitable for a mosquito to live for extended periods, where environmental temperatures are high enough, particularly after rain. For that reason, 90% of malaria cases occur in Africa, and cases are also frequent in humid areas, such as Asia and Latin America [2–4]. If the disease is not treated at the early stages, this may even lead to death. The usual process for detecting malaria starts with collecting blood samples and counting the parasites and red blood cells (RBCs). Figure 1 shows images of RBCs both uninfected and infected by the malaria parasite. This process needs medical experts to collect and examine millions of blood samples, which is costly, time-consuming, and error-prone processes [5]. There are two traditional approaches for detecting malaria: one is very time-consuming because it needs to identify at least 5,000 RBCs, and another is an antigen-based fast diagnostic examination that is very costly. To overcome the limitations of the traditional approaches, in the last few years, researchers have focused on solving this problem using several machine learning and deep learning algorithms.

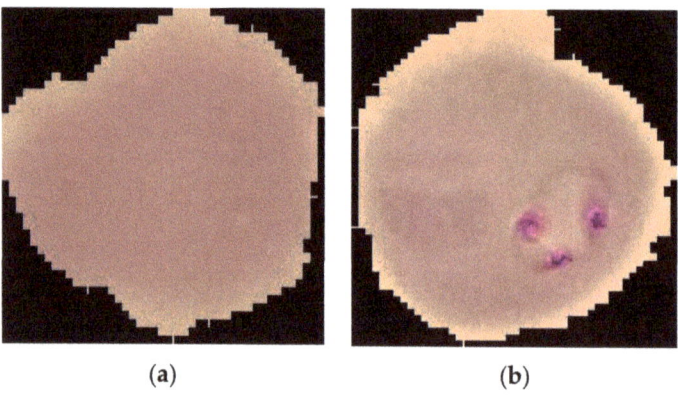

Figure 1. (**a**) Normal and (**b**) malaria-infected RBC images.

A number of studies have been carried out recently to identify malaria using image analysis by artificial intelligence (AI). Bibin et al. proposed a deep belief network (DBN) to detect malaria parasites (MPs) in RBC images [6]. They used 4100 images for training their model and achieved a specificity of 95.92%, a sensitivity of 97.60%, and an F-score of 89.66%. Pandit and Anand detected MPs from the RBC images using an artificial neural network [7] using 24 healthy RBC and 24 infected RBC images in order to train their model and obtained an accuracy of between 90% and 100%. Jain et al. used a CNN model to detect MPs from RBC images [8] without using GPU and preprocessing techniques while providing a low-cost detection algorithm, which achieved an accuracy of 97%. Rajaraman et al. pretrained CNN models for extracting the features from 27,558 RBC cell images to detect MPs and achieved an accuracy of 92.7% [5]. Alqudah et al. developed a lightweight CNN to accurately detect MPs using RBC images [9]. They trained their model using 19,290 images with 4134 test data and achieved an accuracy of 98.85%. Sriporn et al. used six transfer learning models (TL): Xception, Inception-V3, ResNet-50, NasNetMobile, VGG-16, and AlexNet to detect MPs [10]. Several combinations of activation function and optimizer were employed to improve the model's effectiveness. A combined accuracy of 99.28% was achieved by their models trained with 7000 images. Fuhad et al. proposed an automated CNN model to detect MPs from RBC images [11] and performed three training techniques—general, distillation, and autoencoder training—to improve model accuracy after correctly labeling the incorrectly labeled images. Masud et al. proposed leveraging the CNN model to detect MPs using a mobile application [12] and a cyclical stochastic gradient descent optimizer and achieved an accuracy of 97.30%. Maqsood et al. developed a customized CNN model to detect MPs [13] with the assistance of bilateral filtering (BF) and image augmentation methods and achieved an accuracy of 96.82%. Umer et al. developed a stacked CNN model to predict MPs from thin RBC images and achieved an outstanding performance with an accuracy of 99.98%, precision of 100%, and recall of 99.9% [14]. Hung and Carpenter proposed a region-based CNN to detect the object from the RBC images [15]. The total accuracy using one-stage classification and two-stage classification was 59% and 98%, respectively. Pattanaik et al. suggested a methodology for detecting malaria from cell images using computer-aided diagnosis (CAD) [16]. They employed an artificial neural network with a functional link and sparse stacking to pretrain the system's parameters and achieved an accuracy of 89.10% and a sensitivity of 93.90% to detect malaria from a private dataset of 2565 RCB pictures gathered from the University of Alabama at Birmingham. Olugboja et al. used a support vector machine (SVM) and CNN [17] to obtain accuracies of 95% and 91.66%, respectively. Gopakumar et al. created a custom CNN based on a stack of images [18]. A two-level segmentation technique was introduced after the cell counting problem was reinterpreted as a segmentation problem. An accuracy of 98.77%, a sensitivity of 99.14%, and a specificity of 99.62% were achieved from the CNN focus stack model.

Khan et al. used three machine learning (ML) models—logistic regression (LR), decision tree (DT), and random forest (RF)—to predict MPs from RBC images [19]. Firstly, they extracted the aggregated features from the cell images and achieved a high recall of 86% using RF. Fatima and Farid developed a computer-aided system (CAD) to detect MPs from RBC images [20] upon removing the noise and enhancing the quality of the images using the BF method. To detect the MPs, they used adaptive thresholding and morphological image processing and achieved an accuracy of 91%. Mohanty et al. used two models, autoencoder (AE) [21] and self-organizing maps (SOM) [22], to detect MPs and found that AE was better than SOM, which achieved an accuracy of 87.5% [23]. Dong et al. proposed three TL models, LeNet [24], AlexNet, and GoogLeNet [25], to detect MPs [26]. SVM was used to make a comparison with the TL models, which achieved an accuracy of 95%, which was more significant than the accuracy of 92% using the support vector machine (SVM). Anggraini et al. proposed a CAD to detect MPs from RBC images [27] with gray-scale preprocessing for stretching the contrast of the images and global thresholding to gain the different blood cell components from the images.

So far, many computerized systems have been proposed; most of them were based on traditional machine learning or conventional deep learning approaches, which provided satisfactory performances, but there is still scope for further improvement. After developing the vision transformer model [28], the attention-based transformer model has shown promising results in medical imaging, bioinformatics, computer vision tasks, etc. compared with the conventional convolution-based deep learning model. However, to date, no attention-based works have been carried out to detect malaria parasites. Again, the interpretability of a deep CNN model is a major issue. More recently, visualizing what a deep learning model has learned has attracted significant attention to the deep learning community. However, most previous works have failed to introduce the interpretability of the model for malaria parasite detection. To overcome these issues, in this work, an explainable transformer-based model is proposed to detect the malaria parasite from the cell image of blood smear images. Various hyperparameters, such as encoder depth, optimizer (Adam and stochastic gradient descent (SGD)), batch size, etc., were experimented with to achieve better performance. Two malaria parasite datasets (original and modified) were taken into consideration to conduct the experiments.

The key contributions of this paper are:

(1) A multiheaded attention transformer-based model was implemented for the detection of malaria parasites for the first time.
(2) The gradient-weighted class activation map (Grad-CAM) technique was applied to interpret and visualize the trained model.
(3) Original and modified datasets of malaria parasites were used for experimental analysis.
(4) The proposed model for malaria parasite detection was compared with SOTA models.

2. Proposed Methodology

Figure 2 shows the overall design of the proposed methodology. Firstly, the raw images were preprocessed, followed by dataset splitting into training and testing sets to build the model. Finally, to visualize the trained model, Grad-CAM was used to show the heatmap image.

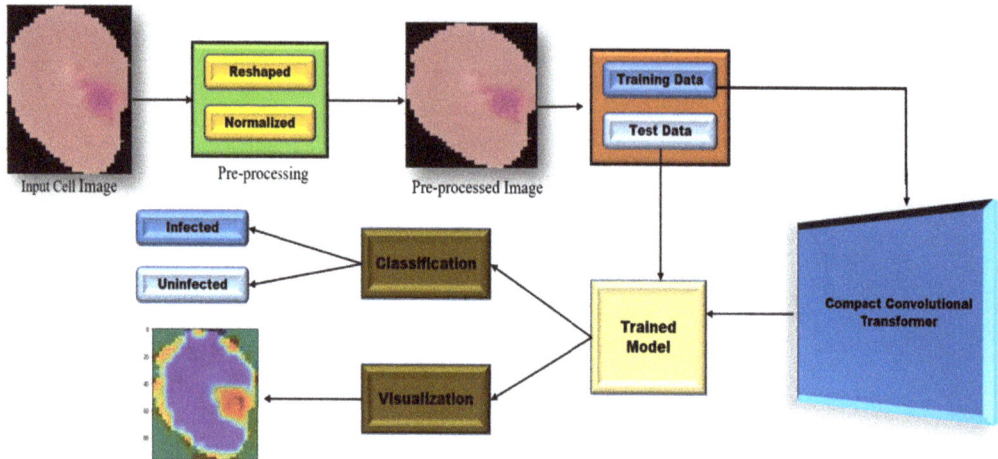

Figure 2. The overall design of the proposed methodology.

2.1. Dataset Description

The dataset for malaria detection contains segmented RBC images. It is archived at the National Library of Medicine and is also openly accessible at "https://lhncbc.nlm.nih.gov/LHC-publications/pubs/MalariaDatasets.html" (accessed on 10 May 2022). Rajaraman et al. [5] developed a segmentation process and implemented it for segmenting RBC images from thin blood smear images. The dataset has a total 27,588 RBC images, among which 13,779 are infected and 13,779 are uninfected images of the malaria parasite. A detailed distribution of the dataset is given in Table 1. This dataset was further studied by a medical expert in the research work conducted by Fuhad et al. [11]. They discovered some suspicious data in the dataset, including data that seemed to be infected but was labeled as uninfected, as well as data that appeared uninfected but was labeled as infected. The data that had been mislabeled was afterward manually annotated. These incorrectly labeled data were simply set aside during annotation, with 647 false infected and suspicious data and 750 false uninfected and suspicious data being eliminated. The updated dataset was uploaded to Google Drive [29], which is open to the public and also taken into account in this work. In both datasets, 20% of the images were used for testing purposes, and 80% of the images were used for training the proposed model.

Table 1. Data distribution in the dataset used in this work.

Dataset	Number of Healthy Images	Number of Infected Images	Total	Total Training Samples (80%)	Total Testing Samples (20%)
Original dataset [5]	13,779	13,779	27,558	22,046	5512
Modified dataset [11]	13,029	13,132	26,161	20,928	5233

2.2. Preprocessing (Resize)

The raw images of the dataset come in a variety of sizes. As the proposed model contains fully connected layers in the classifier layer, the model needs a fixed-sized input image [30]. Therefore, the raw images were resized into $96 \times 96 \times 3$.

2.3. Model Architecture

Various attention-based models have been developed recently. To date, the vision transformer has most attracted researchers for computer vision tasks [28]. The compact convolutional transformer (CCT) is a slightly modified model from the vision transformer introduced in 2021 [31]. CCT with Grad-CAM visualization was implemented in this study to detect the malaria parasite. Figure 3 shows the model architecture of CCT.

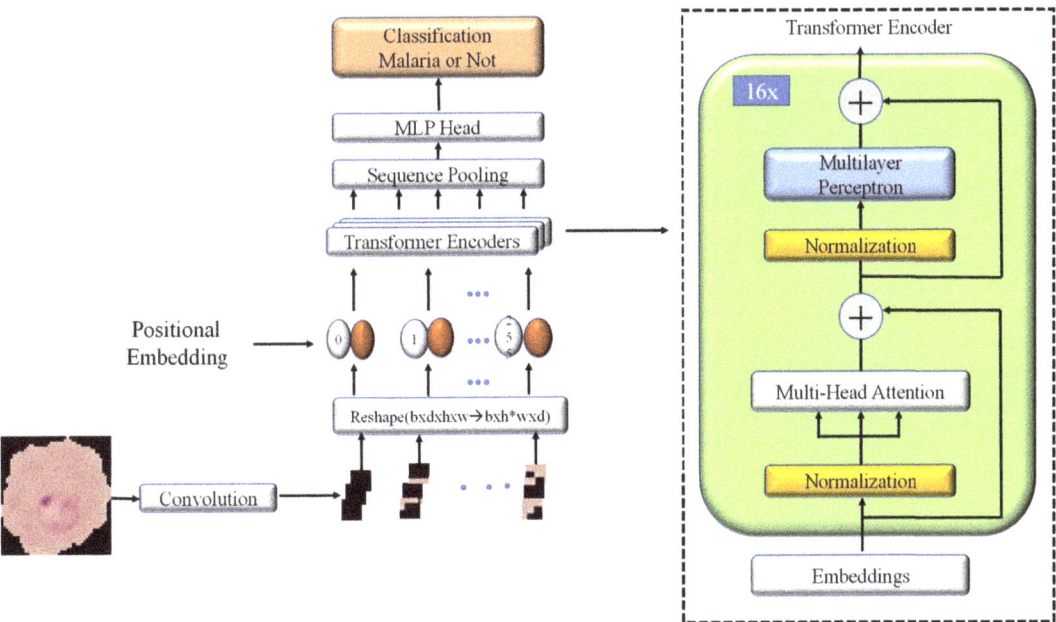

Figure 3. Compact convolutional transformer (CCT) model architecture.

2.3.1. Convolutional Block

The traditional convolutional layer and the ReLU(.) activation function was used. A 3 × 3 kernel size with stride 3 was used to make it a nonoverlapping slide. After that, a maxpool layer was used. Instead of a full input image in the transformer model, the input image is divided into patches/grid images, which are given to the transformer's encoder. In the proposed transformer-based model, the convolution filter was used for patching. Instead of patching images directly, these convolutional blocks took the input images to a latent representation that provides more flexibility than the vision transformer. Filters for the convolutional layer were employed to align with the vision transformer embedding dimension. Given an input image $X \in \mathbb{R}^{H \times W \times C}$

$$X' = MaxPool(ReLU(Conv2d(X))) \in \mathbb{R}^{H' \times W' \times E} \qquad (1)$$

where E is the number of filters = 768.

After the convolutional layer, the output image was reshaped from $\mathbb{R}^{H' \times W' \times E}$ to $\mathbb{R}^{N \times E}$ for converting it to the convolutional patches, where the number of sequences or patches $N \equiv (H'W')$. This convolutional block maintains the locally spatial information. To keep tracking the position or sequence number of each patch, a learnable positional embedding was added.

2.3.2. Multiheaded Attention Mechanism

The main part of the compact convolutional transformer is the multiheaded self-attention (MSA). The whole part of an image is not necessary for extracting valuable information; the attention mechanism focuses on the valuable part. Various attention mechanisms have been developed so far. However, the multiheaded self-attention was first introduced in the vision transformer. Figure 4 shows the scaled dot-product-based multihead attention mechanism [32].

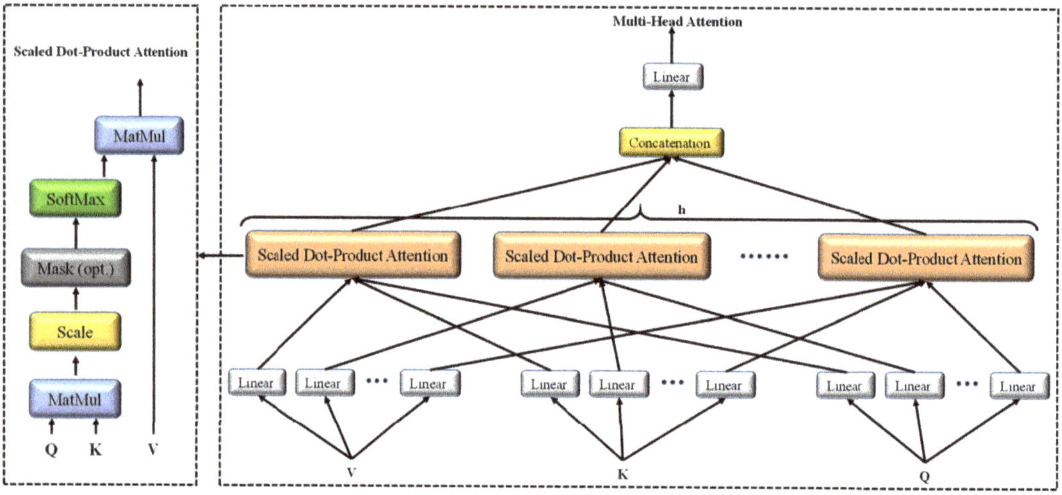

Figure 4. Scaled dot-product-based multihead attention mechanism.

The input is projected to queries, keys, and values using different learnable weights with linear layers in self-attention.

$$Q = (X' \in \mathbb{R}^{N \times E}) \times \left(W_Q \in \mathbb{R}^{E \times d}\right) \qquad (2)$$

$$K = (X' \in \mathbb{R}^{N \times E}) \times \left(W_K \in \mathbb{R}^{E \times d}\right) \qquad (3)$$

$$V = (X' \in \mathbb{R}^{N \times E}) \times \left(W_V \in \mathbb{R}^{E \times d}\right) \qquad (4)$$

Now queries, $Q \in \mathbb{R}^{N \times d}$; keys, $K \in \mathbb{R}^{N \times d}$; and values, $V \in \mathbb{R}^{N \times D}$. In the case of a scaled dot-product form of attention, the dot product is computed between the queries and keys, which is scaled by $\sqrt{d_k}$. After that, nonlinear softmax function is imposed to obtain the attention weights.

$$Z' = \left(Q \in \mathbb{R}^{N \times d}\right) \times \left(K^T \in \mathbb{R}^{d \times N}\right) \qquad (5)$$

$$Z = softmax\left((Z' \in \mathbb{R}^{N \times N})/\sqrt{d_K}\right) \qquad (6)$$

This $Z \in \mathbb{R}^{N \times N}$ is the attention weight among the patches. This attention weight is then multiplied with values to obtain the self-attention weighted output $H \in \mathbb{R}^{N \times d}$

$$H = \left(Z \in \mathbb{R}^{N \times N}\right) x \left(V \in \mathbb{R}^{N \times d}\right) \qquad (7)$$

Therefore, the scaled dot-product attention function can be written in shorted form as below:

$$Attention(Q, K, V) = softmax\left(\frac{QK^T}{\sqrt{d_K}}\right)V \qquad (8)$$

Rather than performing a single attention function with d-dimensional queries, keys, and values, it is advantageous to linearly project queries, keys, and values to d_h, d_k, and d_v dimensions, h times using different learnable weights with linear layers. After that, the scaled dot-product attention function is applied in parallel in all the h heads, resulting in h number of d_v-dimensional values. These attention-weighted values are then concatenated and further projected with linear layers. The multiheaded attention mechanism helps the model attend to different parts from different representation subspaces. In this work, $d_h = d_k = d_v = d$ was applied, and the number of heads used was $h = 8$.

$$head_i = Attention(Q, K, V) \in \mathbb{R}^{N \times d} \tag{9}$$

$$MultiHead(Q, K, V) = Concatenate[head_1, head_2, \ldots head_h]W^0 \tag{10}$$

where $W^0 \in \mathbb{R}^{hd \times d}$.

2.3.3. Transformer Encoder

The image patches come from the convolutional block and are passed through the transformer encoder. Firstly, layer normalization is applied to the image patches that normalize the activations along the feature direction instead of the mini-batch direction in batch normalization. Then multiheaded self-attention is applied to these normalized patches. The results are added with the residual connected original patches, as shown in Figure 4. Further layer normalization and a feed-forward block are imposed along with another residual connection. The feed-forward block has two linear layers along with a dropout layer and a GELU nonlinearity. The first linear layer expands the dimension four times, and the second linear layer reduces the dimension back (feed-forward block).

$$O\prime = Linear\left(A \in \mathbb{R}^{N \times d}\right) \in \mathbb{R}^{N \times 4d} \tag{11}$$

$$O\prime\prime = Dropout(GELU(O\prime)) \in \mathbb{R}^{N \times 4d} \tag{12}$$

$$O = Linear\left(O\prime\prime \in \mathbb{R}^{N \times 4d}\right) \in \mathbb{R}^{N \times d} \tag{13}$$

The outcomes from these two paths are added again. For this work, 16 transformer encoders were implied sequentially.

2.3.4. Sequence Pooling

In the vision transformer, a class token is used to classify the final output. However, in the compact convolutional transformer, instead of using a class token, sequence pooling is used. Sequence pooling pools over the entire sequence of data. Given the output of the last transformer encoder block $X_L \in \mathbb{R}^{b \times N \times d}$, b is the mini-batch size, N is the number of sequences, X_L is sent through a linear layer, and then the output X_L' from this linear layer is multiplied with X_L.

$$X_L' = \in softmax\left(Linear\left(X_L \in \mathbb{R}^{b \times N \times d}\right)\right) \in \mathbb{R}^{b \times N \times 1} \tag{14}$$

$$\begin{aligned} F &= \left(X_L' \in \mathbb{R}^{b \times N \times 1}\right) \cdot \left(X_L \in \mathbb{R}^{b \times N \times d}\right) \\ &= \left(X_L'^T \in \mathbb{R}^{b \times 1 \times N}\right)\left(X_L \in \mathbb{R}^{b \times N \times d}\right) \\ &= Reshape\left(X_L'' \in \mathbb{R}^{b \times 1 \times d}\right) \in \mathbb{R}^{b \times d} \end{aligned} \tag{15}$$

The output F is then sent to the final linear classifier to classify the input data.

3. Grad-CAM Visualization

The gradient-weighted class activation map (Grad-CAM) is a technique to interpret what the model has actually learned [33]. This technique generates a class-specific heatmap using a trained deep learning model for a particular input image. This Grad-CAM approach highlights the input image regions where the model pays much attention to producing discriminative patterns from the last layer before the final classifier, as the last layer contains the most highly semantic features. Grad-CAM uses the feature maps from the last convolutional layer, providing the best discriminative semantics. Let y^c be the class score for class c from the classifier before the SoftMax layer. Grad-CAM has three basic steps:

Step-1: Compute the gradients of class score y^C with respect to the feature maps A^k of the last convolutional layer before the classifier, i.e.,

$$\frac{\partial y^c}{\partial y^k} \in \mathbb{R}^{F \times U \times V}$$

where the feature map is

$$A^k \in \mathbb{R}^{F \times U \times V}$$

Step-2: To obtain the attention weights α^c, global average pool the gradients over the width (indexed by i) and height (indexed by j).

$$\begin{aligned} \alpha_k^c &= \frac{1}{Z} \sum_i \sum_j \frac{\partial y^c}{\partial A_{ij}^k} \\ &= \in \mathbb{R}^{F \times 1 \times 1} \\ &= \in \mathbb{R}^F \, [simplify] \end{aligned} \quad (16)$$

Step-3: Calculate the final Grad-CAM heatmap by the weighted (α^c) sum of feature maps (A^k) and then apply the ReLU (.) function to retain only the positive values and turn all the negative values into zero.

$$\begin{aligned} L_{heatmap}^c &= ReLU\left(\sum_k \alpha_k^c A^k\right) \\ &= \in \mathbb{R}^{U \times V} \end{aligned} \quad (17)$$

Firstly, the proposed model was trained with the training samples from the dataset. After the training phase was completed, the trained model was used for evaluation with the testing parts of the dataset. In addition, to explain what the trained model had actually learned, the Grad-CAM technique explained above was applied. Various test images were selected randomly to generate the corresponding heatmap from the trained model using the Grad-CAM approach. In this case, the multilayer perceptron layer of the last transformer encoder before the final classifier was chosen as the target layer. Features and gradients were extracted from that layer, and a heatmap was generated using the above Grad-CAM formula. Subsequently, the heatmap was resized with nearest-neighbor interpolation as the same size as the input image, and the heatmap was overlaid with the input image. Figure 5 shows the original input images and their corresponding heatmap images. For the heatmap image conversion, a jet color map was used. It can be seen from the overlaid heatmap images that the lesion areas are much more reddish than the other regions of the image. These reddish areas are the main lesions responsible for the malaria parasite [34].

There is no existing segmentation dataset of RBC cell images with the parasite mask on the RBC cell image for quantitative analysis. Annotation made in the dataset used in this work was that normal RBC images come with a clean version without any lesions, but the parasite images come with lesions [34]. Based on the presence of these lesions, the RBC cell images were classified either as normal or parasite. To show explainability of the trained model, the CAM technique was applied to generate heatmap images that showed the actual parts (lesions) the model paid attention to during feature extraction and classification. This technique can bring new insights toward detecting MPs accurately.

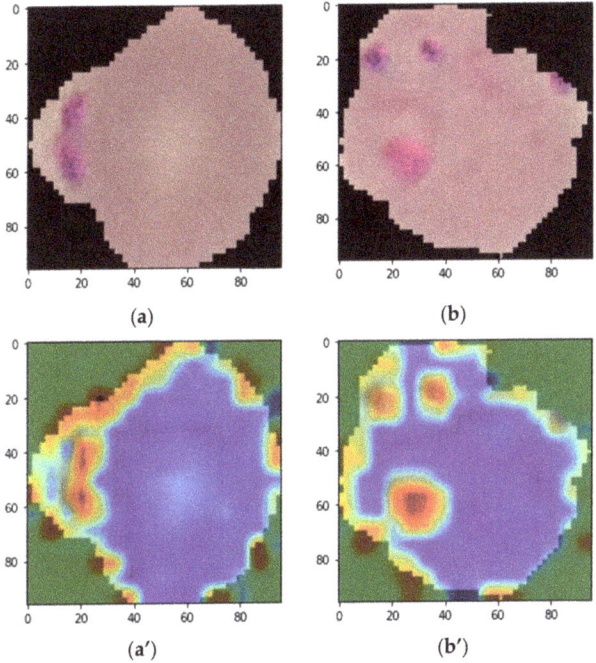

Figure 5. Grad-CAM localization map of the input images (**a**,**b**) and their corresponding overlaid heat map (**a**′,**b**′).

4. Result Analysis

4.1. Performance Evaluation Procedure

Pytorch python framework was used to conduct the whole experiment. The model was run on a highly computing GPU-supported Desktop PC with 11th Generation Intel (R) Core (TM) i9-11900 CPU @2.50GHz, 32 GB RAM, NVIDIA GeForce, and RTX 3090 24 GB GPU running on a 64-bit Windows 10 Pro operating system.

The cell images were preprocessed, then the proposed transformer-based model was trained using original and modified datasets. The performance of the model was tested using 20% of the dataset in both cases. In both cases, the proposed model was trained for 50 epochs, and the learning rate was fixed to 0.001. Various hyperparameters such as optimizers, batch size, and transformer's encoder depth were experimented with for performance analysis.

For measuring the performance of the deep learning model, various evaluation metrics were used. The proposed work was evaluated with confusion matrix (CM), accuracy, precision, recall, f1-score, and area under the curve (AUC) [35,36]:

$$Accuracy = \frac{T_P + T_N}{T_P + T_N + F_P + F_N} \tag{18}$$

$$Recall = \frac{T_P}{T_P + F_N} \tag{19}$$

$$Precision = \frac{T_P}{T_P + F_P} \tag{20}$$

$$AUC = \frac{1}{2}\left(\frac{T_P}{T_P + F_N} + \frac{T_N}{T_N + F_P}\right) \tag{21}$$

where T_P = true positive means that a malaria-infected person is correctly detected as a malaria-infected person, T_N = true negative means that a noninfected person is correctly detected as a noninfected person, F_P = false positive means that a noninfected person is wrongly detected as an infected person, and F_N = false negative means that an infected person is wrongly detected as a noninfected person.

The original and modified datasets were considered for balanced binary classification. To examine the performance of the proposed model, various hyperparameters were considered. Among various optimization methods developed so far for deep learning, "Adam" [37] and "SGD" [38] are the two most used and popular optimization ones. Therefore, to demonstrate their effectiveness in malaria parasite detection, the proposed model was trained using both optimizers. Batch size is also a key factor for the model's learning and obtaining more generalized results. A larger batch size makes a model speed up the training process, whereas a much larger batch size very often provides poor generalization. In this study, the proposed model was also tuned with various batch sizes (8, 16, 32, and 64). Furthermore, different encoder depths (8, 12, and 16) were also experimented with.

4.2. Results Obtained with Original Dataset

4.2.1. Adam Optimizer for Original Dataset

The different performance criteria of the proposed model with the ADAM optimizer, for instance, precision, recall, F1-score, and accuracy, were calculated and are presented in Table 2. The ROCs of the model with the ADAM optimizer for different batch sizes are presented in Figure 6, which shows that the highest AUC of 64.61% was achieved using a batch size of 8. This could be due to the fact that the model trained with larger batch sizes with the Adam optimizer did not show a continuous improvement. Even though the ADAM optimizer produced very high precision, the other results for recall, F1-score, and accuracy were disappointing.

Table 2. Model's performance for various batch sizes with ADAM optimizer and original dataset.

Batch Size	Precision (%)	Recall (%)	F1-Score (%)	Accuracy (%)
8	52.10	**62.03**	56.64	**60.11**
16	63.17	56.05	59.40	56.82
32	100	50	**66.67**	50
64	**99.96**	49.99	66.65	49.98

Note: Bold numbers indicate highest value within a column.

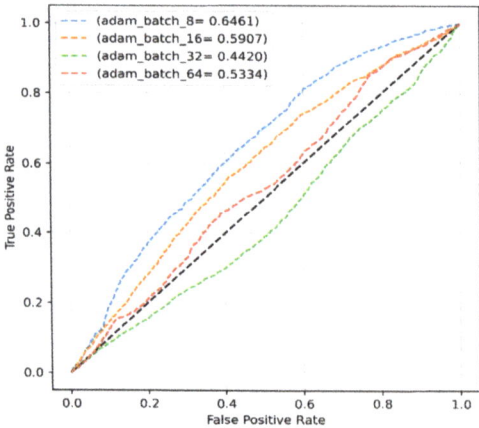

Figure 6. The ROCs of the proposed model obtained with original dataset and ADAM optimizer.

4.2.2. SGD Optimizer for Original Dataset

The effectiveness of the model with the SGD optimizer, along with various batch sizes, is briefly discussed in this section. The model's training and test accuracies for the original dataset are shown in Figure 7, and the training loss curve in Figure 8.

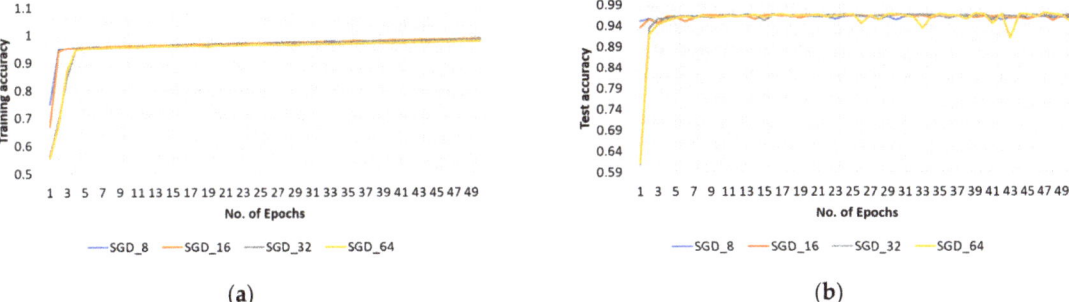

Figure 7. Accuracy curves of (**a**) training and (**b**) test phases of the proposed model obtained with original dataset and SGD optimizer.

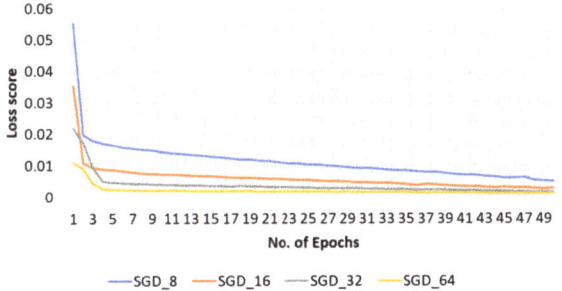

Figure 8. Loss curve of training phase of the proposed model obtained with original dataset and SGD optimizer.

The proposed model's highest training accuracy was 98.61%, which was achieved with a batch size of 8, whereas the highest testing accuracy was 96.86% with a batch size of 64. A batch size of 64 resulted in the shortest training loss of 0.1%. To calculate how well the proposed model with the SGD optimizer detects malaria-infected patients, the same number of cell images as in Adam was used for testing. A number of predicted patients are shown by the CM in Figure 9. The highest accuracy of 96.41% was achieved with the SGD optimizer and a batch size of 32, and the highest recall of 95.88% was achieved for the same batch size (Table 3).

Table 3. Model's performance for various batch sizes with SGD optimizer and original dataset.

Batch Size	Precision (%)	Recall (%)	F1-Score	Accuracy (%)
8	97.06	94.66	95.84	95.79
16	96.73	95.56	96.14	96.12
32	96.99	**95.88**	**96.44**	**96.41**
64	**97.50**	92.53	94.95	94.81

Note: Bold numbers indicate highest value within a column

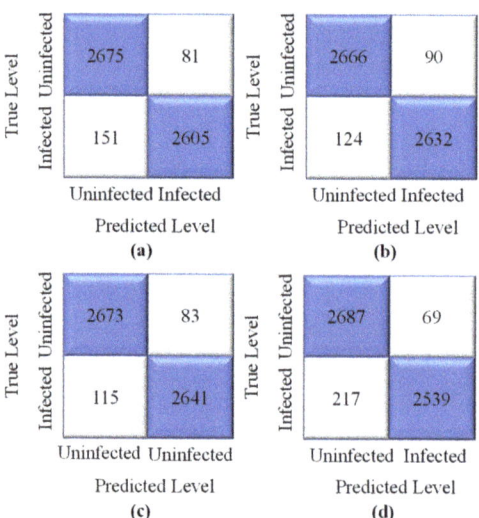

Figure 9. Confusion matrix of the proposed model obtained with original dataset, SGD optimizer, and batch sizes of (**a**) 8 (**b**) 16 (**c**) 32, and (**d**) 64.

Figure 10 shows the ROCs of the proposed model with the SGD optimizer for various batch sizes, and the results indicated the greatest AUC of 99.11% was reached with batch sizes of 16 and 32.

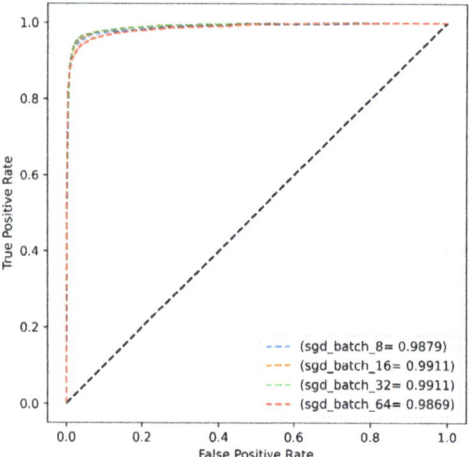

Figure 10. The ROC of the proposed model obtained with original dataset and SGD optimizer.

4.2.3. Encoder Depth for Original Dataset

To show the impact of different depths of the encoder, the SGD optimizer was used, and the batch size was fixed to 16. The highest test accuracy of 96.66% was obtained with an encoder depth of 8 (Figure 11).

ROC scores of ~99% were achieved from the model with all encoder depths of 8, 12, and 16 (Figure 12).

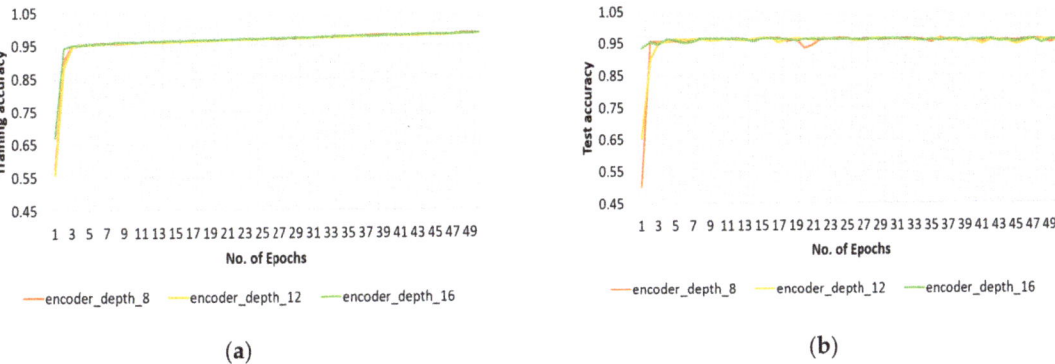

Figure 11. (**a**) Training curves and (**b**) test curves of the proposed model obtained with original dataset and encoder depths of 8, 12, and 16.

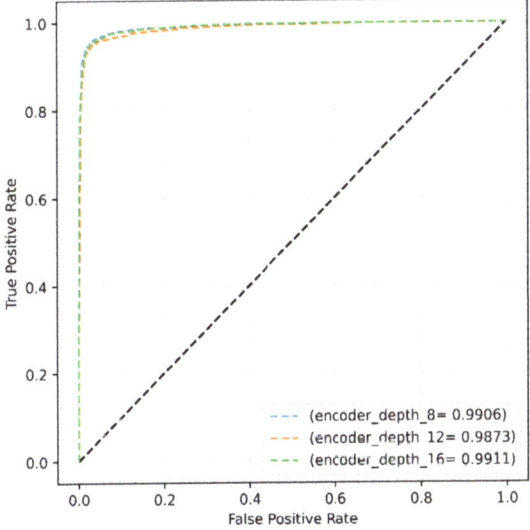

Figure 12. ROC curves of the proposed model obtained with original dataset and encoder depths of 8, 12, and 16.

4.3. Results Obtained with Modified Dataset

4.3.1. Adam Optimizer for Modified Dataset

The proposed transformer-based model's classification performance with the modified dataset and Adam optimizer was evaluated and is presented in Table 4. Again, other than precision, the other performance results were poor. The highest ROC of 59.7% was achieved with a batch size of 16 (Figure 13), indicating no promising results.

Table 4. Model's performance for various batch sizes with ADAM optimizer and modified dataset.

Batch Size	Precision (%)	Recall (%)	F1-Score (%)	Accuracy (%)
8	58.37	53.58	55.87	54.08
16	48.12	**59.69**	53.28	**57.98**
32	100	49.80	66.49	49.80
64	**100**	49.80	**66.49**	49.80

Note: Bold numbers indicate highest value within a column.

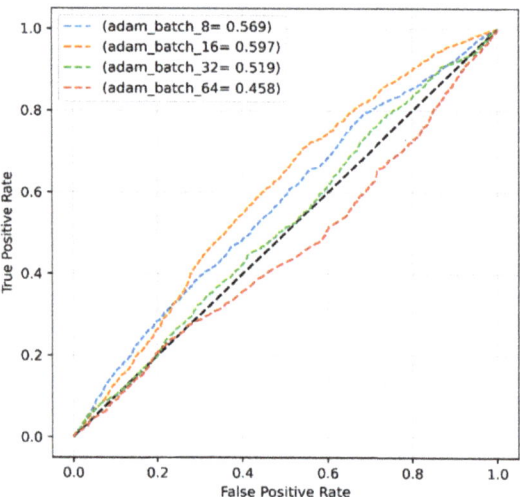

Figure 13. The ROC of the proposed model obtained with modified dataset and ADAM optimizer.

4.3.2. SGD Optimizer for Modified Dataset

Furthermore, the SGD optimizer was used with various batch sizes for the modified dataset, and accuracy and loss curves are shown in Figures 14 and 15, respectively.

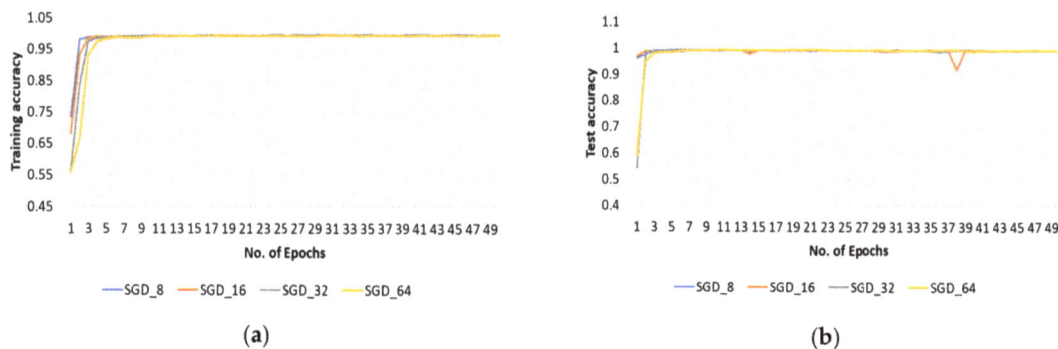

Figure 14. Accuracy curves of (**a**) training and (**b**) test phases of the proposed model obtained with modified dataset and SGD optimizer.

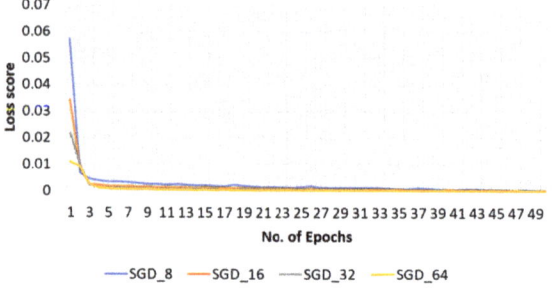

Figure 15. Loss curve of training phase with modified dataset, SGD optimizer, and batch sizes of 8, 16, 32, and 64.

The CMs for the model with the SGD optimizer are shown in Figure 16. Although the highest accuracy of 99.25% and recall of 99.50% were achieved for a batch size of 64 with the SGD optimizer, the results showed insignificant differences between the batch sizes (Table 5). The ROCs of each batch size are demonstrated in Figure 17; the results did not show much difference, with AUC values close to 1.0 in all cases.

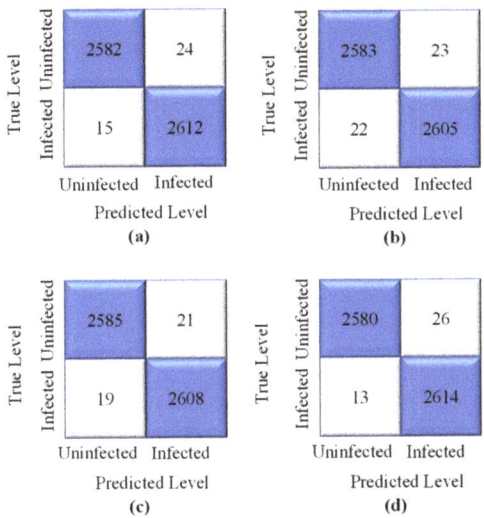

Figure 16. Confusion matrix of the proposed model obtained with modified dataset, SGD optimizer, and batch sizes of (**a**) 8, (**b**) 16, (**c**) 32, and (**d**) 64.

Table 5. Model's performance for various batch sizes with SGD optimizer and modified dataset.

Batch Size	Precision (%)	Recall (%)	F1-Score (%)	Accuracy (%)
8	99.08	99.42	**99.25**	**99.25**
16	99.12	99.16	99.14	99.14
32	**99.19**	99.27	99.23	99.24
64	99.00	**99.50**	**99.25**	**99.25**

Note: Bold numbers indicate highest value within a column.

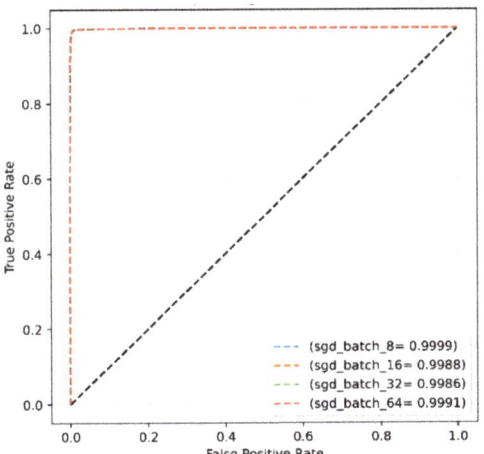

Figure 17. The ROC of the proposed model obtained with SGD optimizer and modified dataset.

4.3.3. Encoder Depth for Modified Dataset

Encoder depth was also finetuned for the modified dataset. The larger model with higher encoder depth showed higher fluctuation, and the highest test accuracy of 99.29% was obtained from the proposed model with encoder depths of 8 and 16. ROC curves in Figure 18 ensured that all models achieved the same high AUC score of approximately 99.9%.

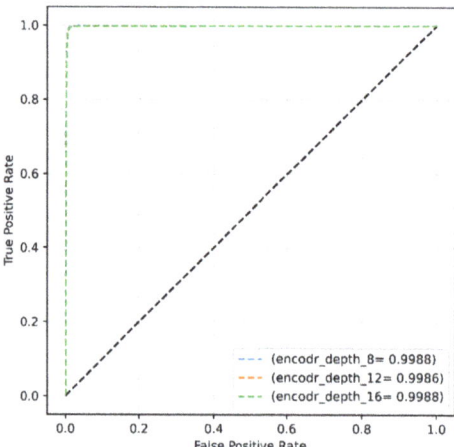

Figure 18. ROC curves of the proposed model obtained with modified dataset and encoder's depth of 8, 12, and 16.

4.4. Performance Comparison between Two Datasets

With smaller batch sizes, the learning process became easier and provided the best evaluation results, but on the other hand, with greater batch sizes, the transformer-based model converged faster and provided much more generalization. After correcting the mislabeled data in the original datasets, the problem became a much easier balanced binary classification. From the above experimental results, it was observed that the Adam optimizer showed poor results in all cases, as it was not guaranteed to converge to the global optimum point. On the other hand, the results obtained using the SGD optimizer were significantly better than that obtained by the Adam optimizer for all performance criteria employed in this work. From Figure 19, it was observed that the classification performance of the model was optimistic with the modified dataset rather than the original dataset. Therefore, the proposed model with the SGD optimizer and modified dataset could be the best combination for accurate MP detection.

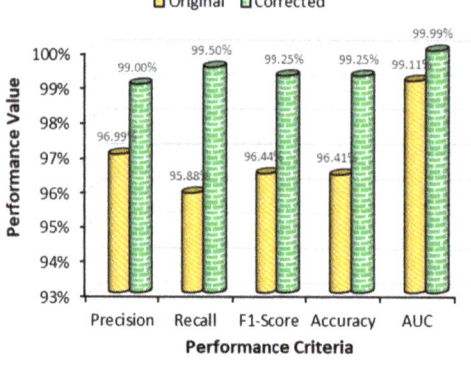

Figure 19. Performance comparison of original and modified datasets for the transformer-based model with SGD optimizer.

However, a similar comparison made between the two datasets using the Adam optimizer showed no improvement with the modified dataset. Furthermore, this indicated that for both datasets, the SGD optimizer could produce optimistic results.

4.5. Performance Comparison with Previous Works

In this section, the performance of the proposed model is compared with several SOTA methods for both datasets. The details of the SOTA methods have been described in the Introduction section.

For the original dataset, the first five rows show the results of the SOTA models in Table 6. It was observed that the highest accuracy score of 91.80% was achieved from the previous work by Fatima and Farid [20]. On the other hand, the proposed transformer-based model achieved a promising accuracy of 96.41% when the SGD optimizer was used with a batch size of 32 and a learning rate of 0.001, almost 5% higher than the other best work reported to date. This suggests that the proposed model could produce even better results than that reported using the same dataset.

Table 6. Performance comparison with previous works.

Reference No	Model Used	Optimizer	Learning Rate	Batch Size	Precision (%)	Recall (%)	AUC (%)	Accuracy (%)
[39]	Custom CNN	Adam	-	-	-	-	-	95
[20]	Image processing	-	-	-	94.66	-	-	91.80
[19]	Random forest	-	-	-	82.00	86.00	-	-
[5]	CNN	SGD	0.0005	-	94.70	**95.90**	**99.90**	-
[16]	Neural network	-	-	-	93.90	-	-	83.10
Proposed work (original dataset)	Transformer	SGD	0.001	32	**96.99**	95.88	99.11	**96.41**
[9]	CNN	Adam	0.001	128	98.79	-	-	98.85
[11]	Custom CNN	SGD	0.01	32	98.92	**99.52**	-	99.23
Proposed work (modified dataset)	Transformer	SGD	0.001	32	**99.00**	99.50	**99.99**	**99.25**

Note: Bold numbers indicate highest value within a column.

For the modified dataset, the highest accuracy of 99.23% was achieved by Fuhad et al. [11]. However, the AUC of the SOTA models did not achieve a satisfactory level, whereas the proposed model showed an optimistic AUC score of 99.99% when the SGD optimizer with a batch size of 64 and a learning rate of 0.001 was employed. The highest AUC of the proposed model proved the highest differentiation capability between malaria-infected and uninfected patients. The table suggested that the proposed transformer-based model achieved a satisfactory classification performance compared with the SOTA models mentioned.

RBC images were collected from an open-source repository. However, the real working procedure starts with segmenting, first, the RBCs in blood smear images that contain various other cells. Afterward, from the affected regions of the RBCs, the malaria parasite is classified. In this work, the segmentation task was ignored, and readymade RBC cell images were used only to classify the malaria parasite. Conventional CNN models show image-specific inductive bias [40], and they are based on a local receptive field. To capture global information, CNN models need larger kernels or very deep network models. However, the transformer models are free from these shortcomings, and, therefore, the transformer-based model proposed for the malaria parasite showed excellent performance. Moreover, the

Grad-CAM visualization demonstrated its explanation visibility. It was also noticed that similar to this study, the application of the SGD optimizer by other studies also produced the highest performance.

5. Conclusions

A multiheaded attention-based transformer model was proposed for malaria parasite detection. In addition, to interpret the trained model, Grad-CAM visualization was used to verify the learning. The proposed work with the transformer model achieved accuracy, precision, recall, and an AUC score of 99.25%, 99.00%, 99.50%, and 99.99%, respectively. Various SOTA works for malaria parasite detection were compared with the proposed model. The model outperformed all the previous works for detecting malaria parasites. Our future work will be focused on segmenting RBCs from blood smear images and classifying malaria parasites from the segmented RBC images.

Author Contributions: All authors have equally contributed to preparing and finalizing the manuscript. Conceptualization, M.R.I., M.N., M.O.F.G. and A.S.; methodology, M.R.I., M.N., M.O.F.G., A.S., M.S.A., M.A. and J.H.; software, M.R.I., M.N., M.O.F.G. and A.S.; validation, M.R.I., M.N., M.O.F.G., A.S., M.S.A., M.A. and J.H.; formal analysis, M.R.I., M.N., M.O.F.G., A.S., M.S.A., M.A. and J.H.; investigation, M.R.I., M.N., M.O.F.G. and A.S.; data curation, M.R.I., M.N., M.O.F.G. and A.S.; writing—original draft preparation, M.R.I., M.N., M.O.F.G. and A.S.; writing—review and editing, M.R.I., M.N., M.O.F.G., A.S., M.S.A., M.A. and J.H.; visualization, M.R.I., M.N., M.O.F.G. and J.H.; supervision, M.S.A., M.A. and J.H.; All authors have read and agreed to the published version of the manuscript.

Funding: This research received no external funding.

Institutional Review Board Statement: Not applicable.

Informed Consent Statement: Not applicable.

Data Availability Statement: The data presented in this study are available in the article.

Acknowledgments: The authors would like to thank the team at the Manchester Met. University and the University of York for supporting this research work and preparing the manuscript.

Conflicts of Interest: The authors declare no conflict of interest.

References

1. World Health Organization. *Malaria Microscopy Quality Assurance Manual-Version 2*; World Health Organization: Geneva, Switzerland, 2016.
2. Caraballo, H.; King, K. Emergency Department Management of Mosquito-Borne Illness: Malaria, Dengue, and West Nile Virus. *Emerg. Med. Pract.* **2014**. Available online: https://europepmc.org/article/med/25207355 (accessed on 10 January 2022).
3. World Health Organization. *Malaria, "Fact Sheet. No,"*; World Health Organization: Geneva, Switzerland, 2014.
4. Wang, H.; Naghavi, M.; Allen, C.; Naghavi, M.; Bhutta, Z.; Carter, A.R.; Casey, D.C.; Charlson, F.J.; Chen, A.; Coates, M.M.; et al. Global, regional, and national life expectancy, all-cause mortality, and cause-specific mortality for 249 causes of death, 1980–2015: A systematic analysis for the Global Burden of Disease Study 2015. *Lancet* **2016**, *388*, 1459–1544. [CrossRef]
5. Rajaraman, S.; Antani, S.K.; Poostchi, M.; Silamut, K.; Hossain, M.A.; Maude, R.J.; Jaeger, S.; Thoma, G.R. Pre-trained convolutional neural networks as feature extractors toward improved malaria parasite detection in thin blood smear images. *PeerJ* **2018**, *6*, e4568. [CrossRef] [PubMed]
6. Bibin, D.; Nair, M.S.; Punitha, P. Malaria parasite detection from peripheral blood smear images using deep belief networks. *IEEE Access* **2017**, *5*, 9099–9108. [CrossRef]
7. Pandit, P.; Anand, A. Artificial neural networks for detection of malaria in RBCs. *arXiv* **2016**, arXiv:1608.06627.
8. Jain, N.; Chauhan, A.; Tripathi, P.; Moosa, S.B.; Aggarwal, P.; Oznacar, B. Cell image analysis for malaria detection using deep convolutional network. *Intell. Decis. Technol.* **2020**, *14*, 55–65. [CrossRef]
9. Alqudah, A.; Alqudah, A.M.; Qazan, S. Lightweight Deep Learning for Malaria Parasite Detection Using Cell-Image of Blood Smear Images. *Rev. d'Intell. Artif.* **2020**, *34*, 571–576. [CrossRef]
10. Sriporn, K.; Tsai, C.-F.; Tsai, C.-E.; Wang, P. Analyzing Malaria Disease Using Effective Deep Learning Approach. *Diagnostics* **2020**, *10*, 744. [CrossRef]
11. Fuhad, K.M.F.; Tuba, J.F.; Sarker, M.R.A.; Momen, S.; Mohammed, N.; Rahman, T. Deep learning based automatic malaria parasite detection from blood smear and its smartphone based application. *Diagnostics* **2020**, *10*, 329. [CrossRef]

12. Masud, M.; Alhumyani, H.; Alshamrani, S.S.; Cheikhrouhou, O.; Ibrahim, S.; Muhammad, G.; Hossain, M.S.; Shorfuzzaman, M. Leveraging deep learning techniques for malaria parasite detection using mobile application. *Wirel. Commun. Mob. Comput.* **2020**, *2020*, 8895429. [CrossRef]
13. Maqsood, A.; Farid, M.S.; Khan, M.H.; Grzegorzek, M. Deep Malaria Parasite Detection in Thin Blood Smear Microscopic Images. *Appl. Sci.* **2021**, *11*, 2284. [CrossRef]
14. Umer, M.; Sadiq, S.; Ahmad, M.; Ullah, S.; Choi, G.S.; Mehmood, A. A novel stacked CNN for malarial parasite detection in thin blood smear images. *IEEE Access* **2020**, *8*, 93782–93792. [CrossRef]
15. Hung, J.; Carpenter, A. Applying faster R-CNN for object detection on malaria images. In Proceedings of the IEEE Conference on Computer Vision and Pattern Recognition Workshops, Honolulu, HI, USA, 21–26 July 2017; pp. 56–61.
16. Pattanaik, P.A.; Mittal, M.; Khan, M.Z. Unsupervised deep learning cad scheme for the detection of malaria in blood smear microscopic images. *IEEE Access* **2020**, *8*, 94936–94946. [CrossRef]
17. Olugboja, A.; Wang, Z. Malaria parasite detection using different machine learning classifier. In Proceedings of the 2017 International Conference on Machine Learning and Cybernetics (ICMLC), Ningbo, China, 9–12 July 2017; Volume 1, pp. 246–250.
18. Gopakumar, G.P.; Swetha, M.; Siva, G.S.; Subrahmanyam, G.R.K.S. Convolutional neural network-based malaria diagnosis from focus stack of blood smear images acquired using custom-built slide scanner. *J. Biophotonics* **2018**, *11*, e201700003. [CrossRef]
19. Khan, A.; Gupta, K.D.; Venugopal, D.; Kumar, N. Cidmp: Completely interpretable detection of malaria parasite in red blood cells using lower-dimensional feature space. In Proceedings of the 2020 International Joint Conference on Neural Networks (IJCNN), Glasgow, UK, 19–24 July 2020; pp. 1–8.
20. Fatima, T.; Farid, M.S. Automatic detection of Plasmodium parasites from microscopic blood images. *J. Parasit. Dis.* **2020**, *44*, 69–78. [CrossRef]
21. Makhzani, A.; Shlens, J.; Jaitly, N.; Goodfellow, I.; Frey, B. Adversarial autoencoders. *arXiv* **2015**, arXiv:1511.05644.
22. Kohonen, T. The self-organizing map. *Proc. IEEE* **1990**, *78*, 1464–1480. [CrossRef]
23. Mohanty, I.; Pattanaik, P.A.; Swarnkar, T. Automatic detection of malaria parasites using unsupervised techniques. In Proceedings of the International Conference on ISMAC in Computational Vision and Bio-Engineering, Palladam, India, 16–17 May 2018; pp. 41–49.
24. El-Sawy, A.; Hazem, E.-B.; Loey, M. CNN for handwritten arabic digits recognition based on LeNet-5. In Proceedings of the International Conference on Advanced Intelligent Systems and Informatics, Cairo, Egypt, 24–26 November 2016; pp. 566–575.
25. Zhong, Z.; Jin, L.; Xie, Z. High performance offline handwritten chinese character recognition using googlenet and directional feature maps. In Proceedings of the 2015 13th International Conference on Document Analysis and Recognition (ICDAR), Tunis, Tunisia, 23–26 August 2015; pp. 846–850.
26. Dong, Y.; Jiang, Z.; Shen, H.; Pan, W.D.; Williams, L.A.; Reddy, V.V.; Benjamin, W.H.; Bryan, A.W. Evaluations of deep convolutional neural networks for automatic identification of malaria infected cells. In Proceedings of the 2017 IEEE EMBS International Conference on Biomedical\Health Informatics (BHI), Orlando, FL, USA, 16–19 February 2017; pp. 101–104.
27. Anggraini, D.; Nugroho, A.S.; Pratama, C.; Rozi, I.E.; Iskandar, A.A.; Hartono, R.N. Automated status identification of microscopic images obtained from malaria thin blood smears. In Proceedings of the 2011 International Conference on Electrical Engineering and Informatics, Bandung, Indonesia, 17–19 July 2011; pp. 1–6.
28. Dosovitskiy, A.; Beyer, L.; Kolesnikov, A.; Weissenborn, D.; Zhai, X.; Unterthiner, T.; Dehghani, M.; Minderer, M.; Heigold, G.; Gelly, S.; et al. An Image is Worth 16 × 16 Words. Transformers for Image Recognition at Scale. October 2020. Available online: https://arxiv.org/abs/2010.11929v2 (accessed on 11 January 2022).
29. Corrected Malaria Data—Google Drive. 2019. Available online: https://drive.google.com/drive/folders/10TXXa6B_D4AKuBV085tX7UudH1hINBRJ?usp=sharing (accessed on 10 January 2022).
30. He, K.; Zhang, X.; Ren, S.; Sun, J. Spatial Pyramid Pooling in Deep Convolutional Networks for Visual Recognition. *Proc. IEEE Trans. Pattern Anal. Mach. Intell.* **2015**, *37*, 1904–1916. [CrossRef]
31. Hassani, A.; Walton, S.; Shah, N.; Abuduweili, A.; Li, J.; Shi, H. Escaping the Big Data Paradigm with Compact Transformers. 2021. Available online: http://arxiv.org/abs/2104.05704 (accessed on 21 February 2022).
32. Vaswani, A.; Shazeer, N.; Parmar, N.; Uszkoreit, J.; Jones, L.; Gomez, A.N.; Kaiser, Ł. Attention is all you need. In Proceedings of the 31st Conference on Neural Information Processing Systems (NIPS 2017), Red Hook, NY, USA, 4–9 December 2017; Volume 30.
33. Selvaraju, R.R.; Cogswell, M.; Das, A.; Vedantam, R.; Parikh, D.; Batra, D. Grad-cam: Visual explanations from deep networks via gradient-based localization. In Proceedings of the IEEE International Conference on Computer Vision, Venice, Italy, 22–29 October 2017; pp. 618–626.
34. Poostchi, M.; Silamut, K.; Maude, R.J.; Jaeger, S.; Thoma, G. Image analysis and machine learning for detecting malaria. *Transl. Res.* **2018**, *194*, 36–55. [CrossRef] [PubMed]
35. Menditto, A.; Patriarca, M.; Magnusson, B. Understanding the meaning of accuracy, trueness and precision. *Accredit. Qual. Assur.* **2007**, *12*, 45–47. [CrossRef]
36. Powers, D.M.W. Evaluation: From Precision, Recall and F-Measure to ROC, Informedness, Markedness and Correlation. October 2020. Available online: http://arxiv.org/abs/2010.16061 (accessed on 10 January 2022).
37. Kingma, D.P.; Ba, J. Adam: A method for stochastic optimization. *arXiv* **2014**, arXiv:1412.6980.
38. Ketkar, N. Stochastic gradient descent. In *Deep Learning with Python*; Springer: Berlin/Heidelberg, Germany, 2017; pp. 113–132.

39. Shah, D.; Kawale, K.; Shah, M.; Randive, S.; Mapari, R. Malaria Parasite Detection Using Deep Learning: (Beneficial to humankind). In Proceedings of the 2020 4th International Conference on Intelligent Computing and Control Systems (ICICCS), Madurai, India, 13–15 May 2020; pp. 984–988.
40. Mondal, A.K.; Bhattacharjee, A.; Singla, P.; Prathosh, A.P. xViTCOS: Explainable Vision Transformer Based COVID-19 Screening Using Radiography. *IEEE J. Transl. Eng. Health Med.* **2021**, *10*, 1–10. [CrossRef] [PubMed]

Article

Hybrid and Deep Learning Approach for Early Diagnosis of Lower Gastrointestinal Diseases

Suliman Mohamed Fati [1,*], Ebrahim Mohammed Senan [2] and Ahmad Taher Azar [1,3]

[1] College of Computer and Information Sciences, Prince Sultan University, Riyadh 11586, Saudi Arabia; aazar@psu.edu.sa

[2] Department of Computer Science & Information Technology, Dr. Babasaheb Ambedkar Marathwada University, Aurangabad 431004, India; senan1710@gmail.com

[3] Faculty of Computers and Artificial Intelligence, Benha University, Benha 13518, Egypt; ahmad.azar@fci.bu.edu.eg

* Correspondence: smfati@yahoo.com

Citation: Fati, S.M.; Senan, E.M.; Azar, A.T. Hybrid and Deep Learning Approach for Early Diagnosis of Lower Gastrointestinal Diseases. *Sensors* 2022, 22, 4079. https://doi.org/10.3390/s22114079

Academic Editors: Mitrea Delia-Alexandrina and Sergiu Nedevschi

Received: 11 April 2022
Accepted: 24 May 2022
Published: 27 May 2022

Publisher's Note: MDPI stays neutral with regard to jurisdictional claims in published maps and institutional affiliations.

Copyright: © 2022 by the authors. Licensee MDPI, Basel, Switzerland. This article is an open access article distributed under the terms and conditions of the Creative Commons Attribution (CC BY) license (https:// creativecommons.org/licenses/by/ 4.0/).

Abstract: Every year, nearly two million people die as a result of gastrointestinal (GI) disorders. Lower gastrointestinal tract tumors are one of the leading causes of death worldwide. Thus, early detection of the type of tumor is of great importance in the survival of patients. Additionally, removing benign tumors in their early stages has more risks than benefits. Video endoscopy technology is essential for imaging the GI tract and identifying disorders such as bleeding, ulcers, polyps, and malignant tumors. Videography generates 5000 frames, which require extensive analysis and take a long time to follow all frames. Thus, artificial intelligence techniques, which have a higher ability to diagnose and assist physicians in making accurate diagnostic decisions, solve these challenges. In this study, many multi-methodologies were developed, where the work was divided into four proposed systems; each system has more than one diagnostic method. The first proposed system utilizes artificial neural networks (ANN) and feed-forward neural networks (FFNN) algorithms based on extracting hybrid features by three algorithms: local binary pattern (LBP), gray level co-occurrence matrix (GLCM), and fuzzy color histogram (FCH) algorithms. The second proposed system uses pre-trained CNN models which are the GoogLeNet and AlexNet based on the extraction of deep feature maps and their classification with high accuracy. The third proposed method uses hybrid techniques consisting of two blocks: the first block of CNN models (GoogLeNet and AlexNet) to extract feature maps; the second block is the support vector machine (SVM) algorithm for classifying deep feature maps. The fourth proposed system uses ANN and FFNN based on the hybrid features between CNN models (GoogLeNet and AlexNet) and LBP, GLCM and FCH algorithms. All the proposed systems achieved superior results in diagnosing endoscopic images for the early detection of lower gastrointestinal diseases. All systems produced promising results; the FFNN classifier based on the hybrid features extracted by GoogLeNet, LBP, GLCM and FCH achieved an accuracy of 99.3%, precision of 99.2%, sensitivity of 99%, specificity of 100%, and AUC of 99.87%.

Keywords: deep learning; hybrid techniques; neural network; gastrointestinal diseases; LBP; GLCM; FCH; endoscope

1. Introduction

Cancer is the greatest threat to human life and is the world's second leading cause of death, following heart disease and atherosclerosis. There are numerous upper and lower gastrointestinal malignancies. Upper gastrointestinal cancers include malignancies of the esophagus and stomach, whereas lower gastrointestinal cancers include colon and rectal cancers. According to a World Health Organization estimate, 3.5 million instances of GI cancer were registered in 2018. The least common type of gastrointestinal cancer is esophageal cancer, but stomach cancer is the fifth most prevalent type of cancer and the third leading cause of death. In contrast, lower-GI tumors are the third most common cancer and the second most common cause of death [1]. Gastrointestinal diseases vary between ulcers, bleeding and polyp, which require early diagnosis otherwise they will develop and be a cause of death. There are many biomarkers to predict health problems in the gastrointestinal tract. However, the high mortality rate shows that there are still possibilities for early diagnosis to receive treatment on time and reduce side effects. If polyps are not diagnosed early, they turn into gastrointestinal cancer [2], which is abnormal cells on the mucous membrane of the colon and stomach. Polyps grow very slowly, and symptoms do not appear until significant [3]. Endoscopy devices that cause pain were used to detect polyps, bleeding and ulcers, and the results of their analysis were inaccurate due to their complex structure [4]. This challenge was overcome in 2000 by using new endoscopic techniques called wireless capsule endoscopy (WCE). WCE is a modern method for detecting and diagnosing diseases of the GI, which has the ability to scan the GI as a whole from the esophagus to the colon in a large video that is divided into frames of thousands of images. Therefore, this massive number of pictures poses a challenge for manual diagnosis because polyps, bleeding, and ulcers may appear in very few frames. In contrast, a massive number of frames look normal. Therefore, all frames must be carefully monitored by experts and doctors, which may take two hours or more to make a proper diagnosis. The WCE technology detects many serious diseases such as polyps, bleeding and ulcers. The large WCE images represent a burden on the gastroenterologist, as it is challenging to track all the images. The morphological features of each disease vary in terms of shape, color, and structure, as well as the etiology of each disease. Each disease has anatomical features that can be distinguished through the endoscope. The endoscope detects the type and location of the disease and gives a brief description and detailed documentation of the most important anatomical landmarks. Additionally, the pathological finding considers abnormal features in the gastrointestinal tract, which can be detected by endoscopy, as a change in the mucous membrane. These signs may be polyps or malignant tumors or persistent disease. Therefore, the early diagnosis of diseases such as polyps and tumors by endoscopy is important in receiving appropriate treatment and survival. Additionally, manual diagnosis is a tedious task that requires tracking all video frames and high experience and clinical knowledge. These challenges can be solved by developing effective computer-aided diagnostic systems represented by machine, deep learning and hybrid techniques. These techniques can assist physicians in deciding on an appropriate diagnosis during the initial stage of the disease [5].

Artificial intelligence systems have shown enormous promise in diagnosing medical images, assisting doctors and specialists in visualizing minute details that the naked eye cannot see [6]. Endoscopic images are used to extract subtle and complicated information using these techniques. They can also distinguish between malignant (neoplastic) and benign tissues. Machine learning approaches can also extract texture, color, form, and edge data with high accuracy and classify all endoscopic pictures according to the type of disease they represent. [7]. Convolutional Neural Networks have proven their great ability to extract feature maps and solve all geometrical feature constraints, which led to accurate diagnosis of GI images. The hybrid technique between machine and deep learning has superior advantages in extracting the deep features by CNN models and sorting them with great speed and accuracy by machine learning algorithms. Thus, artificial intelligence techniques have proven their superiority over the best-specialized experts [8]. ReedT et al. proposed artificial intelligence methods for diagnosing endoscopic images taken from the HyperKvasir dataset. They used grid search to set the best parameters by checking the intersection [9]. Sebastian et al. focused on the detection of polyps in the colon using artificial intelligence techniques that help doctors distinguish the type of tumor [10]. Sharib Ali et al. Due to the presence of artifacts in the internal endoscopy images, which hinder deep learning models for accurate diagnosis. Thus, the researchers focused on removing artifacts, segmentation the lesion area (EAD2020), and detecting the type of disease (EDD2020) [11].

The main contributions of this study are as follows:

- Enhance images using overlapping filters to remove noise, increase contrast, and reveal the edges of the lesion.
- Extracting features by using a hybrid method between three algorithms: LBP, GLCM and FCH.
- Applied hybrid technology consisting of two blocks: the first block is CNN models for extracting feature maps, and the second block is the SVM algorithm for classifying feature maps.
- Fusing features extracted by both CNN models and traditional algorithms (LBP, GLCM, and FCH) to form feature vectors that are more representative of each disease.
- The development of many proposed systems to assist physicians and endoscopy specialists in supporting their diagnostic decisions.

The remainder of the paper is organized as follows: Section 2 reviews the relevant previous studies. Section 3 includes several of methodologies for analyzing and diagnosing the lower gastrointestinal disease dataset. Section 4 presents the experimental results for evaluating the dataset on the proposed systems. Section 5 summarizes the discussion and comparison of the proposed approaches' performance. Section 6 concludes the paper with future directions.

2. Related Work

This section will go over a number of prior studies on the diagnosis of gastrointestinal diseases. In addition, we will develop our technique and vary our methods and materials in order to get superior results in identifying lower GI disorders.

Akshay et al. proposed a colon cancer diagnosis system. The noise was removed to improve the images. Each image was divided into several blocks, and each block was diagnosed using CNN and machine learning algorithms. The system achieved 87% accuracy with the CNN model and 83% accuracy with the KNN algorithm [12]. Tsuyoshi et al. presented a CNN model called Single Shot MultiBox Detector to evaluate the colon dataset. The model was trained on 16,418 images and tested on 7077 images. Each frame was processed for twenty seconds, and the model achieved a sensitivity of 92%, and adenomas were detected with an accuracy of 97% [13]. Alexandr et al. suggested a methodology for early detection of polyps; The system goes through two stages: First, based on the universal features, polyps are classified as having a tumor or not. In the second stage, using CNN models for lesion segmentation, the system achieved a sensitivity and specificity of 93% and 82%, respectively [14]. Ruikai et al. developed a methodology to detect polyps in the colon and rectum. The system worked to identify polyps and predict polyps. The method achieved an accuracy of 87.3% compared to an accuracy of 86.4% by endoscopic specialists [15]. Eduardo et al. presented the CNN model of colonic mucosa diagnosis for colonic polyps. The network extracts and classifies features by exploiting the input pixels and optimizing all noise [16]. Min et al. developed an Asymptotic Laplacian-energy-like invariant of lattices of a system that analyzes the color of the lesions to predict polyps. The dataset was divided into 108 images for training the system and 181 images for testing the system and compared with the results of endoscopy specialists. The system reached an accuracy of 87% during the training phase, and the system achieved an accuracy of 78.4% during the testing phase, compared to an accuracy of 79.6% by endoscopy specialists [17]. Eun et al. Developed an automated computer-aided system to predict the types of diseases in the colon and rectum using CNN models. The system distinguishes between three polyps: serrated polyp (SP), benign adenoma (BA), and deep submucosal cancer (DSMC). The system achieved the highest accuracy of 82.4% [18]. Mehshan et al. presented a method based on deep learning for diagnosing gastrointestinal diseases. The lesion area was segmented by the modified mask and isolated from the rest of the image. ResNet101's pre-trained model is configured to extract the most critical features and categorize them by MSVM [19]. Mustaine et al. discussed improving the way polyps are detected and helping clinicians focus on the most critical areas to diagnose. Colored waveforms and features by a CNN model were extracted to train an SVM classifier. SVM works to see if the images contain a tumor or not [20].

Mahmodul et al. presented a method for polyp diagnosis by fusion of CNN model and contour transformation. Dimensions were reduced and the most important features were combined by Minimum Redundancy Maximum Relevance (MRMR) and Principal Component Analysis (PCA) methods [21]. Chathurika et al. presented a method for integrating the deep features extracted from three CNN models and pooling them into a global average pooling (GAP) layer. The method achieved promising results in the diagnosis of gastrointestinal diseases [22]. Jasper et al. proposed a methodology for evaluating endoscopic images of lesions, including the size and location of the lesion. The method also helps assess the surface pattern and the possibility of excision of the lesion by endoscopic [23]. Roger et al. extracted features by the Global Average Pooling (GAP) to distinguish tumors. Data augmentation technique was applied to balance the dataset; the method reached good results [24]. Şaban et al. provided CNN models to classify the GI dataset; pooling layer features were transferred to the LSTM layer, then all LSTM layers were collected to classify each image [25]. Maghsoudi et al. presented a model for segmentation of polyps of endoscopic images using simple linear iterative clustering (SLIC). The highest sensitivity is found by examining SLIC super-pixels and then classifying them by the SVM classifier, which reached a sensitivity of 91% [26].

Jeph Herrin et al. presented three machine learning algorithms, namely, random survival forests, XGBoost, and RegCox, to predict gastrointestinal bleeding to help clinicians make their decisions. The performance of machine learning algorithms for predicting gastrointestinal bleeding was evaluated using accuracy, sensitivity, and AUC measures. The RegCox algorithm has an AUC of 67%, better than the others [27]. Jayeshkumar et al. presented a random forest algorithm for diagnosing gastrointestinal disorders. The study was conducted on a group of older people with osteoporosis and extract the features. The features were fed to the random forest to diagnose people with gastrointestinal disorders [28]. Hye Jin et al. suggested CAD models for diagnosing gastrointestinal lesions by endoscopic images. Image optimization, noise removal, lesion area segmentation, essential feature extraction, then classification was carried out by machine learning algorithms. The system achieved a sensitivity of 86%, a specificity of 90%, and an AUC of 94% [29].

Previous studies contain drawbacks when applying deep learning models, which are very time-consuming when training the dataset and require costly computers and the inability of the systems to reach the required accuracy. For machine learning algorithms, the drawbacks are that they cannot train a huge dataset. Thus, these obstacles were overcome in this work using hybrid techniques between deep learning to extract feature maps and classify them by the SVM algorithm. Additionally, features from three algorithms were extracted and combined into one feature vector, in addition to the application of a novel method, which is a hybrid technology for extracting features in a hybrid way between deep learning models and GLCM, LBP, and FCH algorithms, then integrating all the features in one vector and classifying them using neural networks, which reached promising results.

3. Materials and Methodology

In this section, the methodology and materials are presented to diagnose endoscopic images for the early detection of lower gastrointestinal diseases, as shown in Figure 1. The first step was to optimize all the images to remove noise and increase the contrast of the edges. Then, the optimized images were fed into four suggested methods. The first method is to classify a dataset using ANN and FFNN based on the hybrid features between LBP, GLCM and FCH algorithms. The second method is to diagnose a dataset by the CNN models GoogLeNet and AlexNet. The third method uses a hybrid technique between CNN models and machine learning (SVM) to diagnose the dataset. The fourth method of dataset diagnosis uses ANN and FFNN based on hybrid features extracted from CNN models and LBP, GLCM and FCH algorithms.

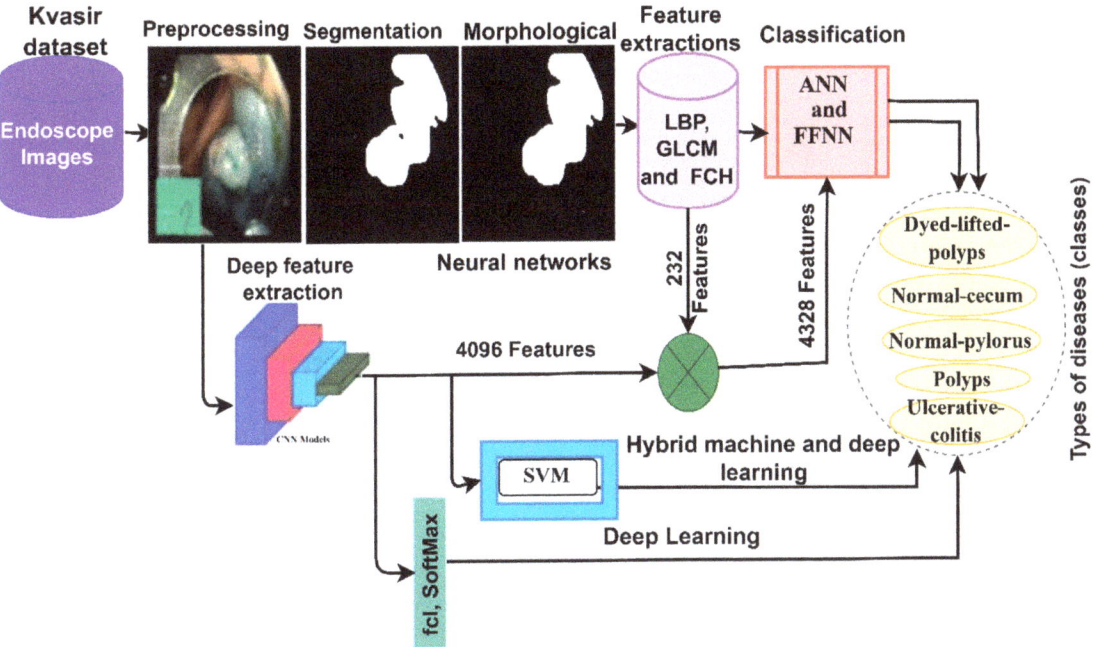

Figure 1. The general methodology for diagnosing lower GI disease dataset.

3.1. Description of the Dataset

The Kvasir dataset was obtained using high-resolution endoscopic cameras by Vestre Viken Health Trust (VV) from the Department of Lower GI, Bærum Hospital, Gjettum, Norway. All images were described by several experts from the Cancer Registry of Norway (CRN) and VV. The CRN is the national body at the University of Oslo Teaching Hospital, responsible for the diagnosis and early detection of cancer to prevent its spread. CRN provides knowledge about cancer and is affiliated with the southeast Norway Health Authority under the supervision of Oslo University Hospital. CRN is responsible for detecting precancerous lesions to prevent death by cancer. All images were described by medical experts, including anatomical landmarks, pathological findings, and regular findings. The dataset contains 5000 images evenly distributed across five classes of diseases: dyed-lifted polyps, normal cecum, normal pylori, polyps, and ulcerative colitis. All images are in RGB color space and have resolutions ranging from 700 × 575 to 1925 × 1075 pixels. Anatomical landmarks include the cecum and pylorus, while pathological findings include polyps and ulcerative colitis, in addition to other images related to removing lesions, such as a lifted polyp. Some image classes contain green images showing the location of the endoscope of the bowel through the electromagnetic imaging system. Figure 2a describes a dataset sample for all the classes contained in the Kvasir dataset [30]. The following link is open source and includes the dataset https://datasets.simula.no/kvasir/#download (accessed on 30 January 2022).

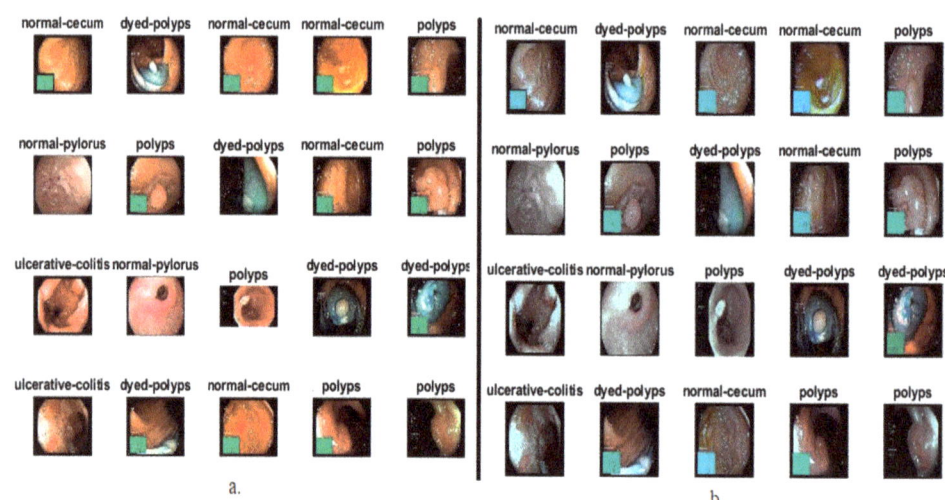

Figure 2. Images of the dataset for all types of diseases (**a**) Before enhancement (**b**) After enhancement.

3.2. Images Enhancement

When using endoscopes to perform interior imaging, several restrictions and obstructions appear as artifacts owing to movement, bubbles, low contrast, bodily fluids, debris, residual feces, and blood. These artifacts impede image classification; hence, all artifacts must be removed to achieve superior outcomes in the subsequent stages of medical image processing. All photos in our study were optimized before being given to the subsequent stages: the dataset images were optimized by averaging RGB colors and passing images through both average and Laplacian filters for removing artifacts, improving image contrast, and showing lesion edges [31]. First, a 5 × 5-pixel averaging filter is defined which removes artefacts and increases image contrast by replacing each central pixel with an average of 24 adjacent pixels. The filter in each iteration works by taking a center pixel and replacing the central pixel's value with the average value of 24 adjacent pixels. Equation (1) describes how the average filter works, and the process continues until the last pixel of the image [32].

$$F(L) = \frac{1}{L} \sum_{i=0}^{L-1} z(L-1) \qquad (1)$$

where $F(L)$ refer to the enhanced image, $z(L-1)$ refers to the input of previous, and L refers the pixel's number in the image.

Secondly, the Laplacian filter is applied which makes the edges of the tumors visible and distinguishes them from the rest of the image. Equation (2) describes the filter's mechanism of action.

$$\nabla^2 f = \frac{d^2 f}{d^2 x} + \frac{d^2 f}{d^2 y} \qquad (2)$$

where x, y are to the 2D matrices and $\nabla^2 f$ is a second-order differential equation.

Finally, the two images are merged together by subtracting the image generated by the average filter from the image generated by the Laplacian filter as shown in Equation (3).

$$O(X) = f(L) - \nabla^2 f \qquad (3)$$

where $O(X)$ represents output image enhanced.

Thus, an improved image is obtained that is fed to all the proposed systems in this study. Figure 2b describes many image samples for all diseases of the dataset after enhancement operations.

3.3. The First Proposed System (Neural Networks)

3.3.1. Segmentation (Active Contour Algorithm)

All lower digestive system endoscopy images contain a specific region affected by a lesion and the rest of the image is healthy. Therefore, analyzing the entire image, including the healthy part, and extracting the features from the complete image will lead to incorrect diagnostic results. Thus, the segmentation technique is necessary to segment the lesion area (area of interest) and isolate it from the healthy part, which will lead to the analysis and extraction of features from the lesion area only to obtain promising diagnostic results.

The lesion area is determined by the curve C defined by the level function $\varnothing: \Omega \to$, where Ω represents the lesion area, and zero is at the first border area in the image of the lower gastrointestinal lesion I. The curve is divided of lesion regions $F_k \subset \Omega$ into sub-regions $F, \overline{F}$ with ϕ, as shown in Equations (4) and (5).

$$\text{inside}(C) = F = \{ x \in F_k : \varnothing(x) > 0 \} \tag{4}$$

$$\text{outside}(C) = \overline{F} = \{ x \in \Omega : \varnothing(x) < 0 \cup x \in \Omega \setminus F_k \} \tag{5}$$

The Active Contour algorithm develops by moving the curve (contour) inward. The first seed is placed at the lesion boundary to map the lesion area (polyp). The curve moves inside to define the polyp subregion when $\phi > 0$ is set. The outer region is determined by subtracting the formerly selected region from the presently selected region as described in Equation (6).

$$\begin{aligned} F_0 &= F_1 + \overline{F}_1 \Rightarrow \overline{F}_1 = F_0 - F_1, \\ F_2 &= F_1 - F_2, \\ \overline{F}_3 &= F_2 - F_3. \end{aligned} \tag{6}$$

Finally, the outer sub-region can be calculated as Equation (7).

$$\overline{F}_k = F_{k-1} - F_k \tag{7}$$

The Active Contour moves toward the lesion boundary by defining the external energy that moves the first point (the zero level) toward the lesion boundary, as described by the functional energy function Equation (8).

$$f_{spz}(\varnothing) = \lambda L_{spz}(\varnothing) + \nu A_{spz}(\varnothing) \tag{8}$$

where ν are constants and $\lambda > 0$. The L_{spz} and A_{spz} are the defined as Equations (9) and (10).

$$L_{spz}(\varnothing) = \int_\Omega spz(I) \delta_\varepsilon(\varnothing) |\nabla \varnothing| \, dx \tag{9}$$

$$A_{spz}(\varnothing) = \int_\Omega spz(I) H_\varepsilon(\varnothing)(-\varnothing) \, dx \tag{10}$$

the term $spz(I)$ will be defined in an equation later, H_ε denotes the Heaviside function and δ_ε is the univariate dirac delta function. The curve is driven from zero point to smooth curve by $L_{spz}(\phi)$ equation. Thus, the small energy $spz(I)$ will speeds up the curve towards the lesion. The coefficient v is a positive or negative value that depends on the position of the curve over the lesion area; If the curve is within the lesion region, v is positive while v is negative for acceleration of the curve at the lesion boundary [33].

We proposed the $E_{proposed}$ for the *SPF* function which previously used. Let $I: \Omega \to R$ be endoscopic image of lower gastrointestinal diseases, C is a closed curve energy functional is defined as by Equation (11).

$$E_{proposed} = \int_\Omega |I(x) - C_1|^2 H_\varepsilon(\varnothing(x)) M^k(x) \, dx + \int_\Omega |I(x) - C_1|^2 (1 - H_\varepsilon(\varnothing(x))) M^k(x) \, dx \tag{11}$$

when the energy $E_{proposed}$ is reduced, ϕ is preserved constant, we get the curve C_1 for F region w and the curve C_1 for area $\overline{F}$ as Equations (12) and (13).

$$C_1(\phi) = \frac{\int_\Omega I(x)\ H_\varepsilon(\varnothing(x))M^k(x)\ dx}{\int_\Omega H_\varepsilon(\varnothing(x))M^k(x)\ dx} \tag{12}$$

$$C_2(\phi) = \frac{\int_\Omega I(x)\ (1 - H_\varepsilon(\varnothing(x)))M^k(x)\ dx}{\int_\Omega (1 - H_\varepsilon(\varnothing(x)))M^k(x)\ dx} \tag{13}$$

M^k refers to the function of characteristic for sub-region as Equation (14).

$$M^k(x) = \phi > 0, \tag{14}$$

$$M^0 : \Omega \to -1,$$

The contour curve continues to move along the edges of the lesion, and when the pixels are similar between two successive contours, the curve will stop, and the algorithm stops by a certain stop value.

If

$$\sum_{i=0}^{row} \sum_{j=0}^{col} M^k_{i,j} < \left(\frac{SV}{100}\right) \sum_{i=0}^{row} \sum_{j=0}^{col} oldM^k_{i,j} \tag{15}$$

where SV represents the stop value when the curve reaches the last point to separate the pest region from the rest of the image.

Then, the contour curve will be stopping moving.

Where $oldM^k_{i,j}$ refers to the last computed mask, and $M^k_{i,j}$ refers to the current mask of the contour curve, col and row are the max number of columns and rows, respectively, in the lesion.

Finally, the polyp lesion area is identified with high accuracy and isolated from the rest of the image, as shown in Figure 3b. After the segmentation process, a binary image of the lesion area is produced and isolated from the rest of the image. Still, some holes appear that do not belong to the lesion, and therefore these holes must be filled to improve the image [34]. Thus, the morphological method was applied, which produces an enhanced binary image, as shown in Figure 3c.

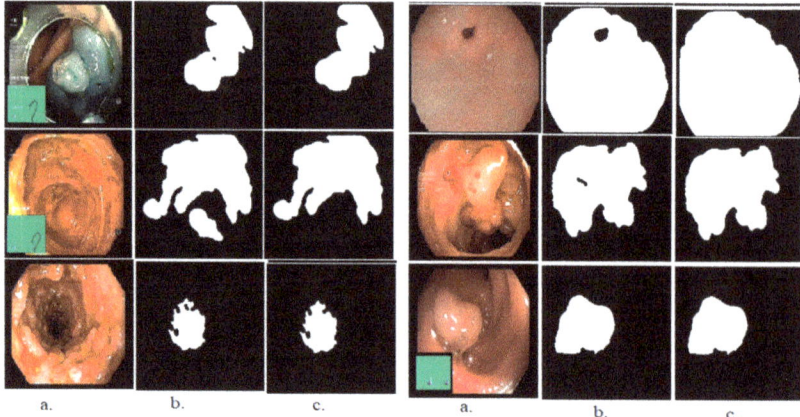

Figure 3. Samples of the dataset (**a**) original image (**b**) After the segmentation (**c**) After the morphological.

3.3.2. Feature Extraction

The feature extraction stage is one of the most critical stages in image processing, which determines the accuracy and efficiency of classification. The image contains thousands of information that is difficult to analyze. Therefore, extracting representative features with high accuracy reduces image dimensions and extracts features from a region of interest (ROI). In this work, the features were extracted by three algorithms: LBP, GLCM and FCH. The features extracted from the three methods are then combined into a single feature vector for each image to obtain more representative features. Combining features is a modern and important method in obtaining more effective and representative features for each lesion.

First, the LBP algorithm extracts the essential features, which is an efficient way to extract the texture features of the binary surface. In this work, an algorithm was tuned on size of 6 × 6; it works to select the central pixel (g_c) and select the 35 pixels adjacent to it (g_p) [35]. According to the algorithm, the central pixel is replaced by neighboring pixels in each iteration. Neighboring pixels are selected according to the radius R. Equations (16) and (17) describe the working mechanism of the LBP algorithm, which replaces the central pixel with adjacent pixels and continues until all pixels of the image are targeted. This algorithm extracted 203 essential features for each image, stored in one attribute vector for each image.

$$LBP(x_c, y_c)_{R,P} = \sum_{P=0}^{P-1} s(g_p - g_c) \, 2^P \qquad (16)$$

$$x(c) = \begin{cases} 0, & c < 0 \\ 1, & c \geq 0 \end{cases} \qquad (17)$$

P is the number of pixels in the image and R is the number of adjacent pixels.

Second, the GLCM algorithm extracted the features, which extracts texture features from the area of interest (tumors).

The algorithm shows multiple levels in the grayscale of an area of interest. The algorithm extracts the statistical features based on its working mechanism. The algorithm relies on distinguishing between smooth and rough pixels through spatial information. An area is rough when its pixel values are far apart, while when the pixel values are close together, the area is smooth. Additionally, spatial information is essential in determining the correlation between pairs of pixels through the distance d between the pixels and the directions θ that determine the location of each pixel from the other [36]. There are four directions for θ, which are 0°, 45°, 90° and 135°; when the angle is vertical θ = 90 or horizontal θ = 0, the distance between one pixel and the other d = 1. When the angle θ between pair pixel θ = 45 or θ = 135, the distance d = $\sqrt{2}$. This algorithm generated 13 statistical features for each image.

Third, the features are extracted from a region of interest by the FCH algorithm. FCH is considered one of the best algorithms for extracting the chromatic features from the pest area. The color is one of the most important features used in medical images to distinguish and diagnose images of gastrointestinal tumors [37]. The lesion region contains many colors, so each color is represented in the histogram bin. Each color in the region of interest represents several histogram bins. Each local color is represented in histogram bin and thus pest colors are distributed over many histogram bins. All two similar colors must be in the same histogram bin, also if the two colors in the histogram bin are different, it means that they are different even if they are almost the same. Thus, the FCH algorithm uses the membership value of each pixel and distributes the pixels based on the total histogram bin. We consider the number of colors in a lesion area containing n pixels as $X(I) = x_1, x_2, \ldots x_i$ where n_i is the image pixels, ith is the all-color bins, $x_{i=ni/n}$ is the probability that any color image belongs to a histogram bin.

$$x_i = \sum_{j=1}^{n} p_{i/j} \; p_j = \frac{1}{n} \sum_{j=1}^{n} p_{i/j} \qquad (18)$$

where p_j is image pixels, conditional probability $p_{i/j}$, means the probability of jth pixel belonging to I histogram color bins using the FCH algorithm.

$$p_{i/j} = \begin{cases} 1, & \text{if } jth \text{ pixels belongs to } ith \text{ histogram color bin} \\ 0, & \text{Otherwise} \end{cases} \quad (19)$$

Finally, the features extracted from the three algorithms are combined into one feature vector for each image. The LBP method produce of 203 features, 13 features produced from the GLCM method, and the FCH method produce 16 features. Therefore, when all the features were hybrid, the Fusion method produced 232 representative features for each image, representing each gastrointestinal polyp's lesion's essential and distinguishing features. Figure 4 describes the fusion method between the three algorithms, LBP, GLCM and FCH.

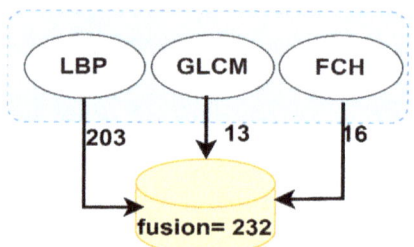

Figure 4. The fusion method between the three algorithms LBP, GLCM and FCH.

3.3.3. ANN and FFNN Algorithms

In this section, the endoscopic dataset for lower GI is evaluated using the ANN and FFNN neural network algorithms. ANN is a type of intelligent neural network that consists of three main layers, each layer having many interconnected neurons. ANN is characterized by its profound ability to solve many complex problems by analyzing, interpreting and transforming complex data into information to solve the required tasks. ANN consists of three layers: the input layer, the hidden layers, and the output layer. Data moves from one layer to another through connections called weights. The first layer of the algorithm is the input layer, which receives information from (the features extracted) and passes it to the hidden layer. In our study, the number of neurons in the input layer is 232 neurons (features extracted). Hidden layers solve complex problems and analyze and interpret inputs across many hidden layers, and each hidden layer has many interconnected neurons. In our study, the hidden layers were set to 30 hidden layers. The output layer receives the output from the last hidden layer and consists of many neurons according to the classes of the dataset. In our study, the output layer produces five neurons: dyed-lifted-polyps, normal-cecum, normal-pylorus, polyps, and ulcerative-colitis. Neurons are interconnected through weights, and information is transmitted from one neuron to another and from one layer to another in each repetition [38]. The algorithm is characterized by many interconnected neurons, the interconnected processing unit, and the activation and bias function associated with each neuron. It is characterized by the basis of learning and calculating the error rate in each iteration. The error rate is calculated between each iteration's actual and predicted output. The process continues, and in each iteration, the weights are updated until the minimum sum squared (MSE) is obtained between the actual and predicted output, as described by Equation (20).

$$\text{MSC} = \frac{1}{n} \sum_{i=1}^{n} (a(x)_i - p(y)_i)^2 \quad (20)$$

where $a(x)_i$ represents the actual values and $p(y)_i$ represents the predicted values.

FFNN is a type of intelligent neural network that consists of three layers, with each layer having many interconnected neurons that forward data. It is distinguished by its superior ability to solve, interpret and analyze many complex problems efficiently [39]. The algorithm consists of three layers: the input, hidden, and output layers. The input layer consists of 232 neurons that receive inputs as features extracted from the feature extraction stage. The algorithm also consists of 30 hidden layers that solve complex problems, analyze, interpret, and send them to the output layer. The output layer consists of five neurons to show the results of classifying each image into its appropriate class as shown in Figure 5. In our study, the algorithm produces five neurons: dyed-lifted-polyps, normal-cecum, normal-pylorus, polyps, and ulcerative-colitis. In this network, information flows between neurons in the forward direction; neurons are interconnected by weights w, and weights are transmitted from one cell to another in the forward direction. Each neuron produces a weight multiplied by the value of the weights of the previous neurons, and the process continues, and with each iteration, the weights are updated. The weights are spoken in each iteration until the minimum sum squared is obtained between the actual and expected output described by the above equation.

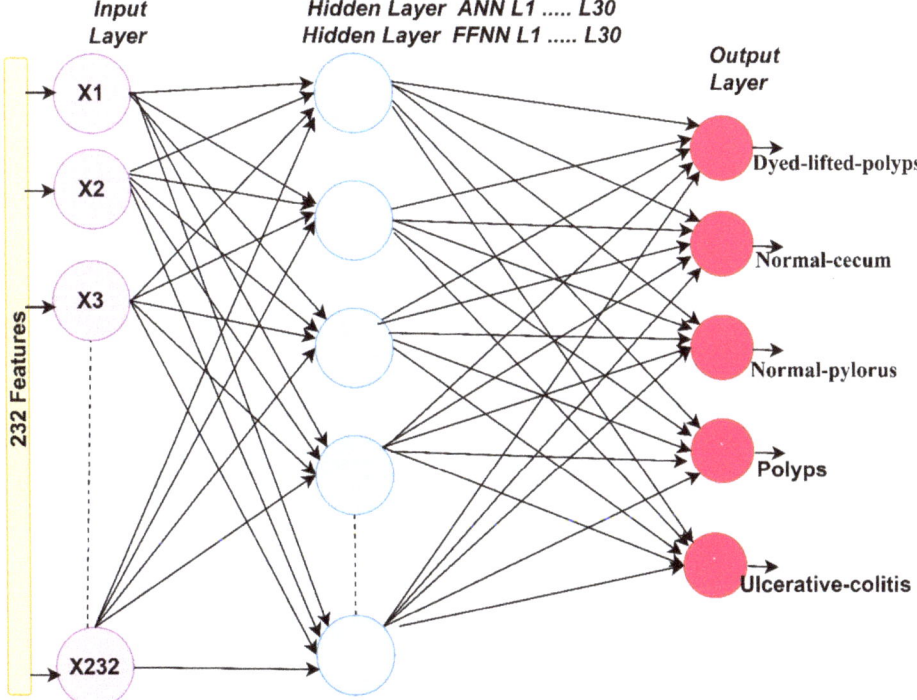

Figure 5. Structure of neural network algorithms for classifying the endoscopy image dataset of lower gastrointestinal diseases.

3.4. The Second Proposed System (CNN Models)

Convolutional neural network (CNN) techniques are compatible with machine learning techniques, and what distinguishes them is the use of many deep layers to analyze and interpret the required tasks with high accuracy and efficiency [40]. CNN models are called deep learning due to the use of deep convolutional layers and the essence of their work is to obtain a representation of many levels to represent the problem starting from simple modular levels that transform the required tasks (endoscopic images of lower GI) from levels to more deeper levels to extract features deeper and more abstract. CNNs are used in classification, regression, texture modelling, image recognition, natural language processing, biomedical image classification, robotics, and other tasks. In classification tasks, networks amplify representation layers to extract the essential features to distinguish each image into its appropriate class and suppress irrelevant differences. CNN architecture refers to many non-linear levels that learn something specific in each layer; for example, one layer works to extract geometric features, while another layer focuses on extracting color features, a layer for texture features, another layer for showing edges, and so on. A CNN consists of many layers, the most important of which are the convolutional layers, the pooling layer, the fully connected layer, and many auxiliary layers.

Convolutional layers are one of the most important layers in CNN models and derive their name from convolutional neural networks. Three parameters that control convolutional layers are filter size, zero-padding, and p-step. The filters work on the process of convolution between $w(t)$ and the target images $x(t)$, which is called the process of convolution $(x \times w)(t)$ or $s(t)$ as shown in Equation (21). The larger the filter size, the more wrap around the image. Each filter has a specific function, some work to discover edges, some filters to extract geometry features, and some to extract texture, shape, color features, and so on. For zero-padding, which preserves the size of the original image, the edges of the image are padded with zeros, and the size of zero-padding is determined based on the size of the filter convoluting around the original image [41]. For p-step, determine the number of steps the filter can move around the image.

$$s(t) = (x \times w)(t) = \int_{-\infty}^{\infty} x(a) w(t-a)\, da \qquad (21)$$

where $(x \times w)(t)$ is the process of convolution, $x(a)$ is the image to be processed, and $w(t-a)$ is the filter convolted around the image.

Pooling layers are one of the primary layers that reduce the dimensions of the input image. After convolutional operations, it produces millions of parameters and thus slows down the training process. Therefore, to represent the high-dimensional data space in the low-dimensional space while preserving the most important features and to speed up the training process, Pooling layers work to solve these challenges. Pooling layers work as convolutional layers, where the size of a filter is determined in the layers Pooling and moving on the image and the representation of groups of pixels by one pixel based on the methods Max-Pooling and Average-Pooling. The Max-Pooling mechanism identifies pixels and represents them by the maximum value [42]. Equation (22) describes the mechanism of Max-Pooling. While in the Average-Pooling method, groups of pixels are selected, their average value is calculated, and groups of pixels are represented by their average value. Equation (23) describes how the Average-Pooling method works.

$$P(i;j) = max_{m,n=1....k}\, A[(i-1)p+m;\ (j-1)p+n] \qquad (22)$$

$$P(i;j) = \frac{1}{k^2} \sum_{m,n=1....k} A[(i-1)p+m;\ (j-1)p+n] \qquad (23)$$

where A is filter size, m,n are filtered size dimensions, p is filtered step size, k is capacity of filter.

Fully Connected Layers (FCL) is one of the primary layers in CNN and is responsible for classifying input images into their appropriate classes. The FCL comprises millions of neurons connected by junctions called weights. FCL converts the dimensions of deep feature maps from 2D to 1D. Some CNN models have many FCL layers. Finally, the FCL layer sends its output to the Softmax activation function, producing neurons with the number of classes in the dataset. In our study, the Softmax layer produces five neurons: dyed-lifted-polyps, normal-cecum, normal-pylorus, polyps, and ulcerative-colitis. Equation (24) describes how the Softmax activation function works.

$$y(x_i) = \frac{\exp x_i}{\sum_{j=1}^{n} \exp x_j} \qquad (24)$$

$y(x)$ between $0 \leq y(x) \leq 1$.

There are also many auxiliary layers such as Rectified Linear Unit (ReLU), dropout layer, and others. ReLU follows convolutional layers to process the output of convolutional layers. This layer passes only positive outputs, while converting negative outputs to zeros. Equation (25) describes the working mechanism of the ReLU layer.

$$\text{ReLU}(x) = \max(0, x) = \begin{cases} x, & x \geq 0 \\ 0, & x < 0 \end{cases} \quad (25)$$

Convolutional layers produce millions of operands, and thus networks experience overfitting. Therefore, CNNs provide a dropout layer to solve these challenges. The dropout layer passes a set number of neurons on each iteration. In this study, CNNs models pass 50% of the neurons in each iteration, but this layer will double the training time of the dataset.

This study will focus on two models, GoogLeNet and AlexNet.

3.4.1. GoogLeNet Model

GoogLeNet is a convolutional neural network used in many applications for pattern recognition and classification purposes, including biomedical image processing. GoogLeNet consists of 27 layers. This model is distinguished from other models in that it contains layers that can significantly reduce the dimensions while preserving the essential features. The network has a convolutional layer with a 7×7 filter that quickly extracts feature maps. The network also contains three 3×3 pooling layers, which can reduce the dimensions and reduce the width and height of the image; Additionally, the network has a pooling layer of 7×7 size which greatly reduces the dimensions and is effective while preserving the essential features [41]. All layers of the network produce seven million parameters. Figure 6 shows the GoogLeNet architecture for classifying endoscopy images of the lower GI dataset.

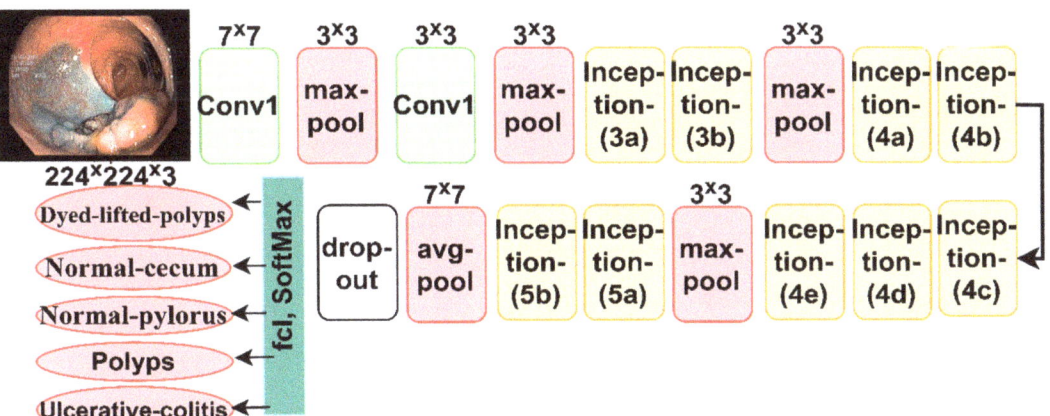

Figure 6. General structure of the GogLeNet model for classifying the lower GI endoscopy image dataset.

3.4.2. AlexNet Model

AlexNet is a type of convolutional neural network that contains many deep layers. AlexNet consists of 25 layers divided between convolutional, pooling, fully connected, and helper layers. The first layer is the input layer, which receives the endoscopy images of the lower GI dataset and adjusts the size of all images to a uniform size of $227 \times 227 \times 3$. There are five convolutional layers for feature map extraction, ReLU layers for further feature processing, and three pooling layers from the max-pooling layers that reduce dimensions; two dropout layers to overcome the problem of overfitting; and three fully connected layers with softmax activation function [43]. AlexNet produces 62 million parameters and 650,000 neurons interconnected by 630 million connections. Figure 7 shows the structure of the AlexNet model to classify the lower GI dataset.

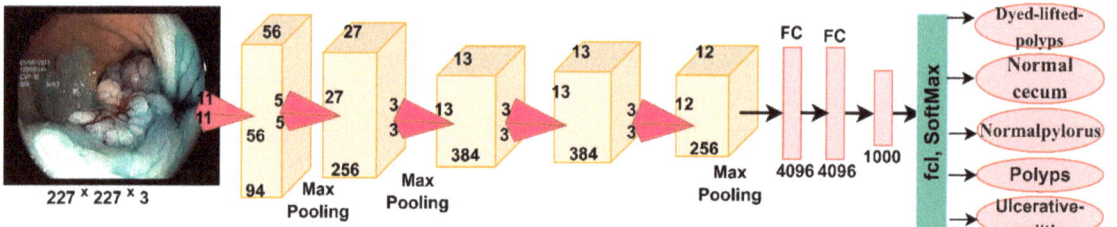

Figure 7. General structure of the AlexNet model for classifying the lower GI endoscopy image dataset.

3.5. The Third Proposed System (Hybrid of Deep Learning and Machine Learning)

In this section, a new technique is introduced, a hybrid between CNN modelling techniques and a machine learning algorithm for classifying endoscopic images of the lower GI dataset. Since CNN models require high-resource and expensive computer specifications and take a long time to train the dataset, these hybrid techniques solve these challenges [44]. Therefore, in this study, a hybrid technique is presented between (GoogLeNet and AlexNet) models and the SVM machine learning algorithm, which requires medium computer resources and is fast to train the data and produces high performance. The basic idea in this technique consists of two blocks; the first block is CNN models that extract the maps of the deep features of the endoscopic images and send them to the second block. The second block is an SVM algorithm that receives deep feature maps and classifies them with high accuracy and efficiency. Figure 8a,b shows the basic structure of the hybrid technique, where it is noted that the technique contains two blocks are CNN models (GoogLeNet and AlexNet). The second block is the SVM algorithm, and therefore the technique is called GoogLeNet+SVM and AlexNet+SVM. It is worth noting that the fully connected layer was removed from the CNN models and replaced with the SVM algorithm.

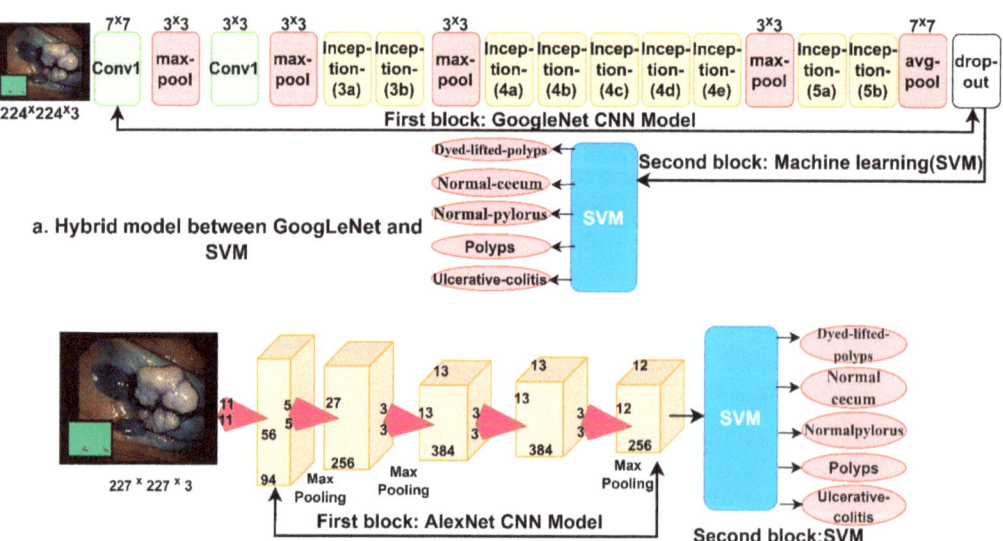

Figure 8. The general structure of the hybrid method between CNN models and SVM (**a**) GoogleNet+SVM and (**b**) AlxNet+SVM.

3.6. The Fourth Proposed System (Hybrid Features)

This section presents modern methods using hybrid techniques between deep feature maps extracted by CNN models and features extracted by traditional algorithms. CNN models require high specification and expensive computer resources and take a long time to train the dataset. Therefore, the hybrid features are classified using the ANN and FFNN algorithms to solve these challenges [45]. The main idea of this technique is as follows: First, extract the deep feature maps from GoogLeNet and AlexNet models and store them in feature vectors, where 4096 features are extracted for each image from each model. Second, combine the features extracted from the LBP, GLCM and FCH algorithms, producing 232 features after fusing them. Third, the deep feature maps were converted into a unified data format. Fourth, the deep feature maps extracted from CNN models (the third step) are combined with the features extracted by traditional algorithms (the second step), so after fusion, it produced 4328 features representing each image (vector). Fifth, these features are fed to the ANN and FFNN algorithms to classify them. Figure 9 shows the methodology for extracting deep feature maps by GoogLeNet and AlexNet and combining them with the features extracted by LBP, GLCM and FCH and then fed to ANN and FFNN algorithms.

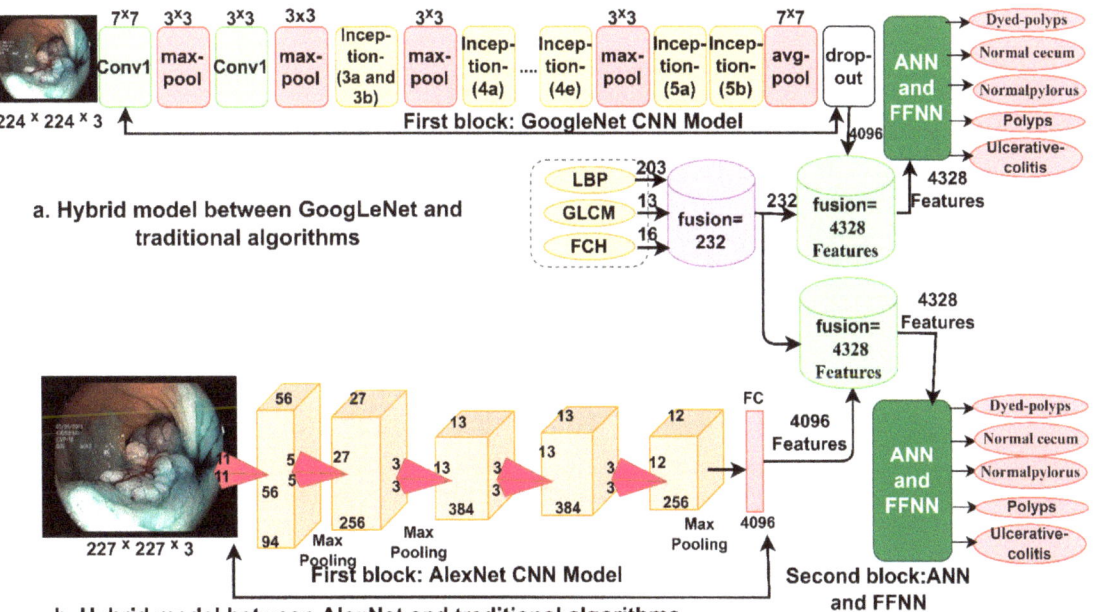

Figure 9. The general structure of feature fusion between CNN models and LBP, GLCM, and FCH algorithms.

4. Experimental Results

4.1. Splitting Dataset

In this study, endoscopic images of the lower gastrointestinal disease dataset were evaluated by four proposed systems, each with more than one algorithm. There are various methods of dataset analysis using neural network algorithms, CNN models, SVM, and hybrid methods for classifying and extracting hybrid features. The dataset contains 5000 images equally divided into five types of diseases: dyed-lifted-polyps, normal-cecum, normal-pylorus, polyps, and ulcerative-colitis. The dataset was divided into 80% for training and validation (80:20) and 20% for testing, as described in Table 1. All the proposed systems in this study were implemented using MATLAB 2018b environment and with Intel® i5 processor specifications, RAM 12GB and GPU 4GB.

Table 1. Splitting of the endoscopy image of lower GI dataset for training and testing.

Phase	80% for Training and Validation (80:20%)		Testing (20%)
Classes	Training (80%)	Validation (20%)	
dyed-lifted-polyps	640	160	200
normal-cecum	640	160	200
normal-pylorus	640	160	200
polyps	640	160	200
ulcerative-colitis	640	160	200

4.2. Evaluation Metrics

In this study, the lower GI dataset was evaluated by several systems proposed, which are neural networks (ANN and FFNN), CNN models (GoogLeNet and AlexNet), hybrid method between CNN models and SVM (GoogLeNet+SVM, AlexNet+SVM), and hybrid features extracted between CNN models (GoogLeNet and AlexNet) and algorithms (LBP, GLCM and FCH) by many statistical measures. The proposed systems were evaluated by using the same scales which are accuracy, precision, sensitivity, specificity, and AUC described in Equations (26)–(30). All the proposed systems produced a confusion matrix that contains all the test samples that are classified as correct and incorrect. Correctly labeled samples are called true positive (TP) for confirmed samples and true negative (TN) for healthy samples. Incorrectly labeled samples are called false negative (FN) and false positive (FP) [46].

$$\text{Accuracy} = \frac{TN + TP}{TN + TP + FN + FP} \times 100\% \tag{26}$$

$$\text{Precision} = \frac{TP}{TP + FP} \times 100\% \tag{27}$$

$$\text{Sensitivity} = \frac{TP}{TP + FN} \times 100\% \tag{28}$$

$$\text{Specificity} = \frac{TN}{TN + FP} \times 100 \tag{29}$$

$$\text{AUC} = \frac{\text{True Positive Rate}}{\text{False Positive Rate}} = \frac{\text{Sensitivity}}{\text{Specificity}} \tag{30}$$

where TP is the number of properly classified GI endoscopy images of as diseases. TN is the number of GI endoscopy images correctly classified as normal. FP is the number of endoscopy images of a normal GI tract but it is classified as diseases. FN is the number of endoscopic images of GI diseases classified as normal.

4.3. Segmentation Performance Evaluation

The segmentation method is one of the most important steps to biomedical image processing, which is widely used in this field to select the affected region and separate it from the rest of the image. In this study, the Active Contour method was applied, which acts as a snake movement and starts with a point and moves along the edges of the lesion until the region of interest (the lesion region) is completely selected, then separated it from the rest of the image. The segmentation method based on Active Contour models was validated by accuracy, precision and recall measures, which reached 99.3%, 99.7%, and 99.6%, respectively. Thus, the lesion region was separated with a promising accuracy and high efficiency, and the region of interest was sent to the feature extraction stage, to extract the most important color, texture, and shape features.

4.4. Results of the First Proposed System (Neural Networks Algorithms)

In many domains, including medical image diagnosis, neural networks are among the most important and successful artificial intelligence networks. The final stage of categorization is neural network algorithms, which are based on their efficiency in previous stages of image processing (pre-processing, segmentation and feature extraction). Endoscopic images of lower gastrointestinal illnesses were classified in this work using the algorithms ANN and FFNN, which were fed the features derived by the hybrid technique. The dataset was separated into two parts: 80 percent for training and validation and 20% for testing. Figure 10 shows the process of training the ANN and FFNN networks, where it is noted that the network consists of an input layer with 232 neurons, 30 hidden layers for both ANN and FFNN, and an output layer with five neurons; each neuron represents one of the classes of the dataset.

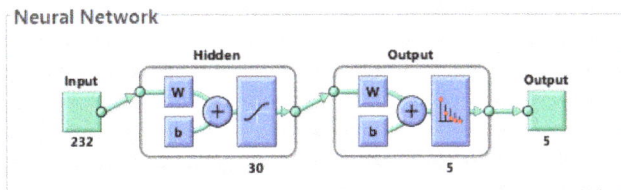

Figure 10. Display performance of the ANN and FFNN algorithms for training a low GI dataset.

4.4.1. Gradient

There are many methods for assessing the lower GI dataset using the ANN and FFNN algorithms, and one of these scales is the gradient values. The gradient value measures the error rate between actual and predicted values. Figure 11a,b shows the gradient values and validation check for the performance of the ANN and FFNN algorithms for evaluating the lower GI dataset. It can be seen from Figure 11a that the dataset was evaluated by ANN, which found the best gradient at a value of 0.012073 at epoch 47 and validated at a value of 6 during the same epoch. It can also be seen from Figure 11b that the dataset was evaluated by the FFNN algorithm, which reached the best gradation at a value of 0.076519 at epoch 12 and was validated at a value of 6 during the same epoch.

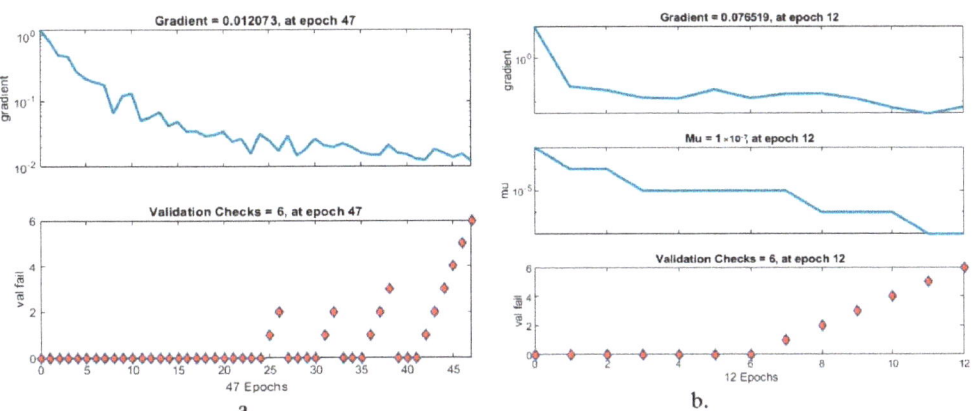

Figure 11. Displays gradient and validation value of lower GI dataset using (**a**) ANN (**b**) FFNN.

4.4.2. Performance Analysis

Cross-entropy loss is one of the performance measures of ANN and FFNN, which measures mean squared error between actual and predicted values. Figure 12 shows the performance of the ANN and FFNN networks for assessing the lower GI dataset during the training, validation and testing phases. It is noticed from the figure that the crossed lines represent the best point reached by the algorithms, and the blue line is for the training stage, green for the validation stage and red for the testing stage. The ANN algorithm achieved the best performance with a value of 0.068474 at epoch 41, as shown in Figure 12a.

The FFNN algorithm achieved the best performance with a value of 0.047785 at epoch 6, as shown in Figure 12b. When the validation stage reaches optimal performance, the network parameters are adjusted, and the network stops training.

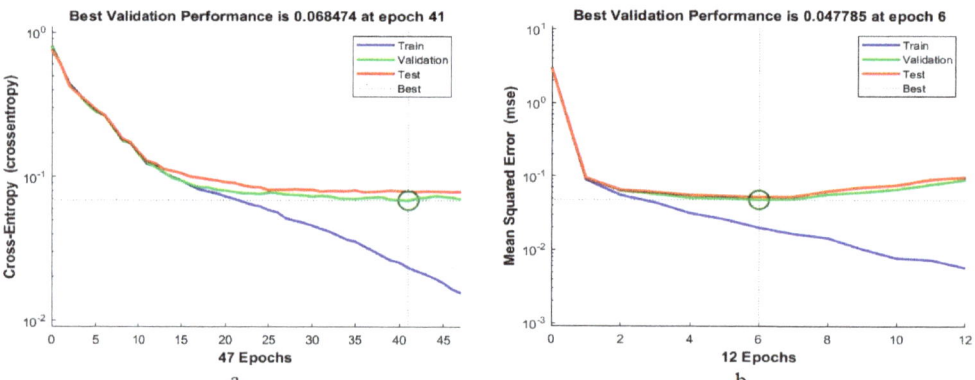

Figure 12. Displays performance for evaluating a lower GI dataset using (**a**) ANN (**b**) FFNN.

4.4.3. Receiver Operating Characteristic (ROC)

ROC is one of the performance measures of neural networks to evaluate their performance on the lower GI dataset. ROC is measured by measuring the positive and false samples rate ratio during the training, validation, and testing phases. Figure 13 describes the performance of the ANN network for assessing the lower GI dataset, where the x-axis represents samples for the false positive rate (FPR) with specificity. In contrast, the y-axis represents samples for the true positive rate (TPR) labelled with sensitivity. It is noted from the figure that there are five colors; each color represents a class in the dataset. ANN achieved an overall ROC of 98.82% for all five types of diseases in the dataset.

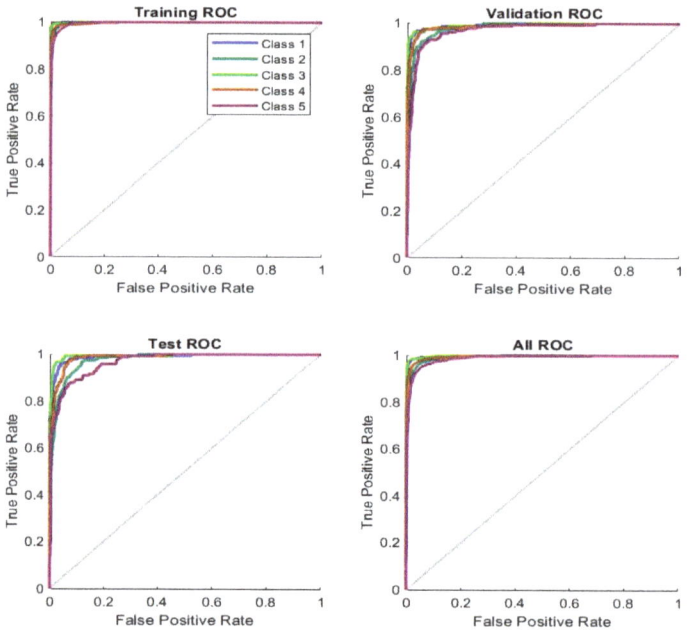

Figure 13. Displays the performance of ANN through ROC for endoscopic image diagnosis.

4.4.4. Regression

Regression is one of the performance measures of neural networks for evaluating a dataset. The network finds the regression value by predicting continuous variables through other available variables by measuring the error rate between the target and output values. The network finds the best value for the regression when it approaches the value 1, which means that the error rate between the target and output values is zero. Figure 14 shows the performance of the FFNN for evaluating the regression of the dataset and predicting continuous values according to the available values. FFNN reached a regression of 93.55% during the training phase and 84% during the validation phase, and during the testing phase, it reached 82.71% and achieved an overall regression of 90.13%.

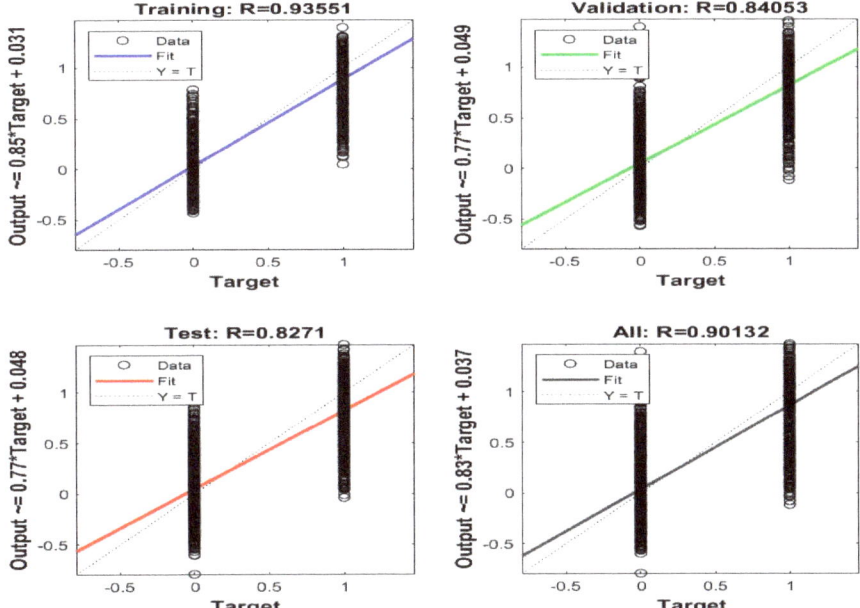

Figure 14. Displays the performance of FFNN through regression values for endoscopic image diagnosis.

4.4.5. Error Histogram

The error histogram measures the performance of the ANN and FFNN algorithms on the lower GI dataset. The ANN and FFNN algorithms find the minimum error between the actual and predicted output during the dataset's training, validation, and testing phase. Figure 15 describes the performance of the ANN and FFNN algorithms on the dataset, where the blue histogram bin is represented during the training phase, the green histogram bin is during the validation phase, the red histogram bin is during the test phase of the dataset, and the orange line is the zero line between the actual and predicted values. Figure 15a shows the performance of the ANN algorithm, which reached the minimum error between the actual and predicted values at 20 bin between the values −0.9494 and 0.95. While Figure 15b shows the performance of the FFNN algorithm, which reached the minimum error between the actual and predicted values at 20 bin between the values −1.333 and 1.054.

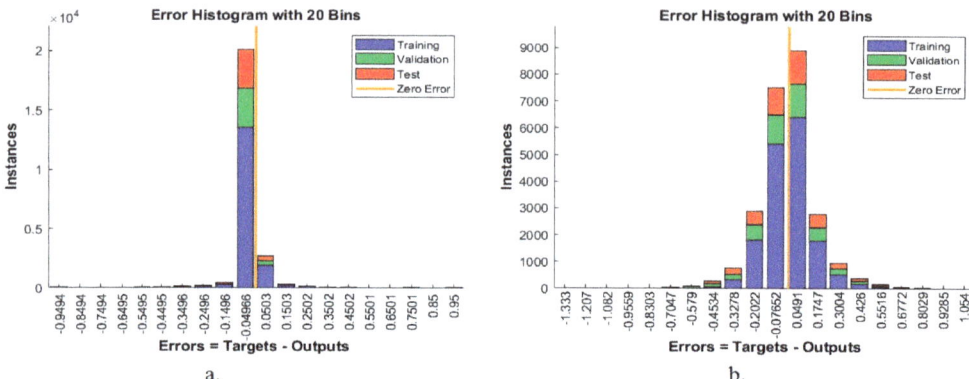

Figure 15. Displays error histogram bin for evaluating a lower GI dataset using (a) ANN (b) FFNN.

4.4.6. Confusion Matrix

The confusion matrix is the comprehensive and most important measure for evaluating networks on a dataset. It is a matrix-like form in which a row and a column represent each class (disease) of the dataset. The rows represent the (actual) output images, while the columns represent the predicted images. The confusion matrix contains all dataset samples that are correctly and incorrectly classified. Correctly classified samples are called true positive (TP) and true negative (TN); incorrectly labelled samples are called false positive (FP) and false negative (FN). In this study, endoscopic images of the lower GI dataset were evaluated by ANN and FFNN during the training, validation and testing phase. Figure 16 shows the resulting confusion matrix from ANN and FFNN algorithms representing the evaluation of the dataset for five diseases as follows: class 1 represents dyed-lifted-polyps, class 2 represents normal-cecum, class 3 represents normal-pylorus, class 4 represents polyps, and class 5 represents ulcerative-colitis. Figure 16a shows the performance of the ANN algorithm, which reached an overall accuracy of 97.4%. Figure 16b also shows the performance of the FFNN algorithm, which reached an overall accuracy of 97.6%.

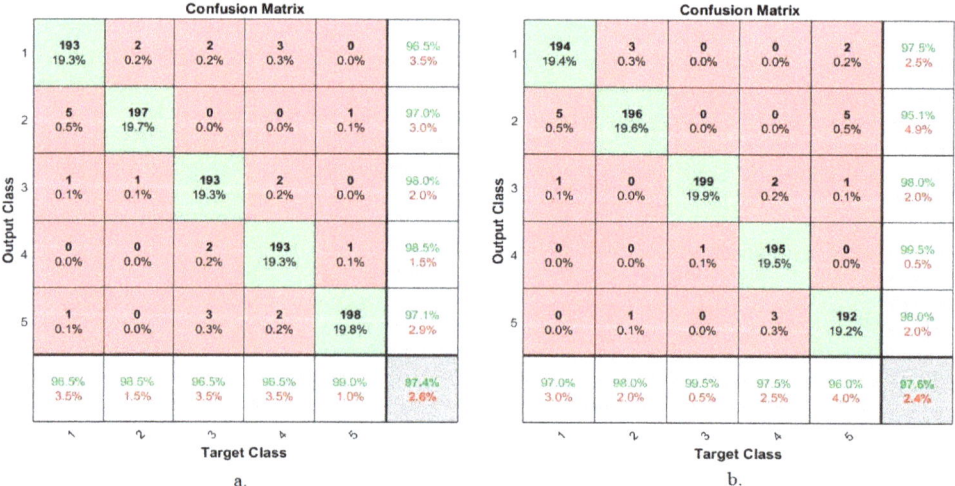

Figure 16. Confusion matrix for the low GI dataset generated by using (a) ANN (b) FFNN.

Table 2 summarizes the results of the evaluation of the ANN and FFNN algorithms on endoscopic images for the early diagnosis of lower gastrointestinal disease. It is noted that the FFNN algorithm is superior to the ANN algorithm. The ANN algorithm achieved an accuracy of 97.4%, a precision of 97.25%, a sensitivity of 96.5%, a specificity of 99.25%, and an AUC of 98.82%. In contrast, the FFNN algorithm achieved an accuracy of 97.6%, precision of 97.25%, the sensitivity of 97.75%, specificity of 99.3%, and AUC of 98.25%. Figure 17 presents the performance of the ANN and FFNN algorithms for evaluating the lower GI dataset.

Table 2. The results of the ANN and FFNN algorithms on the gastroenterology dataset.

Measure	ANN	FFNN
Accuracy %	97.4	97.6
Precision %	97.25	97.25
Sensitivity %	96.5	97.75
Specificity %	99.25	99.3
AUC %	98.82	98.25

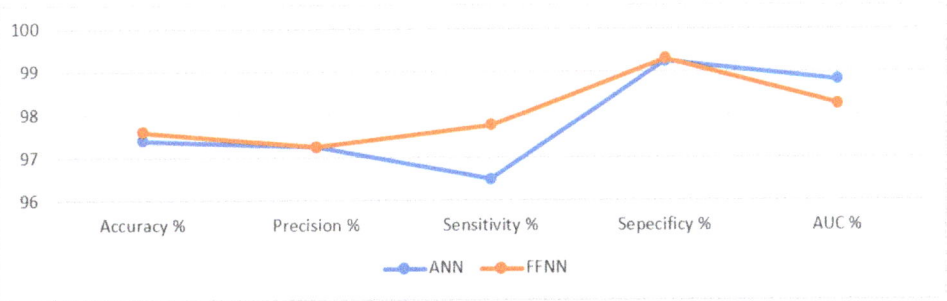

Figure 17. Display of the performance of the ANN and FFNN algorithms for diagnosing a low GI dataset.

4.5. Results of Second Proposed System (CNN Models)

In this section, the endoscopic image of the lower GI dataset is evaluated using the pre-trained CNN models, GoogLeNet and AlexNet. Transfer learning method is pre-trained CNN models on more than one million images to produce more than a thousand classes. Thus, the performance of the experience of the previously trained models is transferred to perform new tasks, as in this study, where the experience of the CNN models for diagnosing lower GI dataset is transferred. One of the challenges facing CNN models is the overfitting problem during the training phase of the dataset. Thus, CNN models introduce the data augmentation technique to overcome this challenge, which artificially augments dataset images. Table 3 summarizes the lower GI dataset before and after using the data augmentation during the training phase. Images are artificially augmented through many operations such as rotation, flipping, shifting, and others. Each image was incremented seven times for all classes equally.

Table 3. Data augmentation method during the training phase.

Phase	Training Phase 80%				
Class name	dyed-lifted-polyps	normal-cecum	normal-pylorus	polyps	ulcerative-colitis
No images before augmentation	640	640	640	640	640
No images after augmentation	**5120**	**5120**	**5120**	**5120**	**5120**

Table 4 summarizes the tuning of the CNN GoogLeNet and AlexNet models, where the adam optimizer and Mini Batch Size, Mini Batch Size, Initial Learn Rate, dataset training time for each model and Validation Frequency were set.

Table 4. Seting parameter options for GoogLeNet and AlexNet models.

Options	GoogleNet	AlexNet
training Options	adam	adam
Mini Batch Size	18	120
Max Epochs	6	10
Initial Learn Rate	0.0003	0.0001
Validation Frequency	3	50
Training time (min)	301 min 23 s	144 min 38 s
Execution Environment	GPU	GPU

The GoogLeNet and AlexNet models achieved superior results for diagnosing endoscopic images of the gastro-intestinal disease dataset. Table 5 describes the evaluation results of the GoogLeNet and AlexNet models, where it is noted that the GoogLeNet model is superior to the AlexNet model. The GoogLeNet model achieved an accuracy of 96%, a precision of 96.2%, a sensitivity of 96%, a specificity of 99.2%, and an AUC of 96%. In contrast, the AlexNet model achieved an accuracy of 91.5%, a precision of 91.8%, a sensitivity of 91.4%, a specificity of 98%, and an AUC of 99.53%.

Table 5. The results of the GoogLeNet and AlexNet models on the lower GI dataset.

Measure	GoogLeNet	AlexNet
Accuracy %	96	91.5
Precision %	96.2	91.8
Sensitivity %	96	91.4
Specificity %	99.2	98
AUC %	96	99.53

Figure 18 presents the evaluation results of the performance of the GoogLeNet and AlexNet models on the lower GI dataset in a graph.

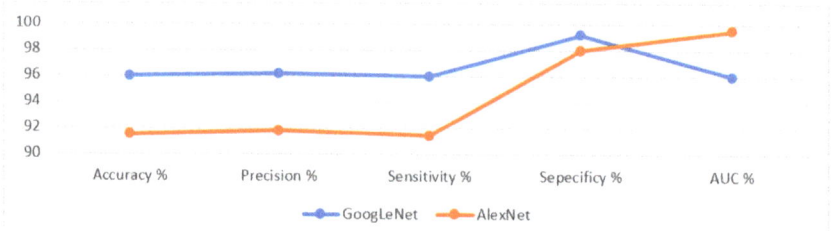

Figure 18. Display results of the GoogLeNet and AlexNet models for diagnosing a low GI dataset.

Figure 19 describes the confusion matrix generated by the CNN models, GoogLeNet and AlexNet for the early diagnosis of lower GI disease. In contrast, the confusion matrix describes all dataset samples that are correctly or incorrectly categorized. It also describes the diagnostic accuracy reached by the models for each class. The figure shows that dyed-lifted-polyps was diagnosed with 98% and 94% accuracy for GoogLeNet and AlexNet, respectively. Normal-cecum was diagnosed with 100% and 94% accuracy for GoogLeNet and AlexNet, respectively. Normal-pylorus was diagnosed with 99.5% and 99% accuracy for GoogLeNet and AlexNet, respectively. Polyps were diagnosed with an accuracy of 92% and 86.5% for GoogLeNet and AlexNet, respectively. Ulcerative colitis was diagnosed with an accuracy of 90.5% and 84% for GoogLeNet and AlexNet, respectively.

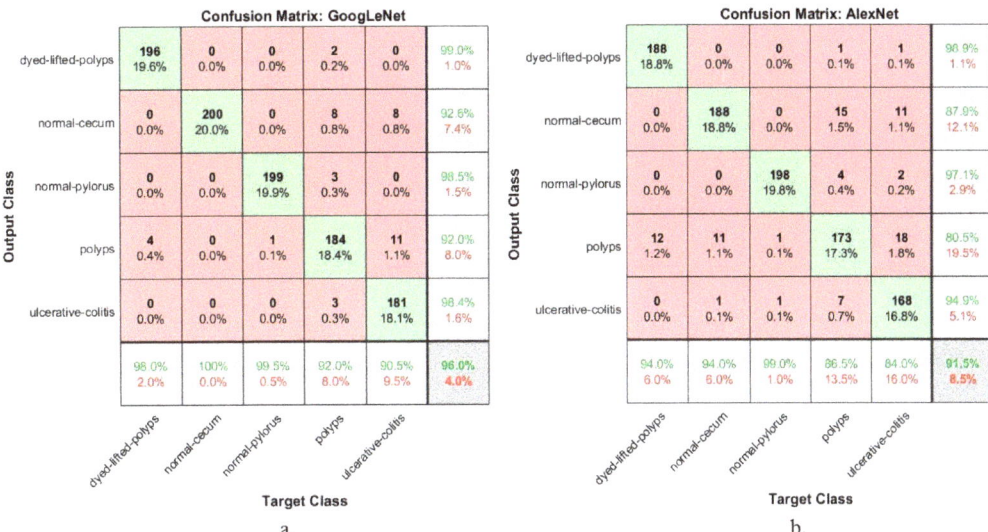

Figure 19. Confusion matrix for the lower GI dataset generated by using (**a**) GoogLeNet (**b**) AlexNet.

4.6. Results of Third Proposed System (Hybrid CNN with SVM)

This section presents the findings of the hybrid techniques between CNN models (GoogLeNet and AlexNet) and the SVM algorithm. The technique consists of two blocks: the first is CNN models for extracting feature maps, and the second block is the SVM algorithm for classifying feature maps. One of the most important reasons for using this technique is that it requires medium-specification computer resources, speed in training the dataset, and high accuracy in diagnosis. Table 6 summarizes the assessment of the lower gastrointestinal diseases dataset by hybrid GoogLeNet+SVM and AlexNet+SVM technique for early diagnosis of gastrointestinal tumors and ulcers. The GoogLeNet+SVM hybrid technique is superior to AlexNet+SVM. The GoogLeNet+SVM achieved an accuracy of 96.7%, a precision of 96.8%, a sensitivity of 96.8%, a specificity of 99%, and an AUC of 99.1%. In contrast, the AlexNet+SVM model achieved an accuracy of 94.7%, a precision of 94.8%, a sensitivity of 94.8%, a specificity of 98.6%, and an AUC of 99.6%.

Table 6. The results of the hybrid method on the lower GI dataset.

Measure	GoogLeNet+SVM	AlexNet+SVM
Accuracy %	96.7	94.7
Precision %	96.8	94.8
Sensitivity %	96.8	94.8
Specificity %	99	98.6
AUC %	99.1	99.62

Figure 20 displays the evaluation results of the GoogLeNet+SVM and AlexNet+SVM techniques on the lower GI dataset in a graph.

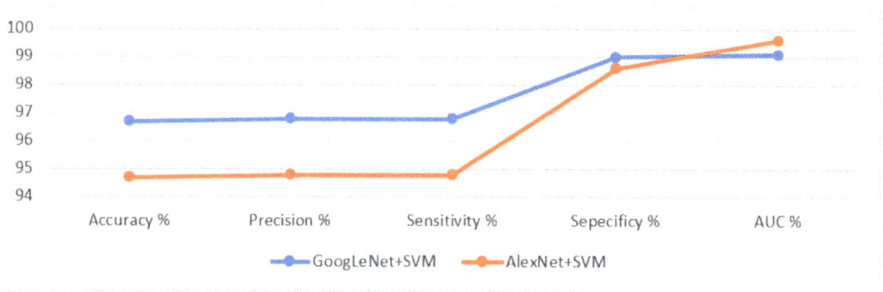

Figure 20. Display results of the GoogLeNet+SVM and AlexNet+SVM techniques for diagnosing a low GI dataset.

Figure 21 shows the performance of the hybrid techniques GoogLeNet+SVM and AlexNet+SVM for diagnosing lower gastrointestinal disease dataset in the form of a confusion matrix. The hybrid methods produced a confusion matrix that describes all samples of the dataset correctly labelled represented in the primary diameter and all samples incorrectly classified and distributed over the rest of the matrix cells. The figure shows the performance of hybrid techniques for diagnosing each disease and the overall accuracy. The figure shows that dyed-lifted-polyps was diagnosed with 98.5% and 95% accuracy for GoogLeNet+SVM and AlexNet+SVM, respectively. Normal-cecum was diagnosed with 100% and 95% accuracy for GoogLeNet+SVM and AlexNet+SVM, respectively. Normal-pylorus was diagnosed with 100% and 100% accuracy for GoogLeNet+SVM and AlexNet+SVM, respectively. Polyps were diagnosed with an accuracy of 93.5% and 90% for GoogLeNet+SVM and AlexNet+SVM, respectively. Ulcerative colitis was diagnosed with an accuracy of 91.5% and 93.5% for GoogLeNet+SVM and AlexNet+SVM, respectively.

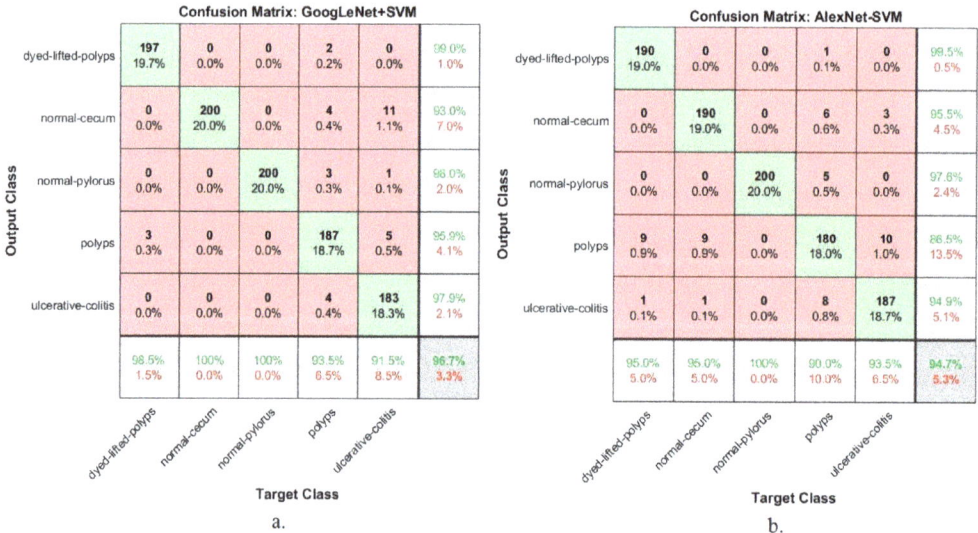

Figure 21. Confusion matrix for evaluating of lower GI dataset using (**a**) GoogLeNet+SVM and (**b**) AlexNet+SVM.

4.7. Results of Fourth Proposed System (Hybrid Features CNN and Traditional Algorithms)

This section presents the evaluation results of hybrid feature techniques between CNN models (GoogLeNet and AlexNet) and features extracted by traditional algorithms (LBP, GLCM and FCH); after fusion, all the features are classified by ANN and FFNN algorithms.

These techniques require low-resource computer specifications, execution speed, and high accuracy in diagnosing endoscopic images of the lower GI dataset. Table 7 summarizes the evaluation results of the performance of the ANN algorithm. When using the hybrid features extracted by CNN models and traditional algorithms (LBP, GLCM and FCH), the systems reached superior results in diagnosing the lower GI dataset. All features are fused into a single feature vector for each image, where each feature vector contains 4328 features fed into the ANN and FFNN classifiers.

Table 7. Results of dataset evaluation by ANN and FFNN based on hybrid features.

Classifiers	ANN		FFNN	
Hybrid Features	GoogLeNet Feature + LBP, GLCM and FCH	AlexNet Feature + LBP, GLCM and FCH	GoogLeNet Feature + LBP, GLCM and FCH	AlexNet Feature + LBP, GLCM and FCH
Accuracy %	98	99.1	98.5	99.3
Precision %	98.25	99	98.6	99.2
Sensitivity %	97.8	98.8	98.2	99
Specificity %	99.4	99.8	99.75	100
AUC %	98.69	99.76	98.83	99.87

First, when diagnosing by ANN algorithm based on the combined features of GoogLeNet and traditional algorithms (LBP, GLCM and FCH), the system reached accuracy, precision, sensitivity, specificity and AUC of 98%, 98.25%, 97.8%, 99.4% and 98.69%, respectively. When using the hybrid features between AlexNet and traditional algorithms (LBP, GLCM and FCH), the system reached accuracy, precision, sensitivity, specificity and AUC with a percentage of 99.1%, 99%, 98.8%, 99.8% and 99.76%, respectively.

Second, when diagnosing by FFNN algorithm based on the combined features of GoogLeNet and traditional algorithms (LBP, GLCM and FCH), the system reached accuracy, precision, sensitivity, specificity and AUC of 98.8%, 98.6%, 98.2%, 99.75% and 98.83%, respectively. When using the hybrid features between AlexNet and traditional algorithms (LBP, GLCM and FCH), the system reached accuracy, precision, sensitivity, specificity and AUC with a percentage of 99.3%, 99.2%, 99%, 100% and 99.87%, respectively.

Figure 22 displays the evaluation of the ANN and FFNN algorithms based on the fusion of features between CNN models and traditional algorithms to classify the lower GI dataset accurately.

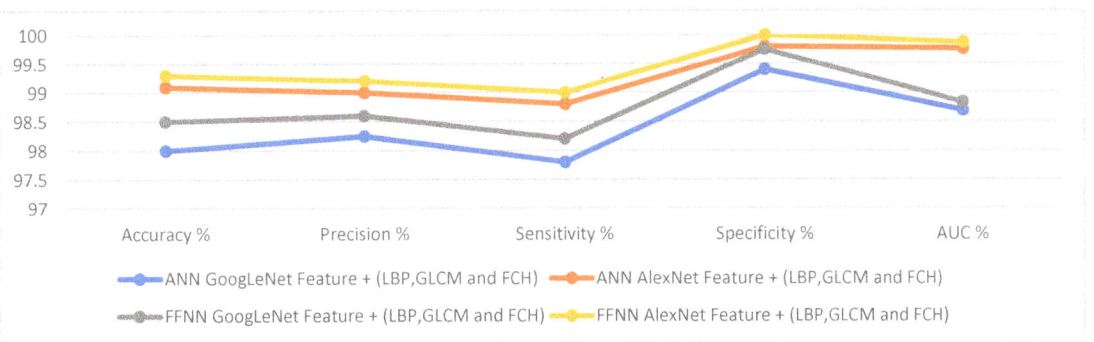

Figure 22. Display of the performance of the ANN and FFNN based on hybrid features for diagnosing a low GI dataset.

Figure 23 shows the evaluation results of the ANN algorithm based on the hybrid features between CNN models (GoogLeNet and AlexNet) with the features extracted by LBP, GLCM and FCH methods for early diagnosis of lower gastrointestinal diseases. The figure summarizes all samples of the correctly classified and incorrectly classified dataset and displays the diagnostic accuracy of each class (disease) in the dataset. First, when using hybrid features extracted from GoogLeNet and conventional, ANN reached an accuracy of 95.4%, 98.5%, 99.5%, 99.5%, and 100% for diagnosing dyed-lifted-polyps, normal-cecum, normal-pylorus, polyps, and ulcerative-colitis, respectively. Second, when using the hybrid features extracted from AlexNet and conventional, ANN reached an accuracy of 97.2%, 99%, 100%, 99.4%, and 100% for diagnosing dyed-lifted-polyps, normal-cecum, normal-pylorus, polyps, and ulcerative-colitis, respectively.

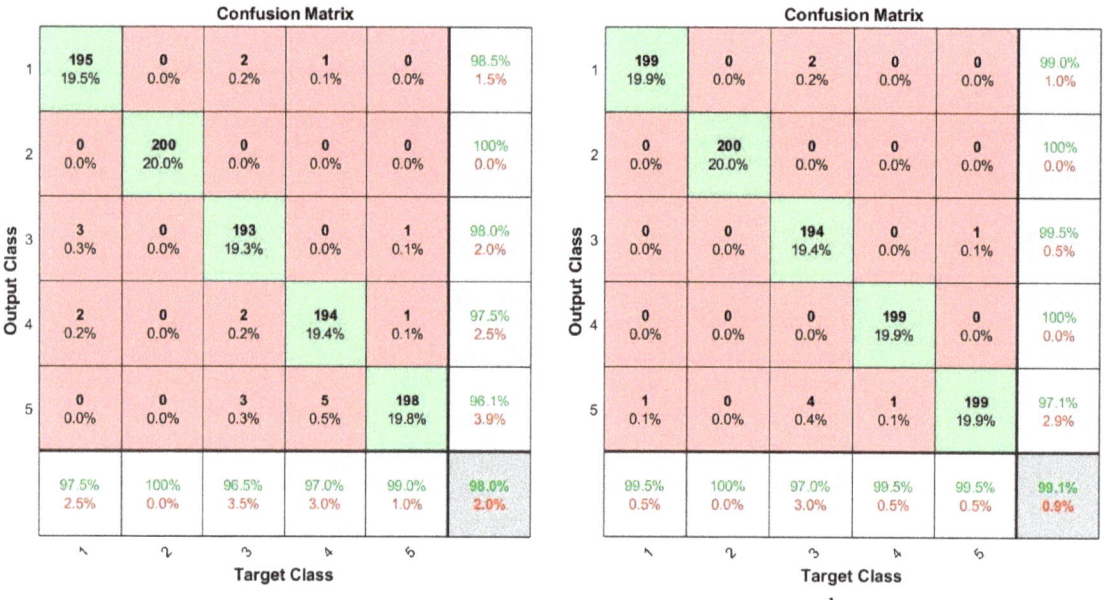

Figure 23. Performance results of the ANN for diagnosing lower gastrointestinal diseases based on hybrid features (**a**) GoogLeNet with traditional algorithm and (**b**) AlexNet with traditional algorithm.

Figure 24 shows the confusion matrix produced by the FFNN algorithm based on the hybrid features between CNN models (GoogLeNet and AlexNet) with the features extracted by LBP, GLCM and FCH methods for early diagnosis of lower gastrointestinal diseases. The figure summarizes all samples of the correctly classified and incorrectly classified dataset and displays the diagnostic accuracy of each class (disease) in the dataset. First, when using hybrid features extracted from GoogLeNet and conventional, FFNN reached an accuracy of 99.5%, 98%, 98.5%, 97%, and 99.5% for diagnosing dyed-lifted-polyps, normal-cecum, normal-pylorus, polyps, and ulcerative-colitis, respectively. Second, when using the hybrid features extracted from AlexNet and conventional, FFNN reached an accuracy of 99.5%, 99%, 99.5%, 99.5%, and 99% for diagnosing dyed-lifted-polyps, normal-cecum, normal-pylorus, polyps, and ulcerative-colitis, respectively.

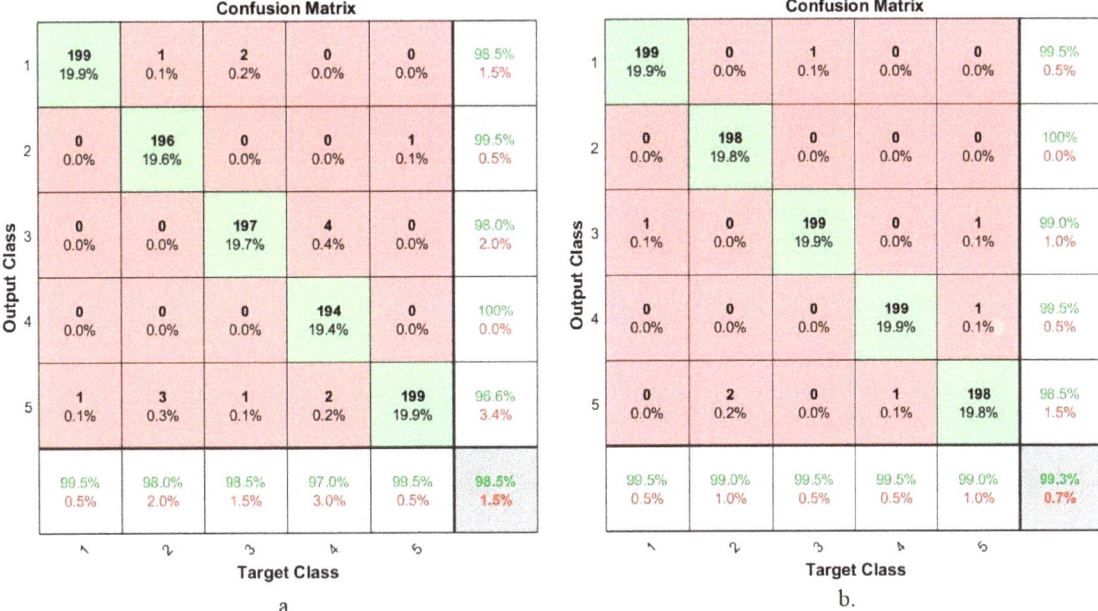

Figure 24. Results of the FFNN for diagnosing lower gastrointestinal diseases based on hybrid features (**a**) GoogLeNet with traditional algorithm and (**b**) AlexNet with traditional algorithm.

5. Discussion and Comparative Analysis

This study presented many methods of artificial intelligence techniques that vary between neural network algorithms, CNN models, hybrid techniques between CNN models, SVM algorithm, and feature merging techniques for early detection of lower GI diseases, which includes 5000 images. Moreover, proposed systems detect and diagnose lower gastrointestinal diseases with a high performance, thus helping to treat and reduce treatment that benefits patients. Clinicians must apply artificial intelligence techniques to diagnose patients and support their diagnostic decisions. The process of data collection and image acquisition from several devices and under different conditions; the influence of external factors such as light reflection; some noise; and internal factors such as mucous membranes and some traces of stool have a negative impact on the diagnostic process, so the average is applied to the three RGB channels. In addition, average and Laplacian filters were applied to enhance the images. Due to the scarcity of medical images, CNN models augment training images by applying the method of data augmentation through flipping, zooming, zooming, and rotating.

Table 8 describes the performance of all proposed systems for diagnosing endoscopic images of the lower gastrointestinal disease dataset. First, for dyed lifted polyps, the ANN algorithm based on fusion features (AlexNet and traditional algorithms) and the FFNN algorithm based on fusion features achieved the best performance for diagnosing this disease with an accuracy of 99.5%.

Table 8. The accuracy achieved by all the proposed systems for evaluating the gastroenterology dataset.

	Diseases		Dyed-Lifted-Polyps	Normal-Cecum	Normal-Pylorus	Polyps	Ulcerative-Colitis	Accuracy %
Neural Networks		ANN	96.5	98.5	96.5	96.5	99	97.4
		FFNN	97	98	99.5	97.5	96	97.6
Deep learning		GoogLeNet	96	100	99.5	92	90.5	96
		AlexNet	94	94	99	86.5	84	91.5
Hybrid		GoogLeNet+SVM	98.5	100	100	93.5	91.5	96.7
		AlexNet+SVM	95	95	100	90	93.5	94.7
Hybrid Features	ANN	GoogLeNet and traditional	97.5	100	96.5	97	99	98
		AlexNet and traditional	99.5	100	97	99.5	99.5	99.1
	FFNN	GoogLeNet and traditional	99.5	98	98.5	97	99.5	98.5
		AlexNet and traditional	99.5	99	99.5	99.5	99	99.3

Second, for Normal-cecum, GoogLeNet, the hybrid technique between GoogLeNet with SVM and the ANN algorithm based on fusion features (GoogLeNet with traditional algorithms and AlexNet with traditional algorithms) achieved the best performance for diagnosing this disease with 100% accuracy. Third, for normal pylorus, the GoogLeNet+SVM and AlexNet+SVM achieved the best performance for diagnosing this disease with 100% accuracy. Fourth, for polyps, the ANN algorithm based on fusion features (AlexNet with traditional algorithms) and the FFNN algorithm based on fusion features (AlexNet with traditional algorithms) achieved the best performance for diagnosing this disease with an accuracy of 99.5%. Fifthly, for ulcerative colitis, the ANN algorithm based on fusion features (AlexNet with traditional algorithms) and the FFNN algorithm based on fusion features (GoogLeNet with traditional algorithms) achieved the best performance for diagnosing this disease with an accuracy of 99.5%.

Figure 25 presents the performance of all proposed systems for diagnosing endoscopic images for early detection of lower gastrointestinal diseases in graphic form.

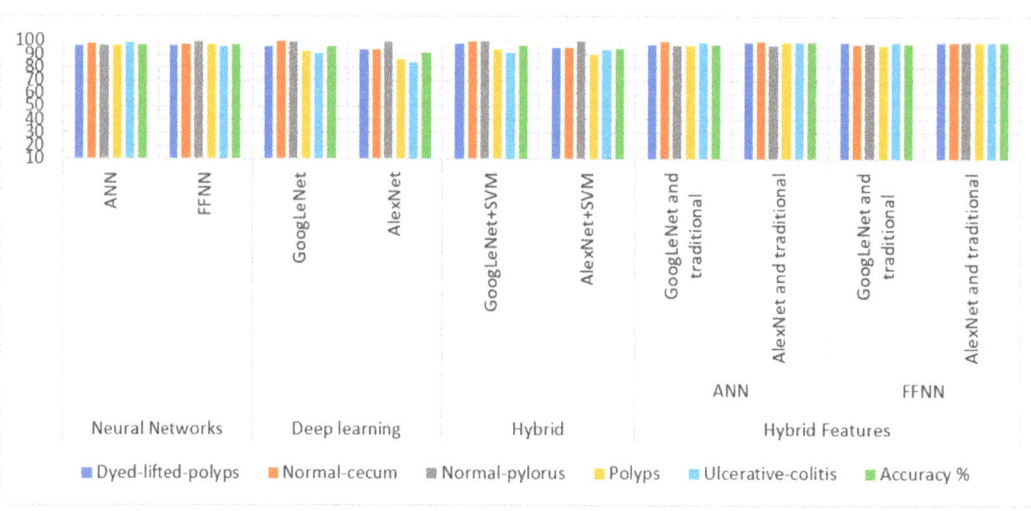

Figure 25. Display of results comparison of all the proposed methods for diagnosing a low GI dataset.

6. Conclusions

The increase in deaths due to lower GI diseases, especially tumors, results from the lack of manual diagnosis due to the difficulty in tracking all the frames. Therefore, this study presents a set of proposed multi-method systems for the early diagnosis of endoscopic images of a lower GI dataset. The first proposed system uses the neural networks ANN and FFNN, which are based on segmentation of the region of interest and feature extraction by LBP, GLCM and FCH algorithms and merging them into one feature vector for each image. The second proposed system uses the CNN models GoogLeNet and AlexNet, which are based on extracting deep feature maps and classifying them accurately. The third proposed system uses hybrid techniques based on CNN models (GoogLeNet and AlexNet) to extract deep feature maps and classify them by the SVM algorithm. The fourth proposed system using ANN and FFNN neural network algorithms is based on fused features extracted by CNN models (GoogLeNet and AlexNet) and LBP, GLCM and FCH algorithms. All the proposed systems achieved highly accurate diagnostic results in diagnosing endoscopic images of the lower gastrointestinal disease dataset with high efficiency.

Future work will apply the principal component analysis (PCA) algorithm to reduce the dimensions of deep feature maps extracted by CNN models, in addition to integrating deep feature maps from more than one CNN model and reducing their dimensions by the PCA algorithm.

Author Contributions: Conceptualization, S.M.F., E.M.S. and A.T.A.; methodology, E.M.S., S.M.F. and A.T.A.; validation, S.M.F., A.T.A. and E.M.S.; formal analysis, A.T.A., S.M.F. and E.M.S.; investigation, E.M.S., S.M.F. and A.T.A.; resources, S.M.F., E.M.S. and A.T.A.; data curation E.M.S., S.M.F. and A.T.A.; writing—original draft preparation, E.M.S.; writing—review and editing, S.M.F. and A.T.A.; visualization, A.T.A., S.M.F. and E.M.S.; supervision, S.M.F., E.M.S. and A.T.A.; project administration, S.M.F. and A.T.A.; funding acquisition, S.M.F. and A.T.A. All authors have read and agreed to the published version of the manuscript.

Funding: This research was funded by Prince Sultan University.

Data Availability Statement: In this study, the data supporting all the proposed systems were collected by the Kvasir dataset available at this link: https://datasets.simula.no/kvasir/#download (accessed on 30 January 2022).

Acknowledgments: The authors would like to acknowledge the support of Prince Sultan University for paying the Article Processing Charges (APC) of this publication. Special acknowledgement to Automated Systems & Soft Computing Lab (ASSCL), Prince Sultan University, Riyadh, Saudi Arabia.

Conflicts of Interest: The authors declare no conflict of interest.

References

1. Bray, F.; Ferlay, J.; Soerjomataram, I.; Siegel, R.L.; Torre, L.A.; Jemal, A. Global cancer statistics 2018: GLOBOCAN estimates of incidence and mortality worldwide for 36 cancers in 185 countries. *CA A Cancer J. Clin.* **2018**, *68*, 394–424. [CrossRef] [PubMed]
2. Liu, J.B.; Cao, J.; Alofi, A.; Abdullah, A.M.; Elaiw, A. Applications of Laplacian spectra for n-prism networks. *Neurocomputing* **2016**, *198*, 69–73. Available online: https://www.sciencedirect.com/science/article/pii/S0925231216003088 (accessed on 1 February 2022). [CrossRef]
3. Liu, J.B.; Pan, X.F. Minimizing Kirchhoff index among graphs with a given vertex bipartiteness. *Appl. Math. Comput.* **2016**, *291*, 84–88. [CrossRef]
4. Lan, L.; Ye, C.; Wang, C.; Zhou, S. Deep convolutional neural networks for WCE abnormality detection: CNN architecture, region proposal and transfer learning. *IEEE Access* **2019**, *7*, 30017–30032. Available online: https://ieeexplore.ieee.org/abstract/document/8651510/ (accessed on 1 February 2022). [CrossRef]
5. Owais, M.; Arsalan, M.; Choi, J.; Mahmood, T.; Park, K.R. Artificial intelligence-based classification of multiple gastrointestinal diseases using endoscopy videos for clinical diagnosis. *J. Clin. Med.* **2019**, *8*, 986. Available online: https://www.mdpi.com/492780 (accessed on 1 February 2022). [CrossRef]
6. Liu, J.B.; Pan, X.F.; Hu, F.T.; Hu, F.F. Asymptotic Laplacian-energy-like invariant of lattices. *Appl. Math. Comput.* **2015**, *253*, 205–214. Available online: https://www.sciencedirect.com/science/article/pii/S0096300314016890 (accessed on 1 February 2022). [CrossRef]
7. Liu, J.B.; Pan, X.F.; Yu, L.; Li, D. Complete characterization of bicyclic graphs with minimal Kirchhoff index. *Discret. Appl. Math.* **2016**, *200*, 95–107. Available online: https://www.sciencedirect.com/science/article/pii/S0166218X15003236 (accessed on 1 February 2022). [CrossRef]

8. Ueyama, H.; Kato, Y.; Akazawa, Y.; Yatagai, N.; Komori, H.; Takeda, T.; Matsumoto, K.; Ueda, K.; Matsumoto, K.; Hojo, M.; et al. Application of artificial intelligence using a convolutional neural network for diagnosis of early gastric cancer based on magnifying endoscopy with narrow-band imaging. *J. Gastroenterol. Hepatol.* **2021**, *36*, 482–489. [CrossRef]
9. Sutton, R.T.; Zaïane, O.R.; Goebel, R.; Baumgart, D.C. Artificial intelligence enabled automated diagnosis and grading of ulcerative colitis endoscopy images. *Sci. Rep.* **2022**, *12*, 2748. [CrossRef]
10. Milluzzo, S.M.; Cesaro, P.; Grazioli, L.M.; Olivari, N.; Spada, C. Artificial intelligence in lower gastrointestinal endoscopy: The current status and future perspective. *Clin. Endosc.* **2021**, *54*, 329. [CrossRef]
11. Ali, S.; Dmitrieva, M.; Ghatwary, N.; Bano, S.; Polat, G.; Temizel, A.; Rittscher, J. Deep learning for detection and segmentation of artefact and disease instances in gastrointestinal endoscopy. *Med. Image Anal.* **2021**, *70*, 102002. [CrossRef]
12. Godkhindi, A.M.; Gowda, R.M. Automated detection of polyps in CT colonography images using deep learning algorithms in colon cancer diagnosis. In Proceedings of the 2017 International Conference on Energy, Communication, Data Analytics and Soft Computing (ICECDS), Chennai, India, 1–2 August 2017; pp. 1722–1728. Available online: https://ieeexplore.ieee.org/abstract/document/8389744/ (accessed on 1 February 2022).
13. Ozawa, T.; Ishihara, S.; Fujishiro, M.; Kumagai, Y.; Shichijo, S.; Tada, T. Automated endoscopic detection and classification of colorectal polyps using convolutional neural networks. *Ther. Adv. Gastroenterol.* **2020**, *13*, 1756284820910659. [CrossRef]
14. Pozdeev, A.A.; Obukhova, N.A.; Motyko, A.A. Automatic analysis of endoscopic images for polyps detection and segmentation. In Proceedings of the 2019 IEEE Conference of Russian Young Researchers in Electrical and Electronic Engineering (EIConRus), Saint Petersburg and Moscow, Russia, 28–31 January 2019; pp. 1216–1220. Available online: https://ieeexplore.ieee.org/abstract/document/8657018/ (accessed on 1 February 2022).
15. Zhang, R.; Zheng, Y.; Mak, T.W.C.; Yu, R.; Wong, S.H.; Lau, J.Y.; Poon, C.C. Automatic detection and classification of colorectal polyps by transferring low-level CNN features from nonmedical domain. *IEEE J. Biomed. Health Inform.* **2016**, *21*, 41–47. [CrossRef]
16. Ribeiro, E.; Uhl, A.; Häfner, M. Colonic polyp classification with convolutional neural networks. In Proceedings of the 2016 IEEE 29th International Symposium on Computer-Based Medical Systems (CBMS), Belfast and Dublin, Ireland, 20–24 June 2016; pp. 253–258. [CrossRef]
17. Min, M.; Su, S.; He, W.; Bi, Y.; Ma, Z.; Liu, Y. Computer-aided diagnosis of colorectal polyps using linked color imaging colonoscopy to predict histology. *Sci. Rep.* **2019**, *9*, 2881. Available online: https://www.nature.com/articles/s41598-019-39416-7 (accessed on 1 February 2022). [CrossRef]
18. Song, E.M.; Park, B.; Ha, C.; Hwang, S.W.; Park, S.H.; Yang, D.H.; Byeon, J.S. Endoscopic diagnosis and treatment planning for colorectal polyps using a deep-learning model. *Sci. Rep.* **2020**, *10*, 30. Available online: https://www.nature.com/articles/s41598-019-56697-0 (accessed on 1 February 2022). [CrossRef]
19. Khan, M.A.; Khan, M.A.; Ahmed, F.; Mittal, M.; Goyal, L.M.; Hemanth, D.J.; Satapathy, S.C. Gastrointestinal diseases segmentation and classification based on duo-deep architectures. *Pattern Recognit. Lett.* **2020**, *131*, 193–204. Available online: https://www.sciencedirect.com/science/article/pii/S016786551930399X (accessed on 1 February 2022). [CrossRef]
20. Billah, M.; Waheed, S. Gastrointestinal polyp detection in endoscopic images using an improved feature extraction method. *Biomed. Eng. Lett.* **2018**, *8*, 69–75. [CrossRef]
21. Hasan, M.M.; Islam, N.; Rahman, M.M. Gastrointestinal polyp detection through a fusion of contourlet transform and Neural features. *J. King Saud Univ. Comput. Inf. Sci.* **2020**, *11*, 2022. Available online: https://www.sciencedirect.com/science/article/pii/S1319157819313151 (accessed on 1 February 2022). [CrossRef]
22. Gamage, C.; Wijesinghe, I.; Chitraranjan, C.; Perera, I. GI-Net: Anomalies classification in gastrointestinal tract through endoscopic imagery with deep learning. In Proceedings of the 2019 Moratuwa Engineering Research Conference (MERCon), Moratuwa, Sri Lanka, 3–5 July 2019; pp. 66–71. Available online: https://ieeexplore.ieee.org/abstract/document/8818929/ (accessed on 1 February 2022).
23. Vleugels, J.L.; Hazewinkel, Y.; Dekker, E. Morphological classifications of gastrointestinal lesions. *Best Pract. Res. Clin. Gastroenterol.* **2017**, *31*, 359–367. Available online: https://www.sciencedirect.com/science/article/pii/S1521691817300434 (accessed on 1 February 2022). [CrossRef]
24. Fonollá, R.; van der Sommen, F.; Schreuder, R.M.; Schoon, E.J.; de With, P.H. Multi-modal classification of polyp malignancy using CNN features with balanced class augmentation. In Proceedings of the 2019 IEEE 16th International Symposium on Biomedical Imaging (ISBI 2019), Venice, Italy, 8–11 April 2019; pp. 74–78. Available online: https://ieeexplore.ieee.org/abstract/document/8759320/ (accessed on 1 February 2022).
25. Öztürk, Ş.; Özkaya, U. Residual LSTM layered CNN for classification of gastrointestinal tract diseases. *J. Biomed. Inform.* **2021**, *113*, 103638. Available online: https://www.sciencedirect.com/science/article/pii/S1532046420302677 (accessed on 1 February 2022). [CrossRef]
26. Maghsoudi, O.H. Superpixel based segmentation and classification of polyps in wireless capsule endoscopy. In Proceedings of the 2017 IEEE Signal Processing in Medicine and Biology Symposium (SPMB), Philadelphia, PA, USA, 2 December 2017; Volume 2018, pp. 1–4. [CrossRef]
27. Herrin, J.; Abraham, N.S.; Yao, X.; Noseworthy, P.A.; Inselman, J.; Shah, N.D.; Ngufor, C. Comparative effectiveness of machine learning approaches for predicting gastrointestinal bleeds in patients receiving antithrombotic treatment. *JAMA Netw. Open* **2021**, *4*, e2110703. [CrossRef]

28. Patel, J.; Ladani, A.; Sambamoorthi, N.; LeMasters, T.; Dwibedi, N.; Sambamoorthi, U. Predictors of Co-occurring Cardiovascular and Gastrointestinal Disorders among Elderly with Osteoarthritis. *Osteoarthr. Cartil. Open* **2021**, *3*, 100148. [CrossRef]
29. Kim, H.J.; Gong, E.J.; Bang, C.S.; Lee, J.J.; Suk, K.T.; Baik, G.H. Computer-Aided Diagnosis of Gastrointestinal Protruded Lesions Using Wireless Capsule Endoscopy: A Systematic Review and Diagnostic Test Accuracy Meta-Analysis. *J. Pers. Med.* **2022**, *12*, 644. [CrossRef]
30. Pogorelov, K.; Randel, K.R.; Griwodz, C.; Eskeland, S.L.; de Lange, T.; Johansen, D.; Halvorsen, P. Kvasir: A multi-class image dataset for computer aided gastrointestinal disease detection. In Proceedings of the 8th ACM on Multimedia Systems Conference, Taipei, Taiwan, 20–23 June 2017; pp. 164–169. [CrossRef]
31. Budak, Ü.; Cömert, Z.; Rashid, Z.N.; Şengür, A.; Çıbuk, M. Computer-aided diagnosis system combining FCN and Bi-LSTM model for efficient breast cancer detection from histopathological images. *Appl. Soft Comput.* **2019**, *85*, 105765. Available online: https://www.sciencedirect.com/science/article/pii/S1568494619305460 (accessed on 4 March 2022). [CrossRef]
32. Mohammed, B.A.; Senan, E.M.; Rassem, T.H.; Makbol, N.M.; Alanazi, A.A.; Al-Mekhlafi, Z.G.; Ghaleb, F.A. Multi-Method Analysis of Medical Records and MRI Images for Early Diagnosis of Dementia and Alzheimer's Disease Based on Deep Learning and Hybrid Methods. *Electronics* **2021**, *10*, 2860. [CrossRef]
33. Saeidifar, M.; Yazdi, M.; Zolghadrasli, A. Performance Improvement in Brain Tumor Detection in MRI Images Using a Combination of Evolutionary Algorithms and Active Contour Method. *J. Digit. Imaging* **2021**, *34*, 1209–1224. [CrossRef] [PubMed]
34. Senan, E.M.; Jadhav, M.E.; Kadam, A. Classification of PH2 images for early detection of skin diseases. In Proceedings of the 2021 6th International Conference for Convergence in Technology (I2CT), Maharashtra, India, 2–4 April 2021; pp. 1–7. [CrossRef]
35. Senan, E.M.; Jadhav, M.E. Techniques for the Detection of Skin Lesions in PH 2 Dermoscopy Images Using Local Binary Pattern (LBP). In Proceedings of the International Conference on Recent Trends in Image Processing and Pattern Recognition, Aurangabad, India, 3–4 January 2020; Springer: Singapore, 2021; pp. 14–25. [CrossRef]
36. Senan, E.M.; Jadhav, M.E. Diagnosis of Dermoscopy Images for the Detection of Skin Lesions Using SVM and KNN. In Proceedings of the Third International Conference on Sustainable Computing, Laguna Hills, CA, USA, 9–13 December 2020; pp. 125–134. [CrossRef]
37. Chaki, J.; Dey, N. Histogram-Based Image Color Features. In *Image Color Feature Extraction Techniques*; Springer: Singapore, 2021; pp. 29–41. [CrossRef]
38. Nudel, J.; Bishara, A.M.; de Geus, S.W.; Patil, P.; Srinivasan, J.; Hess, D.T.; Woodson, J. Development and validation of machine learning models to predict gastrointestinal leak and venous thromboembolism after weight loss surgery: An analysis of the MBSAQIP database. *Surg. Endosc.* **2021**, *35*, 182–191. [CrossRef]
39. Roy, R.B.; Rokonuzzaman, M.; Amin, N.; Mishu, M.K.; Alahakoon, S.; Rahman, S.; Pasupuleti, J. A comparative performance analysis of ANN algorithms for MPPT energy harvesting in solar PV system. *IEEE Access* **2021**, *9*, 102137–102152. [CrossRef]
40. Ezzat, D.; Afify, H.M.; Taha, M.H.N.; Hassanien, A.E. Convolutional Neural Network with Batch Normalization for Classification of Endoscopic Gastrointestinal Diseases. In *Machine Learning and Big Data Analytics Paradigms: Analysis, Applications and Challenges*; Springer: Cham, Switzerland, 2021; pp. 113–128. [CrossRef]
41. Hmoud Al-Adhaileh, M.; Mohammed Senan, E.; Alsaade, W.; Aldhyani, T.H.; Alsharif, N.; Abdullah Alqarni, A.; Jadhav, M.E. Deep Learning Algorithms for Detection and Classification of Gastrointestinal Diseases. *Complexity* **2021**, *2021*, 6170416. [CrossRef]
42. Senan, E.M.; Alzahrani, A.; Alzahrani, M.Y.; Alsharif, N.; Aldhyani, T.H. Automated Diagnosis of Chest X-Ray for Early Detection of COVID-19 Disease. *Comput. Math. Methods Med.* **2021**, *2021*, 6919483. Available online: https://www.hindawi.com/journals/cmmm/2021/6919483/ (accessed on 6 February 2022). [CrossRef]
43. Senan, E.M.; Alsaade, F.W.; Al-Mashhadani, M.I.A.; Theyazn, H.H.; Al-Adhaileh, M.H. Classification of histopathological images for early detection of breast cancer using deep learning. *J. Appl. Sci. Eng.* **2021**, *24*, 323–329. [CrossRef]
44. Liaqat, S.; Dashtipour, K.; Arshad, K.; Assaleh, K.; Ramzan, N. A hybrid posture detection framework: Integrating machine learning and deep neural networks. *IEEE Sens. J.* **2021**, *21*, 9515–9522. [CrossRef]
45. Liu, M.; Lu, Y.; Long, S.; Bai, J.; Lian, W. An attention-based CNN-BiLSTM hybrid neural network enhanced with features of discrete wavelet transformation for fetal acidosis classification. *Expert Syst. Appl.* **2021**, *186*, 115714. [CrossRef]
46. Senan, E.M.; Abunadi, I.; Jadhav, M.E.; Fati, S.M. Score and Correlation Coefficient-Based Feature Selection for Predicting Heart Failure Diagnosis by Using Machine Learning Algorithms. *Comput. Math. Methods Med.* **2021**, *2021*, 8500314. [CrossRef]

Article

Nuclei-Guided Network for Breast Cancer Grading in HE-Stained Pathological Images [†]

Rui Yan [1,2], Fei Ren [1], Jintao Li [1], Xiaosong Rao [3,4], Zhilong Lv [1], Chunhou Zheng [5] and Fa Zhang [1,*]

[1] High Performance Computer Research Center, Institute of Computing Technology, Chinese Academy of Sciences, Beijing 100045, China; yanrui20b@ict.ac.cn (R.Y.); renfei@ict.ac.cn (F.R.); jtli@ict.ac.cn (J.L.); lvzhilong17g@ict.ac.cn (Z.L.)
[2] University of Chinese Academy of Sciences, Beijing 101408, China
[3] Department of Pathology, Boao Evergrande International Hospital, Qionghai 571435, China; raoxiaosong2006@126.com
[4] Department of Pathology, Peking University International Hospital, Beijing 100084, China
[5] College of Computer Science and Technology, Anhui University, Hefei 230093, China; zhengch99@ahu.edu.cn
* Correspondence: zhangfa@ict.ac.cn
[†] This paper is an extension version of the conference paper: Yan, R.; Li, J.; Rao, X.; Lv, Z.; Zheng, C.; Dou, J.; Wang, X.; Ren, F.; Zhang, F. NANet: Nuclei-Aware Network for Grading of Breast Cancer in HE Stained Pathological Images. In Proceedings of the 2020 IEEE International Conference on Bioinformatics and Biomedicine (BIBM), Seoul, Korea, 16–19 December 2020.

Citation: Yan, R.; Ren, F.; Li, J.; Rao, X.; Lv, Z.; Zheng, C.; Zhang, F. Nuclei-Guided Network for Breast Cancer Grading in HE-Stained Pathological Images. *Sensors* 2022, 22, 4061. https://doi.org/10.3390/s22114061

Academic Editors: Sergiu Nedevschi and Mitrea Delia-Alexandrina

Received: 20 April 2022
Accepted: 24 May 2022
Published: 27 May 2022

Publisher's Note: MDPI stays neutral with regard to jurisdictional claims in published maps and institutional affiliations.

Copyright: © 2022 by the authors. Licensee MDPI, Basel, Switzerland. This article is an open access article distributed under the terms and conditions of the Creative Commons Attribution (CC BY) license (https://creativecommons.org/licenses/by/4.0/).

Abstract: Breast cancer grading methods based on hematoxylin-eosin (HE) stained pathological images can be summarized into two categories. The first category is to directly extract the pathological image features for breast cancer grading. However, unlike the coarse-grained problem of breast cancer classification, breast cancer grading is a fine-grained classification problem, so general methods cannot achieve satisfactory results. The second category is to apply the three evaluation criteria of the Nottingham Grading System (NGS) separately, and then integrate the results of the three criteria to obtain the final grading result. However, NGS is only a semiquantitative evaluation method, and there may be far more image features related to breast cancer grading. In this paper, we proposed a Nuclei-Guided Network (NGNet) for breast invasive ductal carcinoma (IDC) grading in pathological images. The proposed nuclei-guided attention module plays the role of nucleus attention, so as to learn more nuclei-related feature representations for breast IDC grading. In addition, the proposed nuclei-guided fusion module in the fusion process of different branches can further enable the network to focus on learning nuclei-related features. Overall, under the guidance of nuclei-related features, the entire NGNet can learn more fine-grained features for breast IDC grading. The experimental results show that the performance of the proposed method is better than that of state-of-the-art method. In addition, we released a well-labeled dataset with 3644 pathological images for breast IDC grading. This dataset is currently the largest publicly available breast IDC grading dataset and can serve as a benchmark to facilitate a broader study of breast IDC grading.

Keywords: breast cancer grading; histopathological image; nuclei segmentation; convolutional neural network; attention mechanism

1. Introduction

Breast invasive ductal carcinoma (IDC) is the most widespread type of breast cancer, making up approximately 80% of all diagnosed cases. Histological grading has direct guiding significance for the prognostic evaluation of IDC. The most popular grading scheme is the Nottingham Grading System (NGS) [1] which gives a more objective assessment than previous grading systems. NGS includes three semi-quantitative criteria: mitotic count, nucleus atypia, and tubular formation. However, in clinical practice, the burden of pathological diagnosis is very heavy, and many pathologists cannot accurately grasp

NGS, which will greatly weaken the guiding significance of histological grading for clinical prognosis evaluation, and even mislead the clinical judgment of prognoses. Therefore, there is an urgent need for an automatic and accurate pathological grading method.

The automatic breast cancer grading methods based on pathological images can be summarized into two categories. The first category is to use machine-learning or deep-learning methods to directly extract the features of the pathological image for breast cancer grading. However, unlike the coarse-grained problem of breast cancer classification, IDC grading is a fine-grained classification problem. Using only general methods cannot classify IDC well because the classification boundaries among intermediate-grade and low- and high-grade IDC pathological images are blurred.

The second category is to compute the three evaluation criteria of NGS separately and then integrate those results to obtain the final IDC grading result. However, NGS is only a semiquantitative evaluation method. The inherent medical motivation of NGS is to classify IDC based on the morphological and texture characteristics of the cell nucleus and the topological structure of the cell population. With the end-to-end advantage of deep learning, not only can the medical goal of emphasizing nuclei-related features be achieved, but more fine-grained feature representations of pathological images that are too abstract for pathologists to understand can also be learned.

In this paper, we propose a Nuclei-Guided Network (NGNet) for IDC grading in hematoxylin-eosin (HE) stained pathological images. Specifically, our network includes two branches. The main branch is used to extract the feature representation of the entire pathological image, and the nuclei branch is used to extract the feature representation of the nuclei image. Then, the nuclei-guided attention module between the two branches plays the role of nucleus attention in end-to-end learning, so that more nuclei-related feature representations for IDC grading can be learned. In addition, the proposed nuclei-guided fusion module in the fusion process of two branches can further enable the network to focus on learning nuclei-related features. Overall, under the guidance of nuclei-related features, the entire NGNet can learn more fine-grained features for breast IDC grading. It should be pointed out that this is different from the general attention mechanism [2–4] that cannot artificially emphasize the region of interest.

Experimental results show that the proposed NGNet significantly outperforms the state-of-the-art method, achieving 93.4% average classification accuracy and 0.93 AUC with our released dataset. In addition, we release a new dataset containing 3644 pathological images with different magnifications (20× and 40×) for evaluating the IDC grading methods. Compared with the previous publicly available breast cancer grading dataset with only 300 images in total, the number of images in our dataset has increased by an order of magnitude. The dataset is publicly available from https://github.com/YANRUI121/Breast-cancer-grading (accessed on 1 April 2022).

2. Related Works

Recently, the application of deep learning has enabled breast cancer pathological image classification to achieve high performance. However, breast cancer classification is not enough for the final medical diagnosis. The classification must be subdivided and accurate to the extent of the pathological grade of the cancer, because the gold standard of the final medical diagnosis, the choice of treatment plan and the prediction of patient outcome are all based on the results of the pathological grade.

The classification boundaries among intermediate-grade and low- and high-grade IDC pathological images are ambiguous; thus, general methods cannot classify the IDC grade well. The current IDC grading methods can be divided into two categories. The first category is to classify the features extracted directly from the pathological image. The second category is to first calculate the three evaluation criteria of NGS (1) mitotic count [5–8], (2) nucleus atypia [9,10], and (3) tubular formation [11–13], and then artificially integrate these three criteria to obtain the final result. Figure 1 is a brief description of NGS. By analyzing the three evaluation criteria of NGS, we observe that nuclei-related features are very important for breast cancer pathological diagnosis. Specifically, mitotic count and nucleus atypia are concerned with the morphological and texture characteristics of the cell nucleus, whereas tubular formation is concerned with the topological structure of the cell population. Because we are primarily concerned with end-to-end breast cancer grading studies, we will only briefly introduce the related works of the first category in the following.

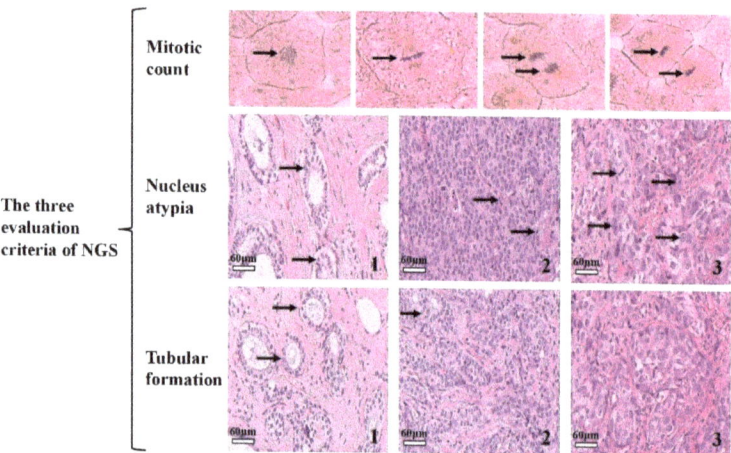

Figure 1. A brief description of the three evaluation criteria of NGS adopted by the World Health Organization. (1) Mitotic count: the images represent prophase, metaphase, anaphase and telophase stages of mitosis from left to right. (2) Nucleus atypia: the nucleus atypia score reflects the variations in the size, shape, and appearance of the cancer cells relative to normal cells. The nuclear atypia score values are 1, 2, and 3 from left to right. (3) Tubular formation: a large number of tubules are formed in the pathological image on the left. As the grade increases, the tubules gradually disappear from left to right.

Before the era of deep learning, research on breast cancer pathological image grading was mainly based on traditional machine-learning methods. For example, Doyle et al. [14] proposed a novel method to classify low- and high-grade of breast cancer histopathological images by using architectural features. Naik et al. [15] classify the low- and high-grade breast cancer by using a combination of low-level, high-level, and domain-specific information. They first segment glands and nuclei. Then, morphological and architectural attributes derived from the segmented gland and nuclei were used to discriminate low-grade from high-grade breast cancer. Basavanhally et al. [16] conducted a multifield-of-view classifier with robust feature selection for classifying ER+ breast cancer pathological images. Their grading system can distinguish low- vs. high-grade patients well, but fails to distinguish low- vs. intermediate-, and intermediate- vs. high-grade patients well.

Deep learning has made great progress in breast cancer pathological image grading. The most representative work was proposed by Wan et al. [17]. They integrated semantic-level features extracted from a convolutional neural network (CNN), pixel-level texture features, and object-level architecture features to classify low-, intermediate-, and high-grade breast cancer pathological images. The method achieved an accuracy of 0.92 for low vs. high, 0.77 for low vs. intermediate, and 0.76 for intermediate vs. high, and an overall accuracy of 0.69 when discriminating all three grades of breast cancer pathological images. Our preliminary work that shows that only using deep learning can help achieve better grading performance was published in BIBM2020 [18]. Compared to the previous work, we put forward new contributions in nuclei-guided branch fusion and further disclosed one of the largest IDC grading datasets.

In the field of computer vision, there are many excellent networks based on attention mechanisms, such as SENet [19], Position and Channel Attention [20], CBAM [4], Criss-Cross Attention [21], and Self-Attention [22,23]. SENet [19] is the abbreviation of Squeeze-and-Excitation Networks. SENet mainly recalibrates the feature responses of channels adaptively by explicitly modeling the interdependence between channels. In other words, the correlation between channels is learned. Convolutional Block Attention Module (CBAM) [4] combines spatial and channel attention mechanism, which can achieve better results than SENet's attention mechanism that only focuses on channels. Because CBAM is a lightweight general module, it can be integrated into any CNN architecture with negligible overhead of this module, and can be trained end-to-end together with the base CNN. Transformer is a deep neural network based on self-attention mechanism, which has been considered as a viable alternative to convolutional and recurrent neural networks. In the field of computer vision, Vision Transformer (ViT) proposed by Dosovitskiy et al. [24] is a pioneering work. Following the paradigm of ViT, a series of ViT variants [25,26] have been proposed to improve the performance. The complexity of the ViT-like model is very high, so it needs a very large training dataset. Therefore, the application of the ViT-like model in the field of pathological images analysis is still few at present, especially for the breast cancer grading tasks that are difficult to manually label. These above-mentioned attention mechanisms are adaptively learned from the data, and are the areas where the algorithm thinks attention should be focused. However, if we need to customize the area where the algorithm focuses attention based on prior knowledge, this is not possible. A more comprehensive review of attention mechanisms can be found in [27,28]. Our proposed network can focus on a specific area. This is different from the general attention mechanism that cannot artificially emphasize the region of interest. This provides a new paradigm for embedding medical prior knowledge into algorithms.

3. Dataset

Deep-learning methods have an important dependence on well-labeled datasets such as BreaKHis dataset [29], the Yan et al. dataset [30], and the BACH dataset [31]. However, due to the difficulty of the IDC grading task, there are few related works. To the best of our knowledge, only Kosmas et al. [32] has released one IDC grading dataset containing 300 pathological images, which is insufficient for deep-learning research. In this work, we cooperated with Peking University International Hospital to release a new benchmark dataset for IDC grading. We conducted experiments on these two datasets to comprehensively verify the effectiveness of our proposed NGNet method. Next, we will introduce these two datasets.

3.1. IDC Pathological Images Dataset

The dataset released by Kosmas et al. [32] includes 300 images (107 Grade1 images, 102 Grade2 images, and 91 Grade3 images). All images were acquired at 40× magnification. Although this released dataset has played a significant role in the IDC grading research, 300 images are not enough for the deep-learning method.

To meet the needs of deep-learning research, we cooperated with Peking University International Hospital to release a new IDC grading dataset. Our annotated HE-stained pathological image dataset consists of 3644 pathological images (1000 × 1000 pixels). Figure 2 is an example of the images and a summary of the dataset. We named it the PathoIDCG dataset, which is an abbreviation of the Pathological Image Dataset for Invasive Ductal Carcinoma Grading. The overall description of the PathoIDCG dataset is shown in Table 1. The preparation procedure used in our research is the standard paraffin process, which is widely used in routine clinical practice. The thickness of pathological sections is 3–5 μm. Each image is labeled Grade1, Grade2, or Grade3 according to the three evaluation criteria of NGS. Image annotation was independently performed by two pathologists in strict accordance with NGS standards, and the images with different annotations were reannotated by a senior pathologist. The Ethics Committee of Peking University International Hospital reviewed and approved the study, and all the related data are anonymous.

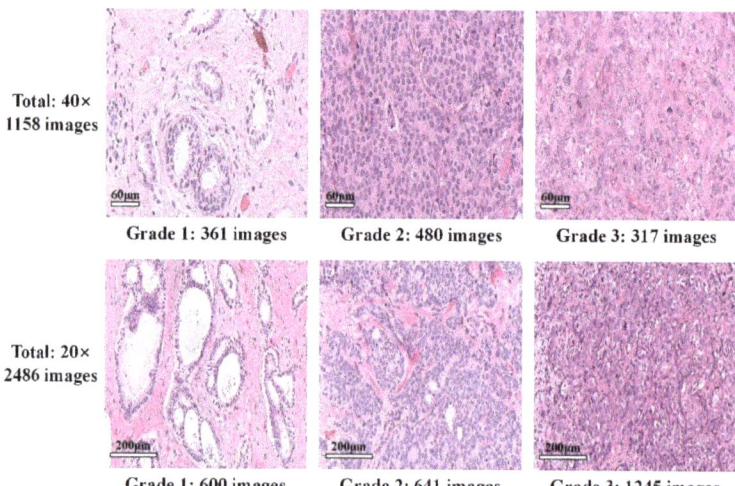

Figure 2. Pathological image examples and quantity statistics of our proposed dataset for IDC grading.

Table 1. The overall description of the PathoIDCG dataset.

Description	Value
No. pathological images (total)	3644
No. pathological images (40×)	1158 (361 G1, 480 G2, 317 G3)
No. pathological images (20×)	2486 (600 G1, 641 G2, 1245 G3)
Size of pathological images	1000 × 1000 pixels
Magnification of pathological images	20×, 40×
Color model of pathological images	R(ed)G(reen)B(lue)
Memory space of pathological images	~1 MB
Type of image label	Image-wise

Our dataset is mainly acquired under a 20× magnified field of view, because the 20× magnified pathological image can contain more information about the topology of the cell population. Another reason is that the commonly available 20× slides are easier to obtain, and the current cell nucleus segmentation technology can also segment pathological images under 20× magnification. At the same time, we also collected pathological images at 40× magnification because a larger magnification can better reflect the texture and morphological characteristics of individual nuclei.

3.2. Nuclei Segmentation Dataset

The dataset released by Kumar et al. [33] included HE-stained pathological images with 21,623 annotated nucleus boundaries, and Figure 3 is an example of this dataset. Kumar et al. [33] downloaded 30 whole slide pathological images of several organs from The Cancer Genomic Atlas (TCGA) [34] and used only one WSI per patient to maximize nuclear appearance variation. In addition, these images come from 18 different hospitals, which makes the dataset sufficiently diverse. It is important to emphasize that although we only segmented the nucleus of breast cancer pathological images, our segmentation model was trained on pathological images of all seven organs: breast, liver, colon, prostate, bladder, kidney, and stomach. For the above reasons, our segmentation model is more robust and generalizable.

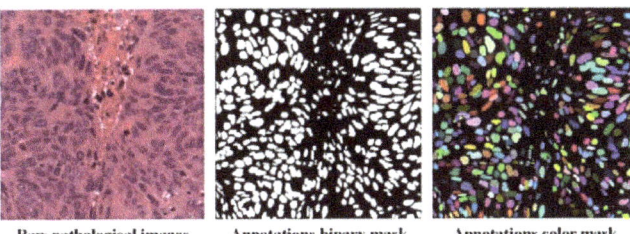

Figure 3. The nuclei segmentation dataset we used for breast cancer grading. We only need binary mask annotations to train the segmentation model. For better visualization, each nucleus is shown in a different color.

4. Methods

The key idea of NGNet is shown in Figure 4. Our method consists of two stages: in the first stage, we segmented the nucleus of each pathological image to obtain all images that only contain the nucleus region. In the second stage, two images (original pathological image and corresponding nuclei image) are input at the same time and sent to the NGNet to obtain the final classification result.

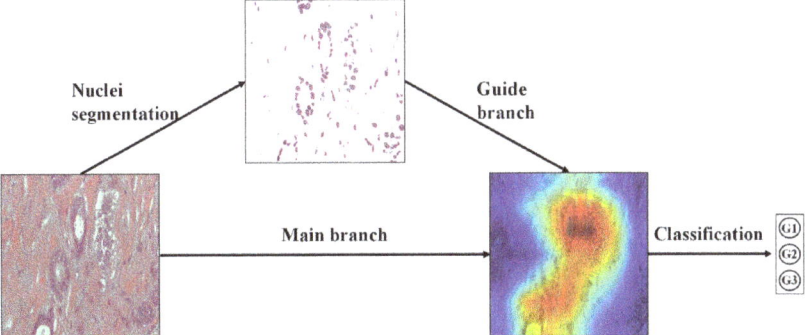

Figure 4. The key idea of the proposed method. NGNet forces the network to focus on learning features related to the nuclei. At the same time, under the guidance of nuclei-related features, the entire network learns more fine-grained features. The visual heat map is obtained through Grad-CAM using our proposed NGNet.

4.1. Nuclei Segmentation

We use DeepLabV3+ [35] as our nuclei segmentation network because it can better address the following challenges. In the HE-stained pathological image, some cell nuclei are very large, whereas some are very small. Moreover, under different magnifications, such as 20× and 40×, the difference in the size of the nucleus is more significant. Therefore, our network is required to be able to use multiscale image features, especially to be able to reconstruct the information of small objects. At the same time, many overlapping nuclei boundaries make nuclei segmentation more difficult, so the segmentation algorithm is required to have the ability to reconstruct nuclei boundaries.

Given a pathological image, the output of DeepLabV3+ is a nuclei segmentation mask. The backbone of the DeepLabV3+ algorithm we applied is Xception [36]. When our training steps are 100,000, we have achieved the best experimental results. The values of atrous rates we used are 6, 12, and 18. We adopt an output stride equal to 16. Here, we denote the output stride as the ratio of input image spatial resolution to the final output resolution.

4.2. NGNet Architecture

The overall network architecture is shown in Figure 5. The proposed NGNet has two inputs $[I_{main}, I_{nuclei}]$. The input to the main branch is the original pathological image I_{main}, and the input to the guide branch is the image I_{nuclei} containing only the nuclei, respectively. The relationships between the two inputs are:

$$I_{nuclei} = S \times I_{main}, \qquad (1)$$

where S is the nuclei segmentation result corresponding to the original pathological image.

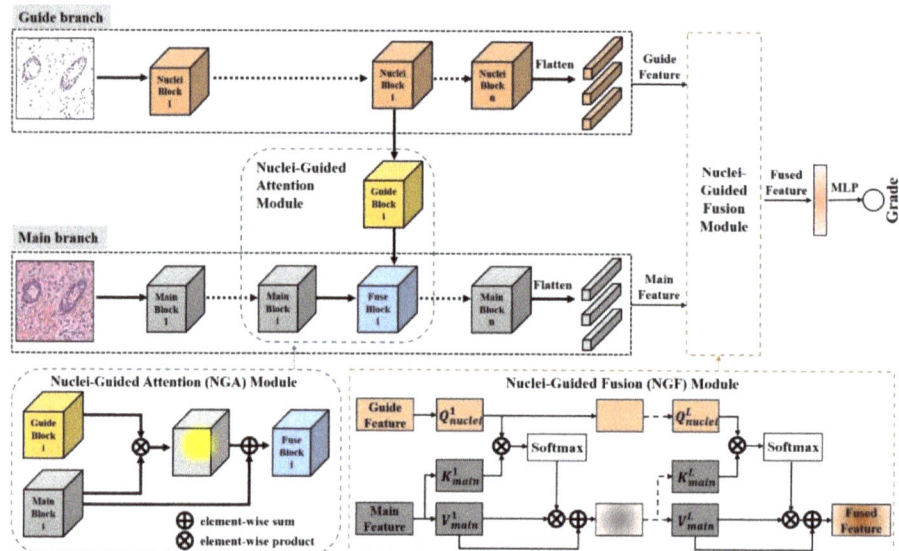

Figure 5. The overall network architecture of NGNet we proposed. The input of NGNet has two corresponding images: one is the original pathological image, and the other is the result of nucleus segmentation corresponding to this original pathological image. The entire NGNet is trained end-to-end.

The guide branch and main branch contain the same number of convolutional layers. Between the corresponding convolutional layers of the two branches, the Nuclei-Guided Attention (NGA) module transfers the nuclei-related features of the guide branch to the main branch. On top of the last convolution layer of each branch, feature maps $F_{main}^M(n)$ and $F_{nuclei}^M(n)$ were flattened to several feature vectors $P_{main}^M(n)$ and $P_{nuclei}^M(n)$, respectively, where M represents the number of convolutional layers of each branch, and n represents the n-th feature map. Then, the feature vectors $P_{main}^M(n)$ and $P_{nuclei}^M(n)$ were passed through the Nuclei-Guided Fusion (NGF) module to obtain fused feature representation. Finally, the grading result is obtained through the multilayer perceptron (MLP) module.

The following is a detailed introduction to the NGA module and NGF module. The specific implementation details of the NGA module can be illustrated by the specific example of the "Guide 21" step in NGNet, as shown in Figure 6. Given a pathological image I, $F_{main}^m(I_{main})$ and $F_{nuclei}^m(I_{nuclei})$ is denoted as the convolutional feature maps from the m-th convolutional layer of the main branch and guide branch, respectively. In each corresponding convolutional layer, the guide branch extracting nuclei features has a guide block $F_{guide}^m(I_{nuclei})$ pointing to the main branch extracting pathological image features.

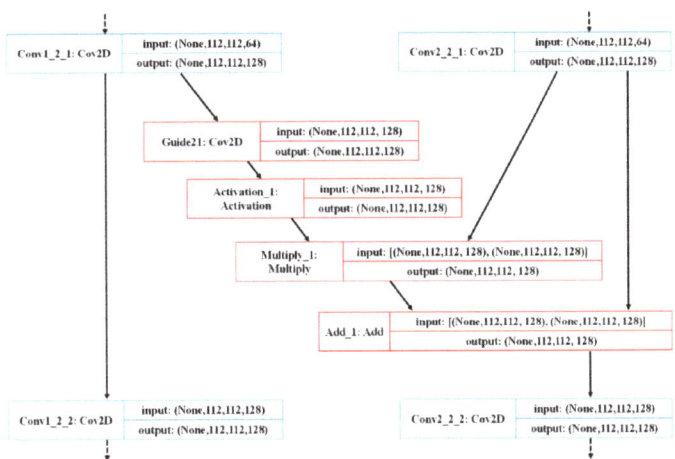

Figure 6. Detailed schematic diagram of the nuclei-guided attention module in the NGNet we proposed; the example comes from the "Guide 21" step.

We first perform a 1×1 convolution on the feature maps of the corresponding nuclei block $F_{nuclei}^{m}(I_{nuclei})$, in which the input and output dimensions are equal. After performing the 1×1 convolution operation on the feature maps of the corresponding nuclei block $F_{nuclei}^{m}(I_{nuclei})$, the Softmax activation function is used to generate the attention map A^m. Thus, the value of the feature map is adjusted to between 0 and 1. Then, we perform elementwise multiplication with the feature map of the corresponding main branch $F_{main}^{m}(I_{main})$, thereby increasing the weight of the important area of the feature map. The purpose of this is to focus on the features related to the nuclei. Specifically, we calculate the attention map A^m and guide block $F_{guide}^{m}(I_{nuclei})$ as follows:

$$A^m = \text{Softmax}(\,(Conv_{1\times 1}(F_{nuclei}^{m}(I_{nuclei}))), \qquad (2)$$

$$F_{guide}^{m}(I_{nuclei}) = F_{main}^{m}(I_{main}) \otimes A^m, \qquad (3)$$

where the Softmax (.) is the Softmax activation function, $Conv_{1\times 1}$ (.) is a 1×1 convolution operation, $\otimes$ represents elementwise multiplication. At the end of each NGA module, an elementwise addition $\oplus$ is performed:

$$F_{fuse}^{m}(I) = F_{main}^{m}(I_{main}) \oplus F_{guide}^{m}(I_{nuclei}), \qquad (4)$$

where $F_{fuse}^{m}(I)$ is the feature maps guided by nuclei-related features from the m-th convolutional layer.

The NGF module (see Figure 5) is inspired by the self-attention mechanism which can capture various dependencies within a sequence (e.g., short-range and long-range dependencies). The self-attention mechanism is implemented via the Query-Key-Value (QKV) model. Given a sequence and its packed matrix representations of Q, K, and V, the scaled dot-product attention is given by

$$Att(Q, K, V) = \text{Softmax}\left(\frac{QK^T}{\sqrt{d_k}}\right)V = AV, \qquad (5)$$

where d_k is the dimension of key, and A is often called the attention matrix which computes the similarity score of the QK pairs. Different from the standard self-attention QKV which comes from the same input sequence, our Q_{nuclei} is the feature vector from the guide branch, and the K_{main}, V_{main} are the feature vectors from the main branch. Therefore, the $Q_{nuclei}K_{main}$ similarity we calculated represents the similarity between the nuclear features and the original pathological image features. The similarity score of $Q_{nuclei}K_{main}$ is then mapped to V_{main}, allowing the network to pay more attention to the nuclei-related features. The $Q^l_{nuclei}K^l_{main}V^l_{main}$ calculation can be performed one or more times (L); here we set $L = 3$. In addition, we also added a residual connection between V^l_{main} and $Att\left(Q^l_{nuclei}, K^l_{main}, V^l_{main}\right)$ to preserve the information of the main branch. At the end of the NGF module, we obtained the fused feature representation of the guide and main branch. Formally, we have

$$P = V^L_{main} + Att\left(Q^L_{nuclei}, K^L_{main}, V^L_{main}\right). \tag{6}$$

To get the final classification result, P is flattened into the vector, and then goes through the fully connected layer. The loss function for NGNet is defined as the cross entropy (CE) loss:

$$L_{CE} = -\frac{1}{m}\sum_{i=1}^{z}\sum_{k=1}^{k} q^z_k \log(p^z_k), \tag{7}$$

where q^z_k and p^z_k indicates the ground truth and prediction probability of the z-th image for k-th class.

It should be emphasized that our method is universal and can be easily generalized to another task that needs to emphasize a certain local area (such as a lesion) in the model. First, determine the image area of interest through prior knowledge and segment this area. Then, our algorithm framework can model this particular part of attention into the algorithm through end-to-end learning. The design of this network structure provides an end-to-end modeling methodology for custom attention.

5. Results and Discussion

In this section, we will evaluate the performance of NGNet. We randomly selected 80% of the dataset to train and validate the model, and the remaining 20% was used for testing. All experiments in this paper are finished on three NVIDIA GPUs by using the Keras framework with TensorFlow backend. We mainly use the average accuracy to evaluate the performance of NGNet. Apart from the average accuracy, the classification performance of an algorithm can be further evaluated by using the sensitivity, specificity, confusion matrix, and AUC. The accuracy, sensitivity, and specificity metrics can be defined as follows:

$$\text{Accuracy} = \frac{TP + TN}{TP + TN + FP + FN},$$

$$\text{Sensitivity} = \frac{TP}{TP + FN},$$

$$\text{Specificity} = \frac{TN}{TN + FP},$$

where TP (TN) represents number of true positive (true negative) classified pathological images, and FP (FN) represents number of false positive (false negative) classified pathological images.

5.1. Comparison of the Accuracy with Previous Methods

To verify the effectiveness of the method, we conduct comprehensive comparative experiments. For the three-class classifications, our method achieved 93.4% average accuracy based on the PathoIDCG dataset (see Table 2). The morphological differences between grade 1 (G1) and grade 2 (G2), as well as grade 2 (G2) and grade 3 (G3), is very subtle, so it is difficult to distinguish. This problem is reflected by our experimental results and previous studies. For this reason, previous studies have only focused on the classification tasks of G1 and G3. We have made comprehensive comparisons with previous state-of-the-art studies and the classic CNN: ResNet50 [37] and Xception [36]; the experimental results are shown in Table 2. It can be seen from the results that our method has achieved good classification accuracy in each category. However, only 94.1% and 93.9% accuracy are achieved on G1 vs. G2 and G2 vs. G3, respectively. Compared with the classification results of these two difficult categories, the classification accuracy of G1 vs. G3 is much better, reaching 97.8%.

Table 2. Comparison of accuracy with previous methods.

Methods	Acc (%) G1 vs. G2	Acc (%) G1 vs. G3	Acc (%) G2 vs. G3	Acc (%) G1 vs. G2 vs. G3
Naik et al. [15]	-	80.5	-	-
Doyle et al. [14]	-	93.0	-	-
Basavanhally et al. [16]	74.0	91.0	75.0	-
Wan et al. [17]	77.0	92.0	76.0	69.0
ResNet50 [37]	87.5	91.0	88.5	87.2
Xception [36]	88.3	92.3	88.6	87.9
NGNet	94.1	97.8	93.9	93.4

5.2. Confusion Matrix and AUC

We conduct experiments on the PathoIDCG dataset to comprehensively evaluate the performance of our method. The confusion matrix of the predictions is presented in Figure 7 by using the proposed NGNet on the test set. Figure 8 shows the mean area under curve (AUC) of 0.93, corresponding to 0.94, 0.91, and 0.93 based on receiver operating characteristic analysis.

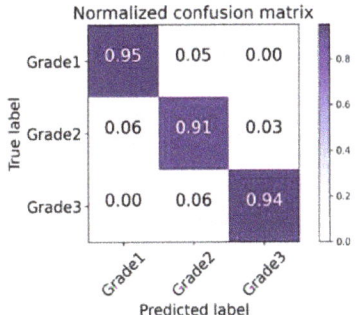

Figure 7. Visualization of normalized confusion matrix.

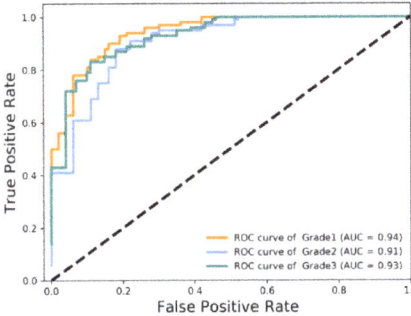

Figure 8. Visualization of receiver operating characteristic curve (ROC) and area under curve (AUC).

As seen from the experimental results in Figures 7 and 8, the results obtained in G1 vs. G2 and G2 vs. G3 are not as good as the classification results of G1 vs. G3. This also further illustrates that the classification bottleneck is to learn more distinguished features for similar categories.

5.3. Nuclei Segmentation Results

To select the suitable method for nucleus segmentation, we compare with three methods: Watershed, UNet [38], and DeepLabV3+ [35]. The watershed is the most representative traditional image processing method, and the version we used in the experiment is Fiji [39]. At the same time, we also conduct experiments on representative deep-learning methods UNet [38] and DeepLabV3+. As can be seen from Figure 9, DeepLabV3+ is suitable for our cell nucleus segmentation task, and achieved satisfactory results.

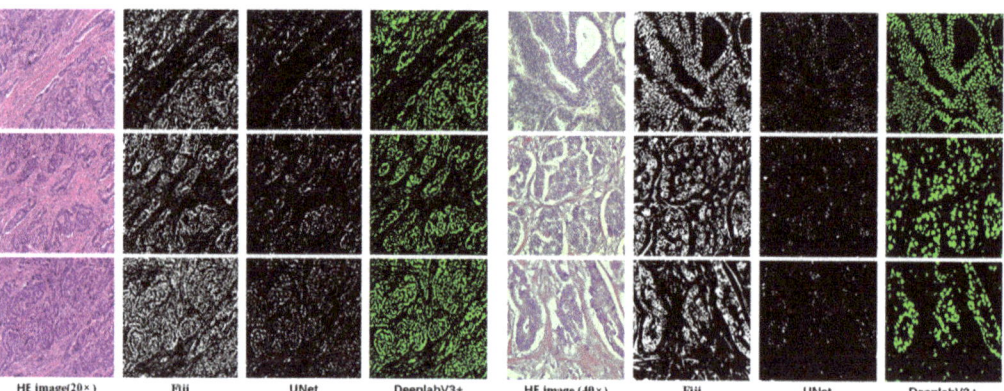

Figure 9. Nuclei segmentation results using Fiji (Watershed), UNet, and DeepLabV3+ (proposed). The left three rows are comparisons of the segmentation results at 20× magnification, and the right three rows are comparisons of the segmentation results at 40× magnification.

We perform a visual qualitative analysis of the segmentation results only. The visual display of the segmentation results is shown in Figure 9. Because we do not have the ground truth of nuclei segmentation for the PathoIDCG dataset, we did not use traditional quantitative indicators such as mean intersection over union (mIOU) to measure the segmentation effect. Our segmentation network is trained on the well-annotated dataset proposed by Kumar et al. [33]. After the segmentation network is well-trained, we directly use this trained segmentation network to segment the IDC grading dataset. Moreover, traditional metrics cannot measure the segmentation results we need. For example, we think that a slightly larger segmentation that includes the edge background of the nuclei may be better. However, the segmentation of nuclei containing a large number of missing nuclei is very poor.

5.4. Grad-CAM Visualization

Gradient-weighted class activation mapping (Grad-CAM) is a method proposed by Selvaraju et al. [40] to produce visual explanations (heat map) of decisions, making CNN-based methods more transparent and explainable. Grad-CAM can generate a rough location map to highlight important areas in the image for prediction. This method only considers the pixels and locations that have a positive impact on the classification result because we only care about the locations that have a positive impact on the classification.

In this section, we use the Grad-CAM method to visualize the pathological image regions that provide support for a particular classification result. We compare the Grad-CAM experimental results of NGNet with VGG16, as shown in Figure 10. From the experimental results, it can be found that the experimental results of NGNet are more focused on the area related to the nuclei. Moreover, NGNet can further refine the nuclei-related feature representations. As shown in the pathological image and the corresponding heat map in Figure 10, attention not only focuses on the nuclei-related area but also focuses on the gland-related nucleus area. This is consistent with the medical knowledge of NGS.

Clinically, breast cancer grading is adopted by pathologists through NGS, and one of the key evaluation criteria is the formation of glands.

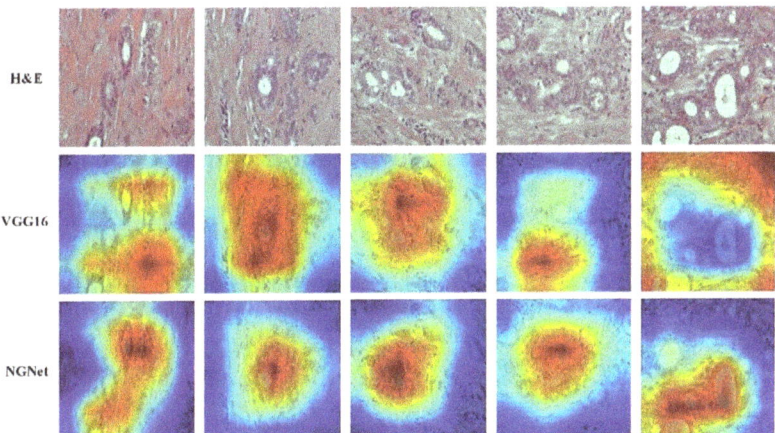

Figure 10. Visualization of class activation maps using Grad-CAM method. Red regions indicate a high score of a certain class. The first line is the pathological image. The second line and third line are the visual heat map using VGG16 and our proposed NGNet, respectively, as the backbone of Grad-CAM. Figure best viewed in color.

5.5. Ablation Study

To evaluate the effectiveness of each component in our proposed method, we conducted an ablation study. The experimental results on the test set are shown in Table 3. The hyperparameters of the experiment include the following: the loss function is categorical cross-entropy, the learning rate is 0.00002, the optimizer is RMSProp, and a total of 300 epoch iterations are performed.

Table 3. Ablation study results with different configurations on the test set.

Methods	Acc.	Sensitivity	Specificity	AUC
VGGNet (pathology image only)	85.1%	86.0%	85.3%	0.87
VGGNet (nuclei image only)	80.6%	81.2%	79.2%	0.79
NGNet (w/o NGA and NGF)	90.6%	89.3%	89.8%	0.89
NGNet (w/o NGF)	92.2%	93.8%	91.1%	0.92
NGNet (w/o NGA)	91.8%	91.6%	90.9%	0.90
NGNet (proposed)	93.4%	95.3%	92.9%	0.93

We conduct comparative experiments on accuracy, sensitivity, specificity and AUC. First, because our single branch network structure is similar to VGG16, we compare the classification performance of NGNet and VGG16. The experimental results show that NGNet has achieved much better results than just using VGG16. Then, we compare the experimental results of NGNet with different experimental configurations. NGNet has achieved better results even with a simple fusion of pathological images and nuclear images; that is NGNet without nuclei-guided attention (NGA) and nuclei-guided fusion (NGF) module. After adding the NGA module and NGF module to NGNet, the best results are achieved. Specifically, compared with NGNet without NGA and NGF module, NGA and NGF module bring an AUC improvement of 0.01 and 0.03 to the network, respectively. When using the NGA and NGF module at the same time—that is, our proposed NGNet—it brings an AUC improvement of 0.04 to the network. The experimental results fully demonstrate the advantages of NGA module and NGF module in NGNet, and also demonstrate that each module is indispensable. The experimental results are shown in Table 3.

6. Conclusions

In this paper, the proposed NGNet can ensure that the network is focused on nuclei-related features, so as to learn fine-grained feature representations for breast IDC grading. Through extensive experimental comparisons, it was shown that NGNet outperforms the state-of-the-art method and has the potential to assist pathologists in breast IDC grading diagnosis. In addition, we released a new dataset containing 3644 pathological images with different magnifications (20× and 40×) for evaluating breast IDC grading methods. Compared with the previous publicly available dataset of breast cancer grading with only 300 images in total, our number of images is an order of magnitude greater. Therefore, the dataset can be used as a benchmark to facilitate a broader study of the breast IDC grading method.

In future work, to further improve the classification performance of breast IDC grading, medical knowledge embedding and semi-supervised learning are two promising directions. Whether in the field of natural image analysis or medical image analysis, the research on the network structure of deep learning has been very comprehensive. Therefore, only by improving the network structure to further improve the classification performance is limited. There are few studies on how to combine medical knowledge with pathological image to further improve classification performance [41]. If we can embed medical knowledge in the end-to-end network learning, the performance of the IDC grading method will be further improved. In terms of pathological image datasets for IDC grading, it is impractical to label a sufficiently large dataset because the cost of labeling pathological images is high. However, the amount of unlabeled pathological image data in each hospital is very large [42]. If a small labeled dataset and a large unlabeled dataset can be used at the same time, the performance of the IDC grading method may be further improved to a level that can be used clinically.

Author Contributions: Methodology, R.Y.; investigation, Z.L. and R.Y.; resources, J.L. and F.Z.; data curation, X.R. and F.R.; writing—original draft preparation, R.Y.; writing—review and editing, C.Z. and F.Z.; funding acquisition, X.R., F.R. and F.Z. All authors have read and agreed to the published version of the manuscript.

Funding: This research was funded by the Strategic Priority Research Program of the Chinese Academy of Sciences (No. XDA16021400), the National Key Research and Development Program of China (No. 2021YFF0704300), and the NSFC projects grants (61932018, 62072441 and 62072280).

Institutional Review Board Statement: Not applicable.

Informed Consent Statement: Not applicable.

Data Availability Statement: The dataset is publicly available from https://github.com/YANRUI121/Breast-cancer-grading (accessed on 1 April 2022).

Conflicts of Interest: The funders had no role in the design of the study; in the collection, analyses, or interpretation of data; in the writing of the manuscript, or in the decision to publish the results.

References

1. Elston, C.W.; Ellis, I.O. Pathological prognostic factors in breast cancer. I. The value of histological grade in breast cancer: Experience from a large study with long–term follow–up. *Histopathology* **1991**, *19*, 403–410. [CrossRef] [PubMed]
2. Wang, F.; Jiang, M.; Qian, C.; Yang, S.; Li, C.; Zhang, H.; Wang, X.; Tang, X. Residual attention network for image classification. In Proceedings of the IEEE Conference on Computer Vision and Pattern Recognition, Honolulu, HI, USA, 21–26 July 2017; pp. 3156–3164.
3. Ma, X.; Guo, J.; Tang, S.; Qiao, Z.; Chen, Q.; Yang, Q.; Fu, S. DCANet: Learning Connected Attentions for Convolutional Neural Networks. *arXiv* **2020**, arXiv:200705099.
4. Woo, S.; Park, J.; Lee, J.-Y.; Kweon, I.S. Cbam: Convolutional block attention module. In Proceedings of the European Conference on Computer Vision (ECCV), Munich, Germany, 8–14 September 2018; pp. 3–19.

5. Huh, S.; Chen, M. Detection of mitosis within a stem cell population of high cell confluence in phase-contrast microscopy images. In Proceedings of the CVPR 2011, Colorado Springs, CO, USA, 20–25 June 2011; IEEE: Piscataway, NJ, USA, 2011; pp. 1033–1040.
6. Tek, F.B. Mitosis detection using generic features and an ensemble of cascade adaboosts. *J. Pathol. Inform.* **2013**, *4*, 12. [CrossRef] [PubMed]
7. Cireşan, D.C.; Giusti, A.; Gambardella, L.M.; Schmidhuber, J. Mitosis detection in breast cancer histology images with deep neural networks. In Proceedings of the International Conference on Medical Image Computing and Computer-Assisted Intervention, Nagoya, Japan, 22–26 September 2013; Springer: Berlin/Heidelberg, Germany, 2013; pp. 411–418.
8. Malon, C.D.; Cosatto, E. Classification of mitotic figures with convolutional neural networks and seeded blob features. *J. Pathol. Informatics* **2013**, *4*, 9. [CrossRef] [PubMed]
9. Khan, A.M.; Sirinukunwattana, K.; Rajpoot, N. A Global Covariance Descriptor for Nuclear Atypia Scoring in Breast Histopathology Images. *IEEE J. Biomed. Health Inform.* **2015**, *19*, 1637–1647. [CrossRef] [PubMed]
10. Lu, C.; Ji, M.; Ma, Z.; Mandal, M. Automated image analysis of nuclear atypia in high-power field histopathological image. *J. Microsc.* **2015**, *258*, 233–240. [CrossRef] [PubMed]
11. BenTaieb, A.; Hamarneh, G. Topology aware fully convolutional networks for histology gland segmentation. In Proceedings of the International Conference on Medical Image Computing and Computer-Assisted Intervention, Athens, Greece, 17–21 October 2016; Springer: Berlin/Heidelberg, Germany, 2016; pp. 460–468.
12. Chen, H.; Qi, X.; Yu, L.; Heng, P.-A. DCAN: Deep contour-aware networks for accurate gland segmentation. In Proceedings of the IEEE conference on Computer Vision and Pattern Recognition, Las Vegas, NV, USA, 27–30 June 2016; pp. 2487–2496.
13. Xu, Y.; Li, Y.; Liu, M.; Wang, Y.; Lai, M.; Eric, I.; Chang, C. Gland instance segmentation by deep multichannel side supervision. In Proceedings of the International Conference on Medical Image Computing and Computer-Assisted Intervention, Athens, Greece, 17–21 October 2016; Springer: Berlin/Heidelberg, Germany, 2016; pp. 496–504.
14. Doyle, S.; Agner, S.; Madabhushi, A.; Feldman, M.; Tomaszewski, J. Automated grading of breast cancer histopathology using spectral clustering with textural and architectural image features. In Proceedings of the 2008 5th IEEE International Symposium on Biomedical Imaging: From Nano to Macro, Paris, France, 14–17 May 2008; IEEE: Piscataway, NJ, USA, 2008; pp. 496–499.
15. Naik, S.; Doyle, S.; Agner, S.; Madabhushi, A.; Feldman, M.; Tomaszewski, J. Automated gland and nuclei segmentation for grading of prostate and breast cancer histopathology. In Proceedings of the 2008 5th IEEE International Symposium on Biomedical Imaging: From Nano to Macro, Paris, France, 14–17 May 2008; IEEE: Piscataway, NJ, USA, 2008; pp. 284–287.
16. Basavanhally, A.; Ganesan, S.; Feldman, M.; Shih, N.; Mies, C.; Tomaszewski, J.; Madabhushi, A. Multi-Field-of-View Framework for Distinguishing Tumor Grade in ER+ Breast Cancer From Entire Histopathology Slides. *IEEE Trans. Biomed. Eng.* **2013**, *60*, 2089–2099. [CrossRef] [PubMed]
17. Wan, T.; Cao, J.; Chen, J.; Qin, Z. Automated grading of breast cancer histopathology using cascaded ensemble with combination of multi-level image features. *Neurocomputing* **2017**, *229*, 34–44. [CrossRef]
18. Yan, R.; Li, J.; Rao, X.; Lv, Z.; Zheng, C.; Dou, J.; Wang, X.; Ren, F.; Zhang, F. NANet: Nuclei-Aware Network for Grading of Breast Cancer in HE Stained Pathological Images. In Proceedings of the 2020 IEEE International Conference on Bioinformatics and Biomedicine (BIBM), Seoul, Korea, 16–19 December 2020; IFEE: Piscataway, NJ, USA, 2020; pp. 865–870.
19. Hu, J.; Shen, L.; Sun, G. Squeeze-and-excitation networks. In Proceedings of the IEEE Conference on Computer Vision and Pattern Recognition, Salt Lake City, UT, USA, 18–23 June 2018; pp. 7132–7141.
20. Fu, J.; Liu, J.; Tian, H.; Li, Y.; Bao, Y.; Fang, Z.; Lu, H. Dual attention network for scene segmentation. In Proceedings of the IEEE/CVF Conference on Computer Vision and Pattern Recognition, Long Beach, CA, USA, 15–20 June 2019; pp. 3146–3154.
21. Huang, Z.; Wang, X.; Huang, L.; Huang, C.; Wei, Y.; Liu, W. Ccnet: Criss-cross attention for semantic segmentation. In Proceedings of the IEEE/CVF International Conference on Computer Vision, Seoul, Korea, 27–28 October 2019; pp. 603–612.
22. Vaswani, A.; Shazeer, N.; Parmar, N.; Uszkoreit, J.; Jones, L.; Gomez, A.N.; Kaiser, Ł.; Polosukhin, I. Attention is all you need. *Adv. Neural Inf. Processing Syst.* **2017**, *30*, 5998–6008.
23. Wang, X.; Girshick, R.; Gupta, A.; He, K. Non-local neural networks. In Proceedings of the IEEE Conference on Computer Vision and Pattern Recognition, Salt Lake City, UT, USA, 18–23 June 2018; pp. 7794–7803.
24. Dosovitskiy, A.; Beyer, L.; Kolesnikov, A.; Weissenborn, D.; Zhai, X.; Unterthiner, T.; Dehghani, M.; Minderer, M.; Heigold, G.; Gelly, S. An image is worth 16x16 words: Transformers for image recognition at scale. *arXiv* **2020**, arXiv:201011929.
25. Liu, Z.; Lin, Y.; Cao, Y.; Hu, H.; Wei, Y.; Zhang, Z.; Lin, S.; Guo, B. Swin transformer: Hierarchical vision transformer using shifted windows. In Proceedings of the IEEE/CVF International Conference on Computer Vision, Montreal, BC, Canada, 11–17 October 2021; pp. 10012–10022.
26. Han, K.; Wang, Y.; Chen, H.; Chen, X.; Guo, J.; Liu, Z.; Tang, Y.; Xiao, A.; Xu, C.; Xu, Y. A survey on vision transformer. In *IEEE Transactions on Pattern Analysis and Machine Intelligence*; IEEE: Piscataway, NJ, USA, 2022; p. 1, Early Access.
27. Guo, M.-H.; Xu, T.-X.; Liu, J.-J.; Liu, Z.-N.; Jiang, P.-T.; Mu, T.-J.; Zhang, S.-H.; Martin, R.R.; Cheng, M.-M.; Hu, S.-M. Attention mechanisms in computer vision: A survey. *Comput. Vis. Media* **2022**, *8*, 331–368. [CrossRef]
28. Hu, D. An introductory survey on attention mechanisms in NLP problems. In Proceedings of the SAI Intelligent Systems Conference, London, UK, 5–6 September 2019; pp. 432–448.
29. Spanhol, F.A.; Oliveira, L.S.; Petitjean, C.; Heutte, L. Breast cancer histopathological image classification using convolutional neural networks. In Proceedings of the International Joint Conference on Neural Networks, Vancouver, BC, Canada, 24–29 July 2016; pp. 717–726.

30. Yan, R.; Ren, F.; Wang, Z.; Wang, L.; Zhang, T.; Liu, Y.; Rao, X.; Zheng, C.; Zhang, F. Breast cancer histopathological image classification using a hybrid deep neural network. *Methods* **2019**, *173*, 52–60. [CrossRef] [PubMed]
31. Aresta, G.; Araújo, T.; Kwok, S.; Chennamsetty, S.S.; Safwan, M.; Alex, V.; Marami, B.; Prastawa, M.; Chan, M.; Donovan, M.; et al. BACH: Grand challenge on breast cancer histology images. *Med Image Anal.* **2019**, *56*, 122–139. [CrossRef] [PubMed]
32. Dimitropoulos, K.; Barmpoutis, P.; Zioga, C.; Kamas, A.; Patsiaoura, K.; Grammalidis, N. Grading of invasive breast carcinoma through Grassmannian VLAD encoding. *PLoS ONE* **2017**, *12*, e0185110.
33. Kumar, N.; Verma, R.; Sharma, S.; Bhargava, S.; Vahadane, A.; Sethi, A. A dataset and a technique for generalized nuclear segmentation for computational pathology. *IEEE Trans. Med. Imaging* **2017**, *36*, 1550–1560. [CrossRef]
34. Hutter, C.; Zenklusen, J.C. The Cancer Genome Atlas: Creating Lasting Value beyond Its Data. *Cell* **2018**, *173*, 283–285. [CrossRef] [PubMed]
35. Chen, L.-C.; Zhu, Y.; Papandreou, G.; Schroff, F.; Adam, H. Encoder-decoder with atrous separable convolution for semantic image segmentation. In Proceedings of the European Conference on Computer Vision (ECCV), Munich, Germany, 8–14 September 2018; pp. 801–818.
36. Chollet, F. Xception: Deep learning with depthwise separable convolutions. In Proceedings of the IEEE Conference on Computer Vision and Pattern Recognition, Honolulu, HI, USA, 21–26 July 2017; pp. 1251–1258.
37. He, K.; Zhang, X.; Ren, S.; Sun, J. Deep residual learning for image recognition. In Proceedings of the IEEE Conference on Computer Vision and Pattern Recognition, Las Vegas, NV, USA, 1–26 July 2016; pp. 770–778.
38. Ronneberger, O.; Fischer, P.; Brox, T. U-net: Convolutional networks for biomedical image segmentation. In Proceedings of the International Conference on Medical Image Computing and Computer-Assisted Intervention, Munich, Germany, 5–9 October 2015; Springer: Berlin/Heidelberg, Germany, 2015; pp. 234–241.
39. Schindelin, J.; Arganda-Carreras, I.; Frise, E.; Kaynig, V.; Longair, M.; Pietzsch, T.; Preibisch, S.; Rueden, C.; Saalfeld, S.; Schmid, B.; et al. Fiji: An open-source platform for biological-image analysis. *Nat. Methods* **2012**, *9*, 676–682. [CrossRef] [PubMed]
40. Selvaraju, R.R.; Cogswell, M.; Das, A.; Vedantam, R.; Parikh, D.; Batra, D. Grad-cam: Visual explanations from deep networks via gradient-based localization. In Proceedings of the IEEE International Conference on Computer Vision, Venice, Italy, 22–29 October 2017; pp. 618–626.
41. Zhou, S.K.; Greenspan, H.; Davatzikos, C.; Duncan, J.S.; Van Ginneken, B.; Madabhushi, A.; Prince, J.L.; Rueckert, D.; Summers, R.M. A Review of Deep Learning in Medical Imaging: Imaging Traits, Technology Trends, Case Studies With Progress Highlights, and Future Promises. *Proc. IEEE* **2021**, *109*, 820–838. [CrossRef]
42. Price, W.N.; Cohen, I.G. Privacy in the age of medical big data. *Nat. Med.* **2019**, *25*, 37–43. [CrossRef] [PubMed]

Article

Accuracy Report on a Handheld 3D Ultrasound Scanner Prototype Based on a Standard Ultrasound Machine and a Spatial Pose Reading Sensor

Radu Chifor [1], Tiberiu Marita [2,*], Tudor Arsenescu [3], Andrei Santoma [2], Alexandru Florin Badea [4], Horatiu Alexandru Colosi [5], Mindra-Eugenia Badea [1] and Ioana Chifor [1]

[1] Department of Preventive Dentistry, University of Medicine and Pharmacy Iuliu Hatieganu, 400083 Cluj-Napoca, Romania; chifor.radu@umfcluj.ro (R.C.); mebadea@umfcluj.ro (M.-E.B.); ioana.chifor@umfcluj.ro (I.C.)
[2] Computer Science Department, Technical University of Cluj-Napoca, 400114 Cluj-Napoca, Romania; santoma.va.andrei@student.utcluj.ro
[3] Chifor Research SRL, 400068 Cluj-Napoca, Romania; tudor.arsenescu@chiforvision.com
[4] Anatomy Department, University of Medicine and Pharmacy, 400006 Cluj-Napoca, Romania; alexandru.badea@umfcluj.ro
[5] Department of Medical Education, Division of Medical Informatics and Biostatistics, Iuliu Hatieganu University of Medicine and Pharmacy, 400349 Cluj-Napoca, Romania; hcolosi@umfcluj.ro
* Correspondence: tiberiu.marita@cs.utcluj.ro

Abstract: The aim of this study was to develop and evaluate a 3D ultrasound scanning method. The main requirements were the freehand architecture of the scanner and high accuracy of the reconstructions. A quantitative evaluation of a freehand 3D ultrasound scanner prototype was performed, comparing the ultrasonographic reconstructions with the CAD (computer-aided design) model of the scanned object, to determine the accuracy of the result. For six consecutive scans, the 3D ultrasonographic reconstructions were scaled and aligned with the model. The mean distance between the 3D objects ranged between 0.019 and 0.05 mm and the standard deviation between 0.287 mm and 0.565 mm. Despite some inherent limitations of our study, the quantitative evaluation of the 3D ultrasonographic reconstructions showed comparable results to other studies performed on smaller areas of the scanned objects, demonstrating the future potential of the developed prototype.

Keywords: 3D ultrasonography; freehand 3D ultrasound scanner prototype; quantitative 3D reconstruction evaluation; 2D image segmentation; pose sensor; coordinate measuring machine

1. Introduction

Ultraportable imaging equipment, such as handheld sonographic machines with wireless systems, shows adequate accuracy, performance and good quality of images compared to high-end sonographic machines [1]. The low cost and the handling of such portable sonographic machines might raise an increased interest among clinicians, especially in emergency medicine departments, but having diagnostic imaging competence may be decisive in driving the correct therapeutic decision. Ultrasound quality is operator dependent and subjective to interpretive error; in order to successfully integrate this technology into their clinical practices, physicians must be familiar with the normal and abnormal appearance of tissues [2]. Conventional two-dimensional (2D) ultrasound imaging is a powerful diagnostic tool in the hands of an experienced user; however, 2D ultrasound remains clinically underutilized and inherently incomplete, with the output being very operator dependent. Providing a simple and inexpensive method of acquiring complete volumetric 3D ultrasound images, with sensed pose information and intuitive feedback displayed to the user, is an important step towards solving the problem of operator dependence. The usefulness of the real-time 3D US was demonstrated by a large variety of clinical

applications, further indicating its role and significance in the fields of medical imaging and diagnosis [3]. Its cost is relatively low in comparison to CT and MRI, no intensive training or radiation protection are required for its operation, and its hardware is movable and can potentially be portable [4]. Previous studies showed that a volume measurement using the 3D US devices has a similar accuracy level to that of CT and MR [5].

The accuracy of ultrasound medical systems seems to depend significantly on settings, as well as on phantom features, probes and investigated parameters. The relative uncertainty due to the influence of probe manipulation on spatial resolution can be very high (i.e., from 10 to more than 30%), and field of view settings must also be taken into account [6]. However, previous studies have shown that an ultrasound scanner was able to scan teeth with an accuracy similar to that of conventional optical scanners when no gingiva was present [7], and ultrasonography is suitable for periodontal imaging [8], even if it requires an extremely high accuracy, due to the size and complexity of the investigated anatomical elements.

Three-dimensional ultrasounds may store volumes describing the whole lesion or organ. A detailed evaluation of the stored data is possible by looking for the features that were not fully appreciated at the time of data collection, or by applying new algorithms for volume rendering, in order to glean important information [9]. Three-dimensional imaging could be an advantage, especially in the education of future surgical generations. Recent studies have shown that the modern 3D technique is superior to 2D, in an experimental setting [10]. The manual guidance of the probe makes reproducible image acquisition almost impossible. Volumetric data offer the distinct advantage of covering entire anatomical structures, and their motion paths can then be used for automated robotic control [11].

The aim of this study was to develop and evaluate a highly accurate 3D ultrasound scanning method. The main requirements imposed on the new scanning method were the free hand architecture of the scanner and the high accuracy of the reconstructions, ranging between the computer tomography reconstructions and optical scans, as well as no movement restrictions during scanning.

The main contributions of this paper are as follows: it documents a quantitative evaluation of a freehand 3D ultrasound scanner prototype, comparing the ultrasonographic reconstructions with the CAD (computer-aided design) model of the scanned object to determine the accuracy of the result; proposes a semi-automatic segmentation method of the raw US images (region growing-based segmentation, followed by morphological filtering and a customized upper contour (envelope) extraction); proposes an evaluation method by comparing the 3D ultrasound reconstructed object with the original 3D CAD model, by computing the mean distance and standard deviation after their alignment.

2. Materials and Methods

A 3D ultrasound scanner prototype based on a 2D standard ultrasound machine and a spatial pose reading sensor was developed using Vinno 6 (Suzhou, China) equipment with a high frequency (10-23 MHz) and a small aperture (12.8 mm) linear transducer (X10-23L) and as a pose reading sensor, an articulated measurement arm (Evo 7, RPS Metrology (Sona/Italy). The articulated measurement arm RPS EVO 7 accuracy was 34 μm. According to the technical specifications, the following information was obtained: "It has no need for calibrations or warm-up time, thanks to its extremely reliable mechanical and electronical design, the automatic temperature compensation and the lightweight structure." The transducer was attached to the articulated arm (coordinate measuring machine, CMM). The spatial and temporal calibration of the employed devices were performed using proprietary algorithms. After calibration, a CAD/CAM manufactured object, used as a phantom, was immersed in a water tank and scanned 6 consecutive times. The CAD/CAM manufactured object was a custom mouth guard, simulating a dental arch, having attached an object with regular contours and planar surfaces, exhibiting both right angles and concave surfaces (Figure 1). The mouth guard was manufactured using

DATRON D5 Linear Scales (Darmstadt, Germany,), with an accuracy of ±5 μm according to the technical specifications and PMMA as the material.

Figure 1. The scanned CAD/CAM manufactured mouth guard, the attachment exhibiting both curved and rectilinear contours and positive and negative relief with regular and irregular multiple concavities and convexities.

Every scanning procedure generated between 452 and 580 bi-dimensional consecutive ultrasound images (Table 1, second column), with a mean scanning time of approximately 14.3 s (the frame rate was 33 frames/second). The ultrasound scanning plane cross-sectioned the object transversally. The scanning procedure started each time at the last molar, going in mesial direction, a 6 teeth area and then backwards, in distal direction, back to the starting point. A total number of 2840 bi-dimensional ultrasound images were acquired and used for the accuracy evaluation of the 3D ultrasound reconstructions.

Table 1. Statistical analysis: mean distance and standard deviation for the 3D ultrasonographic point clouds of six consecutive scans of the same CAD/CAM manufactured object.

Scan	Range of 2D Ultrasound Frames	Segmentation Mode	Scanning Time	Number of 3D Points	Mean Distance	Std Deviation
1	300–752	Segmentation without contour extraction	13.69 s	279,189	0.033 mm	0.387 mm
2	232–611	Segmentation and contour extraction	11.48 s	46,537	0.031 mm	0.287 mm
3	252–703	Segmentation and contour extraction	13.66 s	65,535	0.050 mm	0.350 mm
4	260–779	Segmentation and contour extraction	15.72 s	70,774	0.014 mm	0.352 mm
5	220–674	Segmentation and contour extraction	13.75 s	54,378	0.023 mm	0.372 mm
6	400–979	Segmentation and contour extraction	17.54 s	76,735	0.019 mm	0.565 mm

2.1. Measuring and Verifying the CAD/CAM Manufactured Object, the Mouth Guard, Using a Method with Known and Determined Measurement Error (Intraoral Optical Scanning Method)

The mouth guard was scanned using a TRIOS 3, 3Shape (Denmark) intraoral scanner with an accuracy (trueness) of 6.9 ± 0.9 μm, according to the technical specifications. Using the protocol described below in chapter 2.5 and the CloudCompare open-source software (CCOSS) for evaluating the freehand 3D ultrasound scanner prototype, the mean distance and standard deviation were calculated for the optical scan of the mouth guard aligned with the original STL project, after adjusting the scale of the two 3D objects.

2.2. Semiautomatic Segmentation of the 2D Ultrasound Images

An original semiautomatic segmentation tool for the 2D US images was developed using a customized region growing-based segmentation algorithm (Figure 2). In the process, the user was supposed to click on seed points that were "grown" by iteratively adding neighboring pixels with similar intensities. The algorithm was customized in such a way that the already labeled pixels were not considered. The similarity predicate was controlled by a threshold (T), tunable by the user using a track-bar control and the result (the local grown region) was visible on the fly for any instant position of the track-bar. In general, the initial value for T should be chosen between 2σ and 3σ, where σ is the standard deviation of the Gaussian distribution of the regions of interest (ROIs) computed on some sample image patches and is application dependent. For the current application, the initial value of T was set to $T \approx 15$, since σ was estimated to $\sigma \approx 5$ for a set of samples of whitish ROIs, corresponding to the mouth guard surface regions. There was also available the option of applying morphological-based post-processings (dilation followed by erosion), in order to fill in the holes occurring after the segmentation process. Once the user (which should be a qualified/specialized operator and in this case a dentist specialized in dental ultrasonography) was satisfied with the result (criteria were as follows: maximization of the smoothness and continuity of the upper envelope of the mouth guard in each 2D image/section), the local grown region (local labels matrix) was appended to the global grown region (global labels matrix), which stored the final segmentation result in the form of a binary image.

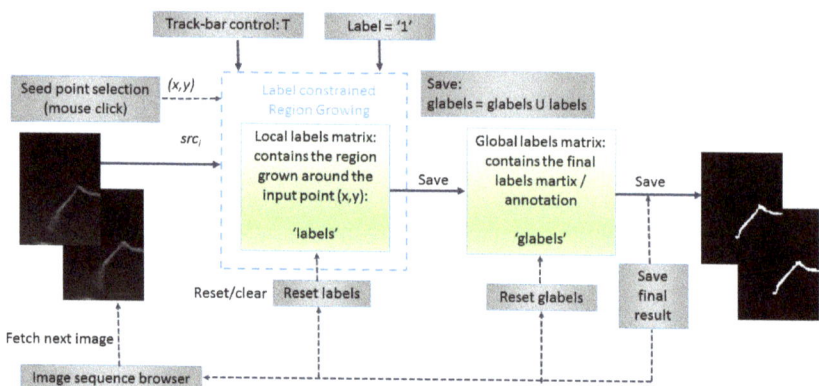

Figure 2. Flowchart of the semiautomatic label constrained region growing (RG)-based segmentation algorithm.

The region growing algorithm is based on the breadth-first search (traversal) algorithm of graphs [12] and uses a queue structure (FIFO list) for optimal implementation. The grown process of each region was constrained to the selected label; therefore, the implementation can be used out-of-the-box for multi-label annotation of more complex anatomical structures in medical imaging. For the current purpose of segmenting the outer surface of the mouth guard, only one label was used (variable label was set to 1 in the segmentation and morphological post-processing algorithms) and a binary result image was obtained. The pseudocode of the proposed labeled constrained region growing-based segmentation algorithm is presented below (Algorithm 1).

Algorithm 1 Label Constrained Region Growing

```
1:   procedure Grow(src; local_labels; label = 1; x; y; T; applyMorph)
2:       h ← src height
3:       w ← src width
4:       Q ← [ ]                                        ▷ Empty queue
5:       W ← 3                                          ▷ Averaging window
6:       d ← W/2                                          size
7:       avgColor ← average(src(y − d: y + d, x − d: x + d))
8:       d ← 1                                          ▷ No. of pixels in the
9:       N ← 1                                            region
10:      Q.append((y,x))
11:      if labels(y,x) = 0 then                        ▷ If pixel is unlabeled
12:          labels(y,x) = label
13:      end if
14:      while Q is not empty do                        ▷ Take out the oldest
15:          oldest ← Q.pop()                             element from the queue
16:          for m ← −d to d, n ← −d to d do
17:              i ← oldest.y + m                       ▷ Search across its
18:              j ← oldest.x + n                         neighbors and add them
19:              if (i,j) inside of src then              to the queue if they are
20:                  color ← src(i,j)                     not labeled and are
21:                  if |color − avgColor| < T and labels(i,j) = 0 and glabels(i,j) = 0 then   similar in terms of color
22:                      labels(i,j) ← label              with the region
23:                      Q.append(i,j)
24:                      avgColor ← (avgColor × N + color)/(N + 1)   ▷ Update the average
25:                      N ← N + 1                        color of the region
26:                  end if
27:              end if
28:          end for
29:      end while                                      ▷ Convert the local
30:      dst ← labels                                     labels matrix into the
31:      if applyMorpho = true then                       destination image
32:          dst ← Dilate(dst,R,label)                  ▷ Post-process the result
33:          dst ← Erode(dst,R,label)                     by morphological
34:      end if                                           operations
         return dst
     end procedure
```

2.3. Morphological Post-Processing

An optional step of the segmentation algorithm was to perform morphological post-processing [13] in order to refine each resulted segment (grown region), mainly for filling up the small holes that occur in the segmented process. For this purpose, a dilation (Algorithm 2), followed by an erosion (Algorithm 3), was applied with a circular structuring element of adjustable radius. The implementation of the algorithms was adapted to the following proposed label constrained paradigm: the foreground pixels were dilated and eroded at the label level. The two complementary morphological operations were applied in pairs with the same structuring element (in terms of size and shape), in order to not alter the area and the overall shape of the segmented regions.

Algorithm 2 Label Constrained Dilation

```
1:   procedure Dilate(src, R, label = 1)
2:       dst ← copy(src)                                          ▷ Clone source image into the destination
3:       h ← img height
4:       w ← img width
5:       for i ← R to h − R − 1, j ← R to w − R − 1 do             ▷ Image scan with safety border
6:           if src(i,j) = label then
7:               for m ← −R to R, n ← −R to R do                   ▷ Apply dilation using a circular structuring
8:                   if R > 2 then                                     element of radius R (R > 2) only on pixels with
9:                       radius ← √(m² + n²)                           the specified label. All pixels in the
10:                      if radius < R and src(i + m, j + n) = 0 then  neighborhood masked by the structuring
11:                          dst(i + m, j + n) ← label                 element are marked with the current label
12:                      end if
13:                  else
14:                      if src(i + m; j + n) = 0 then             ▷ If R ≤ 2, the structuring element has a square
15:                          dst(i + m; j + n) ← label                 shape
16:                      end if
17:                  end if
18:              end for
19:          end if
20:      end for
21:      return dst
     end procedure
```

Algorithm 3 Label Constrained Erosion

```
1:   procedure Erode(src, R, label = 1)
2:       dst ← copy(src)                                          ▷ Clone source image into the destination
3:       h ← img height
4:       w ← img width
5:       for i ← R to h − R − 1, j ← R to w − R − 1 do             ▷ Image scan with safety border
6:           frontier ← false
7:           if src(i,j) = label then
8:               for m ← −R to R, n ← −R to R do                   ▷ Apply erosion using a circular structuring
9:                   if R > 2 then                                     element of radius R (R > 2) only on pixels with
10:                      radius ← √(m² + n²)                           the specified label. If there is a background
11:                      if radius < R and src(i + m, j + n) = 0 then  pixel in the neighborhood masked by the
12:                          frontier ← true                           structuring element, a flag (frontier) is set
13:                      end if
14:                  else
15:                      if src(i + m; j + n) = 0 then             ▷ If R ≤ 2, the structuring element has a square
16:                          frontier ← true                           shape
17:                      end if
18:                  end if
19:              end for
20:          end if
21:          if frontier ← true then
22:              dst(i,j) ← 0
22:      end for                                                   ▷ If a flag is set, the current pixel is removed
22:      return dst                                                  from the result
     end procedure
```

2.4. Upper Envelop/Contour Extraction of the Segmented Objects

Before 3D reconstruction could be applied, the upper/outer envelope of the segmented objects from the 2D binary images had to be extracted in the form of a contour. This contour should correspond to the surface of the mouth guard observed in each 2D US image. The Algorithm 4 is presented below. First, the external contours of the segmented binary objects were detected and drawn as a binary image in the *findExternalContours* function using the following OpenCV [14] methods: *findContours* and *drawContours*. Then, the binary contours image was scanned column by column, from top to bottom and the first vertical sequence of white pixels from each column was stored in the destination image. This approach also dealt with cases of vertical contour segments in the upper envelope.

Algorithm 4 Find Upper Envelope

1:	**procedure** FindEnvelope(*src*)	
2:	*contours* ← *findExternalContours(src)*	▷ Detect external contours in the binary source image and store them (as white pixels) in the *contours* image
3:	h ← img height	
4:	w ← img width	
5:	dst ← (h,w,0)	▷ Create a black destination image
6:	**for** j ← 0 to w − 1 **do**	▷ Scan the binary contour image *contours* on columns
7:	**while** *contours(i,j)* = 0 and i < h − 1 **do**	
8:	i ← I + 1	▷ Skip the first vertical sequence of black pixels
9:	**end while**	
10:	**while** *contours(i,j)* = 255 and i < h − 1	
11:	**do**	▷ Store the first vertical sequence of white (object/contour) pixels in the destination image *dst*
12:	dst(i,j) ← 255	
13:	i ← I + 1	▷ At the end of the sequence, break the for loop (j ← j + 1, i ← 0)
14:	**break**	
15:	**end while**	
16:	**end for**	
	return *dst*	
	end procedure	

The step-by-step results of the segmentation Algorithms 1–4 are shown in Figure 3.

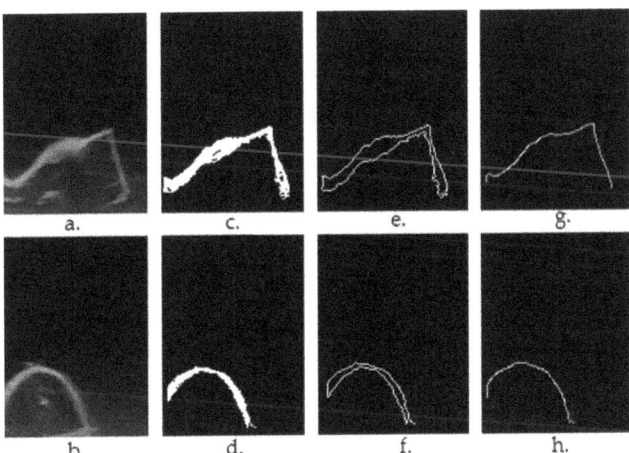

Figure 3. (**a**,**b**)—original ultrasound frames in greyscale ((**a**)—first premolar, regulated shaped object having plane surfaces, right angles; (**b**)—lateral incisor); (**c**,**d**)—results after region growing-based segmentation and morphological post-processing (Algorithms 1–3); (**e**,**f**)—results after contour extraction (*findExternalContours* function in Algorithm 4); (**g**,**h**)—results after upper contour (envelope) extraction (Algorithm 4).

2.5. Generating 3D Ultrasound Reconstructions

Data acquisition and 3D reconstruction were performed using the 3D US scanner prototype and the software developed by Chifor Research's team. After the US data were acquired, each frame was paired or matched with the sensor's readings. The spatial coordinates and orientation of each frame were determined through the time and spatial calibration processes [15].

The 3D reconstruction was performed by introducing the scan planes corresponding to the raw 2D ultrasound images into a tridimensional space. This was carried out by using a series of rotations and translations performed in several stages, as described below. The first step was to create a 3D space to hold the voxels corresponding to the pixels in the image. The next step was to apply a calibration matrix transformation to the 2D frame corresponding to this 3D space. The third step applied the final pose transformation on the output of the previous step, in order to finalize the positioning of the original 2D frame in the corresponding 3D space allocated for it in the beginning. This final transformation represented a bijection between the points in the 2D frame and their 3D correspondents. Finally, the intensity of the original pixel in the 2D frame was assigned to the corresponding voxel in the 3D space. The previous steps were repeated for each acquired 2D ultrasound frame, until all the 2D frames were represented in the 3D space [16].

The segmented envelop/contour points from each 2D US scan were used to mask the 3D points associated with each scan and to generate the 3D point cloud corresponding to the mouth guard's surface, which was further used for the quantitative evaluation of the reconstruction algorithm.

2.6. Evaluating the Accuracy of the 3D Ultrasound Reconstructions

The accuracy of the 3D virtual reconstruction, obtained by the ultrasound scanning of the CAD/CAM manufactured phantom, was evaluated by comparing it with the standard STL project, designed and used for its execution. The alignment, scaling and statistical analysis of the distances between the 3D points of the 3D ultrasound reconstruction point cloud and the reference object, the STL project, were performed using the CloudCompare open-source software (CCOSS). The CCOSS statistically analyzed the distances between the ultrasound point cloud and the STL object after the objects were spatially aligned.

After ultrasound scanning using the developed prototype and generating the 3D reconstruction of the segmented mouth guard's surface, the obtained point cloud was aligned with the CAD project. The mean deviation (distance) of the 3D ultrasound reconstruction from the reference model was calculated, as well as the standard deviation of the distances using CCOSS.

The calculation of the mean distance and the standard deviation was computed during cloud point alignment to the reference point cloud or mesh, as part of the alignment algorithm. This was done in two stages. The first stage was a rough alignment, giving a rough value for the RMS (root mean square) index, representing the square root of the mean value of the squared distances d_i, as described by Equation (1) [17], which is as follows:

$$\text{RMS} = \sqrt{\sum (d_i^2)/n} \tag{1}$$

where d_i^2 is the squared distance between the reconstructed 3D points and corresponding CAD model points, computed over n points.

Once the rough alignment was completed and its corresponding 4 × 4 transformation matrix was calculated so that the two point clouds were moved and scaled into proximity according to at least 3 corresponding points on their surface, the second step of the alignment was performed, providing a fine tuning of the RMS value, by incrementally moving and scaling the two point clouds in order to minimize the RMS. At the end of this process, the corresponding standard deviation and mean distance, which were proportionally correlated with the RMS, were calculated [18,19].

3. Results

The virtual alignment with the reference object (the mouth guard STL project) and the statistical accuracy evaluation were performed on six consecutive 3D ultrasound reconstructions, acquired using the handheld 3D ultrasound scanner prototype.

3.1. Measuring and Verifying the CAD/CAM Manufactured Object, the Mouth Guard, Using a Method with Known and Determined Measurement Error (Intraoral Optical Scanning Method)

The alignment errors measured for the optical scan aligned with the CAD project STL of the mouth guard (Figure 4) are as follows: mean distance of 13,65 µm and standard deviation of 117.14 µm. The measured distances are represented in a Gaussian characteristic symmetric "bell curve" shape and most of the measurements are close to 0, as one can observe in the righthand section of Figure 4.

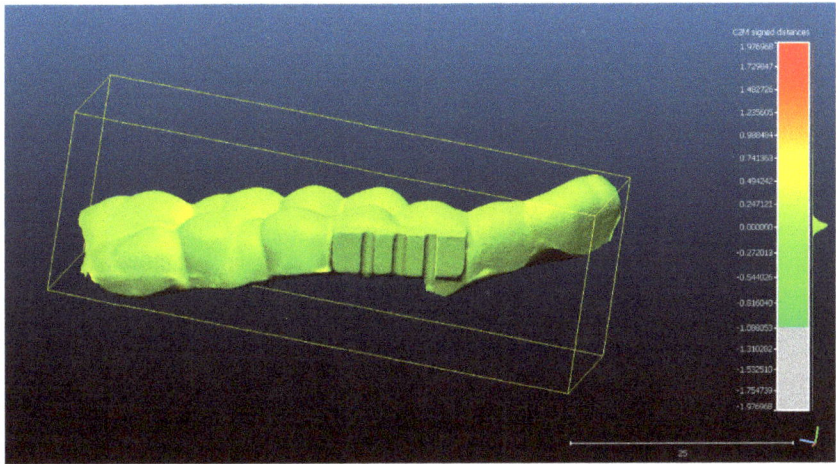

Figure 4. The spatial distribution of the 128,570 measurements between the optical scan of the mouth guard with the reference object (CAD project STL format) after alignment. The distribution of the alignment errors is figured in the form of a colored Gaussian shape on the right side of the color code bar. In light green are the measurements close to 0.

The recalculation of the mean distance and the standard deviation according to the adjusted scale (15 mm = 14.81 mm) can be observed in Figure 5, generated by CCOSS after aligning the two objects with the following errors: mean distance of 13.83 µm and standard deviation of 118.64 µm

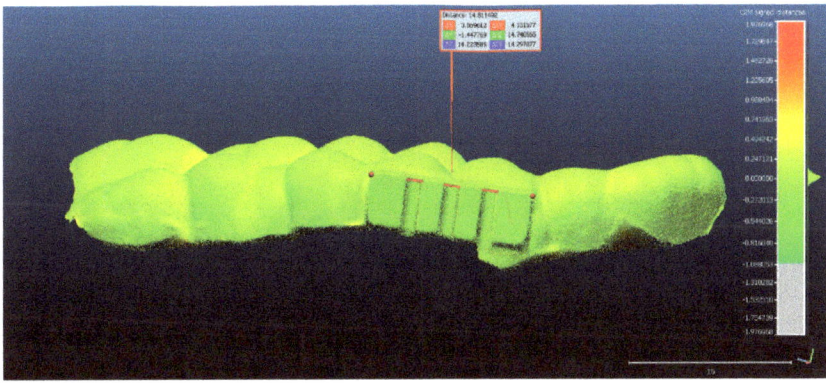

Figure 5. The scaling of the optical scan of the mouth guard, after it had been aligned with the reference object.

3.2. Preparing the 2D Ultrasonographic Images, Performing 3D Ultrasound Reconstructions and Aligning Them with the Reference Object for Statistical Analysis

For scan 1: The original 2D ultrasound images have been segmented without extracting the contours. Subsequently, the 3D reconstruction was performed based on the semi-automatically segmented 2D images and the spatial position reading data related to each 2D frame, resulting in a point cloud of 272,189 3D points.

- The rectangular landmark, used for scaling the objects, measures in real world 15 mm in length. Its length in CCOSS after alignment was 13.939. Thus, 1 mm length in real world equaled 1.07 in CCOSS (Figure 6).

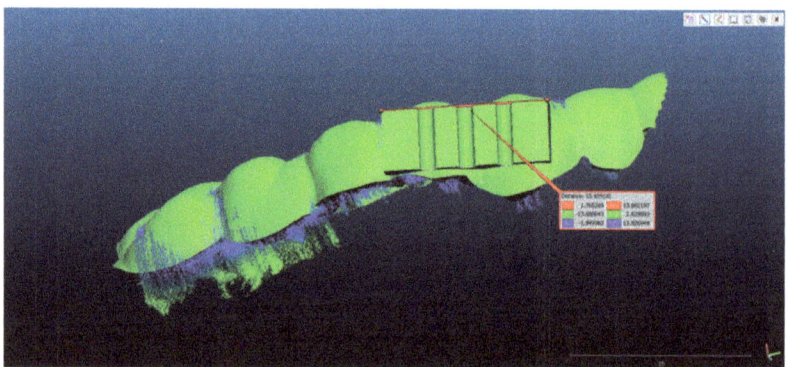

Figure 6. The scaling of the point cloud after it had been aligned with the reference object.

- The spatial distribution of the 272,189 3D ultrasonographic points was compared to the reference object (CAD project in STL format) after alignment. The distance errors of the 3D ultrasonographic points were uniformly distributed, meaning that the reconstruction respected the shape of the scanned object (Figure 7). The mean distance of the 3D ultrasonographic points from the reference object was 0.033 mm and the standard deviation equaled 0.387 mm (Table 1). The deviations were most probably due to the artifacts and to the noise in the 2D ultrasound original frames.

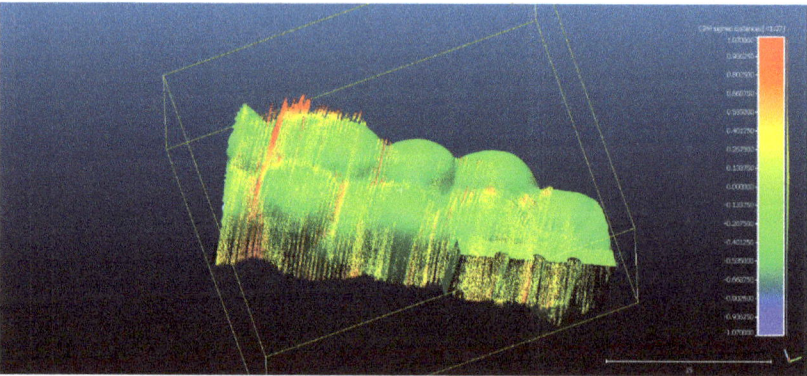

Figure 7. The spatial distribution of the 272,189 3D ultrasonographic points compared to the reference object (CAD project STL format) after alignment. The distribution of the alignment errors is figured in the form of a colored Gaussian shape on the righthand side of the color code bar.

- Scan 2. After segmentation of the 2D images, the contours were extracted, before reconstructing the 3D object. The total number of 3D ultrasonographic points (masked by the segmented contours) was significantly lower (46,537) compared to the scan

1 (279,189 3D points as they were obtained without extracting the contours); the obtained 3D point cloud was aligned with the reference object (STL project), as shown in Figure 8, and the computed mean distance was 0.031 mm and the standard deviation was 0.287 mm (Figure 9a and Table 1). The deviations/errors were isolated to certain areas, probably due to the artifacts in some of the 2D original frames.

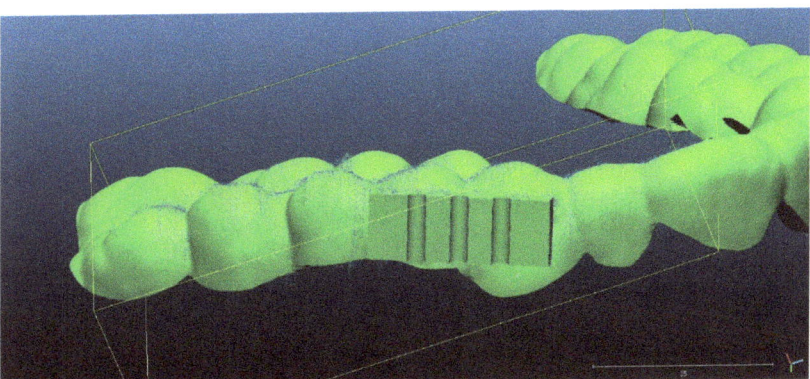

Figure 8. Scan 2 aligned with the reference object using CloudCompare software.

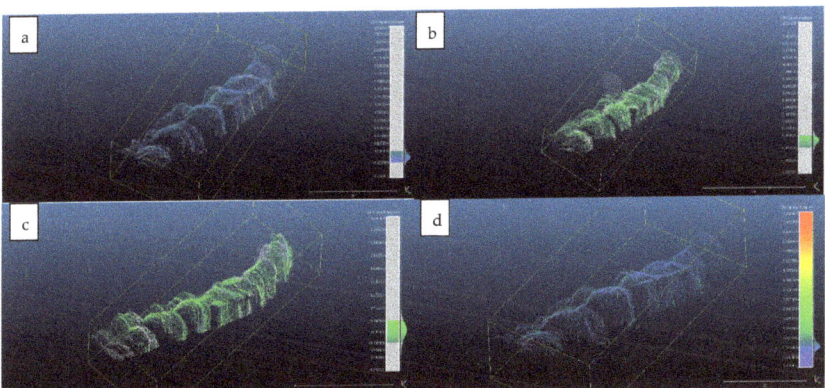

Figure 9. Distribution of the alignment distances/errors of the 3D ultrasonographic points from the scanned reference object after alignment in CCOSS: (**a**) Scan 2—aligned with STL reference object. The farthest points are colored in gray, alignment errors are evenly distributed along the scanned object (observe the blueish distances'/errors' distribution on the righthand side of the figure); (**b**) Scan 3—aligned with the reference STL object (observe the greenish distances'/errors' distribution between the 3D points and reference object on the righthand side of the figure); (**c**) Scan 4 aligned with STL reference object (3D points distances'/errors' spatial distribution in green on the righthand side of the figure); (**d**) Scan 5—aligned with STL reference object (3D points distances'/errors' spatial distribution in blueish on the righthand side of the figure).

- For Scans 3 to 5, the alignment errors are presented in Figure 9b–d. The mean distance between the 3D points of the ultrasonographic reconstructions (obtained by masking the 3D point cloud with the segmented contours) and the reference scanned object varied in the range between 0.019 mm to 0.05 mm (Table 1).

3.3. Quantitative Evaluation by Statistical Error Analysis

The mean and standard deviations of the distances/errors of the 3D ultrasonographic points from the scanned reference object after alignment in CCOSS are presented in Table 1.

The advantage of the method proposed in the present study is that the scanner prototype has six axes freedom of movement during scanning. The accuracy of the reconstruction was not influenced by the length of the reconstructed area. As one can observe in Figure 9a–c, there is a homogeneous distribution of the ultrasonographic 3D points situated at more than 350 microns from the reference, colored in grey, probably due to the artifacts in the 2D ultrasonographic images. If those artifacts had been due to 3D reconstruction errors, the 3D object would have been distorted, with spatially concentrated errors.

In a normal distribution (which can be assumed based on the large number of measurements), approximately 95% of the deviations of the virtual model from the reference model range within the average +/− two standard deviations. In addition, approximately 99% of the deviations of the virtual model from the reference model range within average +/− three standard deviations.

4. Discussion

The aim of the current study has been reached, by developing and evaluating a highly accurate 3D ultrasound scanning method.

The verification of the 3D ultrasound scanned object confirmed that the shape and the size of the scanned object is very close to the technical specification range of the CAD/CAM process, and also confirmed that the method used to appreciate the accuracy of the 3D ultrasound prototype is reliable. Datron D5, used for the manufacturing the mouth guard, has an accuracy of 5 μm and Trios 3 from 3Shape, used to optical scan the mouth guard, has an accuracy of 6.9 μm and 0.9 μm, resulting in a 12.8 μm possible error according to the technical specification of the two devices, because the errors can cumulate.

A previously published low-cost volumetric ultrasound imaging method has been developed by Herickhoff et al. [20], using the augmentation of bidimensional systems generating freehand 3D ultrasounds via probe position tracking. The method allowed a variety of scanning patterns (e.g., linear translation normal to the image plane or panoramic sweep), but it presented the drawback of an expandable, but still limited, field of view, due to its fixture in constraining the probe motion to pivoting about a single axis [20].

Our 3D ultrasound scanning method is based on a closed platform, a standard ultrasound machine. The 3D ultrasound reconstruction is generated from the DICOM files and the data from the coordinate measuring machine (CMM), an articulated arm (as the spatial pose reading sensor). The synchronization method and algorithms of the two data flows were developed in house. This constitutes another advantage compared to other developed methods, because direct data access is typically not enabled by commercial diagnostic systems and, thus, requires the development of open platforms or close collaborations with manufacturers for integration [11].

Other studies [21] measured the difference in the length from the surface of a phantom to the bottom part, using the ultrasound image, and found it to be 6.48 mm. At the same time, the difference in the position data from the ultrasonic sensor was 5.85 mm. The difference between the measured ultrasound image and the position data was only 0.65 mm (9.72%) [21].

A scanner with three translational degrees of freedom was used in another study to scan the teeth from an occlusal direction. One tooth per scan was 3D ultrasound reconstructed. The mean difference between the reconstructed casts and the optical control group was in the range 14–53 μm. The standard deviation was between 21 and 52 μm [22].

Comparing the aforementioned results with the ones obtained in the present study, a similar or better accuracy can be noted in the current study, regarding the mean difference. The overall mean distance ranged between 0.014 mm and 0.050 mm and the standard deviation ranged between 0.287 mm and 0.565 mm. The standard deviation was higher in our case because of the freehand scanning technique, which generated higher artifacts in the acquired 2D ultrasound images. The homogenous distribution of the errors at the scanned area level, the extension of the scanned area, six teeth instead of a single tooth scan and the Gaussian distribution of the errors generate promising premises for the future evaluation and use of a freehand 3D ultrasonographic scanning method in clinical settings.

A study performed by Marotti et al. on extracted teeth covered with porcine gingiva reported mean deviations for 3D ultrasound scans, ranging from between 12.34 to 46.38 µm [23].

The performance of intraoral optical scanners, reported by Winkler et al. in their study about TRIOS 3, displayed slightly higher precision (approximately 10 µm) compared to CS 3600, only after superimposition on the whole dental arch ($p < 0.05$). Both intraoral scanners showed good performance and comparable trueness (median of 0.0154 mm; $p > 0.05$) [24]. Comparing their results with our 3D ultrasound imaging method showed lower precision on our side, mostly due to the artefacts and noise in the acquired 2D ultrasound images. The precision of our pose reading sensor device was a maximum of 25 µm, according to the manufacturer's technical specifications. This also contributed to the accumulated error, which ranged between 14 and 50 µm for the 3D ultrasound reconstructions, compared to the CAD reference object. In addition, the CAM of the reference scanned object has induced errors of at most 7 µm, according to the technical specifications of the device.

The following are the main limitations of the current study: the 3D ultrasound reconstructions were performed only for the surfaces of the scanned object. Comparing the results with the STL CAD project of the object allowed the appreciation of the accuracy (trueness and reproducibility) of the scanning method. Future studies should also evaluate the prototype's scanning accuracy of deep soft tissue and bone surfaces. Another limitation of the current study drew from the fact that only a single object had been scanned. In future studies, we intend to evaluate the proposed prototype by scanning different patients and different types of tissue, so that the segmentation process will also be challenged by the need to correctly identify the anatomical parts or pathological tissues.

5. Conclusions

The quantitative evaluation of the proposed 3D ultrasonographic reconstruction method showed comparable results to other studies performed on smaller areas of scanned objects, thus, demonstrating the future potential of the developed prototype to be used in clinical practice. The freedom of movement during scanning and the accuracy of the 3D reconstructions will have to be exploited in future research to evaluate and monitor the evolution of diseases, by comparing the 3D models. This process can be performed by integrating automatic or semi-automatic methods for the segmentation and alignment of the 3D objects.

Author Contributions: Conceptualization, R.C. and T.M.; methodology, R.C., A.F.B., I.C., M.-E.B. and H.A.C.; software, T.A., T.M. and A.S.; validation, R.C., A.F.B., I.C., H.A.C. and T.A.; formal analysis, R.C., A.F.B., H.A.C. and I.C.; resources, R.C.; data curation, R.C.; writing—original draft preparation, R.C., T.M., T.A., A.S., I.C. and M.-E.B.; writing—review and editing, R.C., T.M. and A.S.; visualization, R.C. and T.A.; supervision, M.-E.B. and I.C.; project administration, R.C.; funding acquisition, R.C. All authors have read and agreed to the published version of the manuscript.

Funding: This study was partially realized with the material, equipment, technology, and logistic support of Chifor Research SRL, through the project Periodontal ultrasonography in diagnosing and monitoring the periodontal disease (Chifor Research SRL, Operational Program Competitivity, Ministry of European Funds from Romania, P_38_930\12.10.2017, Project ID 113124.) This paper was partially funded by the European Social Fund, Human Capital Operational Program 2014-2020, project no. POCU/380/6/13/125171, EIT Health-RIS Innovation Program 2020, project ID 2020 RIS-1001-8253 and "3DentArVis" project no. 70/2018 (research project conducted by the Technical University of Cluj-Napoca and funded by Chifor Research SRL through P_38_930\12.10.2017).

Institutional Review Board Statement: Not applicable.

Informed Consent Statement: Not applicable.

Conflicts of Interest: The authors declare no conflict of interest.

References

1. Zardi, E.M.; Franceschetti, E.; Giorgi, C.; Palumbo, A.; Franceschi, F. Accuracy and Performance of a New Handheld Ultrasound Machine with Wireless System. *Sci. Rep.* **2019**, *9*, 14599. [CrossRef] [PubMed]
2. Stasi, G.; Ruoti, E.M. A Critical Evaluation in the Delivery of the Ultrasound Practice: The Point of View of the Radiologist. *Ital. J. Med.* **2015**, *9*, 5. [CrossRef]
3. Huang, Q.; Zeng, Z. A Review on Real-Time 3D Ultrasound Imaging Technology. *Biomed. Res. Int.* **2017**, *2017*, 6027029. [CrossRef] [PubMed]
4. Huang, Q.H.; Zheng, Y.P.; Lu, M.H.; Chi, Z.R. Development of a Portable 3D Ultrasound Imaging System for Musculoskeletal Tissues. *Ultrasonics* **2005**, *43*, 153–163. [CrossRef] [PubMed]
5. Baek, J.; Huh, J.; Kim, M.; Hyun An, S.; Oh, Y.; Kim, D.; Chung, K.; Cho, S.; Lee, R. Accuracy of Volume Measurement Using 3D Ultrasound and Development of CT-3D US Image Fusion Algorithm for Prostate Cancer Radiotherapy: Volume Measurement and Dual-Modality Image Fusion Using 3D US. *Med. Phys.* **2013**, *40*, 021704. [CrossRef] [PubMed]
6. Scorza, A.; Conforto, S.; D'Anna, C.; Sciuto, S.A. A Comparative Study on the Influence of Probe Placement on Quality Assurance Measurements in B-Mode Ultrasound by Means of Ultrasound Phantoms. *Open Biomed. Eng. J.* **2015**, *9*, 164–178. [CrossRef] [PubMed]
7. Praça, L.; Pekam, F.C.; Rego, R.O.; Radermacher, K.; Wolfart, S.; Marotti, J. Accuracy of Single Crowns Fabricated from Ultrasound Digital Impressions. *Dent. Mater.* **2018**, *34*, e280–e288. [CrossRef] [PubMed]
8. Le, L.H.; Nguyen, K.-C.T.; Kaipatur, N.R.; Major, P.W. Ultrasound for Periodontal Imaging. In *Dental Ultrasound in Periodontology and Implantology*; Chan, H.-L., Kripfgans, O.D., Eds.; Springer International Publishing: Cham, Switzerland, 2021; pp. 115–129. ISBN 978-3-030-51287-3.
9. Slapa, R.; Jakubowski, W.; Slowinska-Srzednicka, J.; Szopinski, K. Advantages and Disadvantages of 3D Ultrasound of Thyroid Nodules Including Thin Slice Volume Rendering. *Thyroid. Res.* **2011**, *4*, 1. [CrossRef] [PubMed]
10. Buia, A.; Stockhausen, F.; Filmann, N.; Hanisch, E. 3D vs. 2D Imaging in Laparoscopic Surgery—An Advantage? Results of Standardised Black Box Training in Laparoscopic Surgery. *Langenbecks Arch. Surg.* **2017**, *402*, 167–171. [CrossRef] [PubMed]
11. von Haxthausen, F.; Böttger, S.; Wulff, D.; Hagenah, J.; García-Vázquez, V.; Ipsen, S. Medical Robotics for Ultrasound Imaging: Current Systems and Future Trends. *Curr. Robot. Rep.* **2021**, *2*, 55–71. [CrossRef] [PubMed]
12. Silvela, J.; Portillo, J. Breadth-First Search and Its Application to Image Processing Problems. *IEEE Trans. Image Process.* **2001**, *10*, 1194–1199. [CrossRef] [PubMed]
13. Shih, F.Y. *Image Processing and Pattern Recognition: Fundamentals and Techniques*; IEEE Press: Piscataway, NJ, USA; Wiley: Hoboken, NJ, USA, 2010; ISBN 978-0-470-40461-4.
14. Open Computer Vision Library. Available online: https://Docs.Opencv.Org/3.4.17/Index.Html (accessed on 20 January 2022).
15. Chifor, R.; Li, M.; Nguyen, K.-C.T.; Arsenescu, T.; Chifor, I.; Badea, A.F.; Badea, M.E.; Hotoleanu, M.; Major, P.W.; Le, L.H. Three-Dimensional Periodontal Investigations Using a Prototype Handheld Ultrasound Scanner with Spatial Positioning Reading Sensor. *Med. Ultrason.* **2021**, *23*, 297–304. [CrossRef] [PubMed]
16. Hsu, P.-W.; Prager, R.W.; Gee, A.H.; Treece, G.M. Freehand 3D Ultrasound Calibration: A Review. In *Advanced Imaging in Biology and Medicine*; Sensen, C.W., Hallgrímsson, B., Eds.; Springer: Berlin/Heidelberg, Germany, 2009; pp. 47–84. ISBN 978-3-540-68992-8.
17. Beddard, T. Registration RMS-CloudCompare ForumRegistration RMS-CloudCompare Forum. Available online: https://www.Danielgm.Net/Cc/Forum/Viewtopic.Php?T=1296 (accessed on 20 January 2022).
18. Align-CloudCompareWiki. Available online: https://www.Cloudcompare.Org/Doc/Wiki/Index.Php?Title=Align (accessed on 20 January 2022).
19. Alignment and Registration-CloudCompareWiki. Available online: https://www.Cloudcompare.Org/Doc/Wiki/Index.Php?Title=Alignment_and_Registration (accessed on 20 January 2022).
20. Herickhoff, C.D.; Morgan, M.R.; Broder, J.S.; Dahl, J.J. Low-Cost Volumetric Ultrasound by Augmentation of 2D Systems: Design and Prototype. *Ultrason. Imaging* **2018**, *40*, 35–48. [CrossRef] [PubMed]
21. Kim, T.; Kang, D.-H.; Shim, S.; Im, M.; Seo, B.K.; Kim, H.; Lee, B.C. Versatile Low-Cost Volumetric 3D Ultrasound Imaging Using Gimbal-Assisted Distance Sensors and an Inertial Measurement Unit. *Sensors* **2020**, *20*, 6613. [CrossRef] [PubMed]
22. Chuembou Pekam, F.; Marotti, J.; Wolfart, S.; Tinschert, J.; Radermacher, K.; Heger, S. High-Frequency Ultrasound as an Option for Scanning of Prepared Teeth: An in Vitro Study. *Ultrasound Med. Biol.* **2015**, *41*, 309–316. [CrossRef] [PubMed]
23. Marotti, J.; Broeckmann, J.; Chuembou Pekam, F.; Praça, L.; Radermacher, K.; Wolfart, S. Impression of Subgingival Dental Preparation Can Be Taken with Ultrasound. *Ultrasound Med. Biol.* **2019**, *45*, 558–567. [CrossRef] [PubMed]
24. Winkler, J.; Gkantidis, N. Trueness and Precision of Intraoral Scanners in the Maxillary Dental Arch: An in Vivo Analysis. *Sci Rep.* **2020**, *10*, 1172. [CrossRef] [PubMed]

Review

Comparison of Diagnostic Test Accuracy of Cone-Beam Breast Computed Tomography and Digital Breast Tomosynthesis for Breast Cancer: A Systematic Review and Meta-Analysis Approach

Temitope Emmanuel Komolafe [1,2,3,†], Cheng Zhang [1,†], Oluwatosin Atinuke Olagbaju [4,5], Gang Yuan [1], Qiang Du [1], Ming Li [1], Jian Zheng [1,*] and Xiaodong Yang [1,*]

[1] Department of Medical Imaging, Suzhou Institute of Biomedical Engineering and Technology, Chinese Academy of Sciences, Suzhou 215163, China; teakomo@mail.ustc.edu.cn or tekomolafe@shanghaitech.edu.cn (T.E.K.); zhangc@sibet.ac.cn (C.Z.); yuangang@sibet.ac.cn (G.Y.); dut@sibet.ac.cn (Q.D.); lim@sibet.ac.cn (M.L.)
[2] School of Biomedical Engineering, ShanghaiTech University, Shanghai 201210, China
[3] Division of Life Sciences and Medicine, School of Biomedical Engineering (Suzhou), University of Science and Technology of China, Hefei 230026, China
[4] Molecular Imaging Research Center, Harbin Medical University, Harbin 150028, China; triplet852002@gmail.com
[5] TOF-PET/CT/MR Center, The Fourth Hospital of Harbin Medical University, Harbin 150028, China
* Correspondence: zhengj@sibet.ac.cn (J.Z.); xiaodong.yang@sibet.ac.cn (X.Y.); Tel.: +86-512-6958813 (J.Z.); +86-512-69588133 (X.Y.)
† These authors contributed equally to this work.

Abstract: Background: Cone-beam breast computed tomography (CBBCT) and digital breast tomosynthesis (DBT) remain the main 3D modalities for X-ray breast imaging. This study aimed to systematically evaluate and meta-analyze the comparison of diagnostic accuracy of CBBCT and DBT to characterize breast cancers. Methods: Two independent reviewers identified screening on diagnostic studies from 1 January 2015 to 30 December 2021, with at least reported sensitivity and specificity for both CBBCT and DBT. A univariate pooled meta-analysis was performed using the random-effects model to estimate the sensitivity and specificity while other diagnostic parameters like the area under the ROC curve (AUC), positive likelihood ratio (LR^+), and negative likelihood ratio (LR^-) were estimated using the bivariate model. Results: The pooled sensitivity specificity, LR^+ and LR^- and AUC at 95% confidence interval are 86.7% (80.3–91.2), 87.0% (79.9–91.8), 6.28 (4.40–8.96), 0.17 (0.12–0.25) and 0.925 for the 17 included studies in DBT arm, respectively, while, 83.7% (54.6–95.7), 71.3% (47.5–87.2), 2.71 (1.39–5.29), 0.20 (0.04–1.05), and 0.831 are the pooled sensitivity specificity, LR^+ and LR^- and AUC for the five studies in the CBBCT arm, respectively. Conclusions: Our study demonstrates that DBT shows improved diagnostic performance over CBBCT regarding all estimated diagnostic parameters; with the statistical improvement in the AUC of DBT over CBBCT. The CBBCT might be a useful modality for breast cancer detection, thus we recommend more prospective studies on CBBCT application.

Keywords: breast cancer; cone-beam computed tomography; digital breast tomosynthesis; meta-analysis; sensitivity; specificity

Citation: Komolafe, T.E.; Zhang, C.; Olagbaju, O.A.; Yuan, G.; Du, Q.; Li, M.; Zheng, J.; Yang, X. Comparison of Diagnostic Test Accuracy of Cone-Beam Breast Computed Tomography and Digital Breast Tomosynthesis for Breast Cancer: A Systematic Review and Meta-Analysis Approach. *Sensors* 2022, 22, 3594. https://doi.org/10.3390/s22093594

Academic Editors: Sylvain Girard and Anna Eva Morabito

Received: 9 March 2022
Accepted: 5 May 2022
Published: 9 May 2022

Publisher's Note: MDPI stays neutral with regard to jurisdictional claims in published maps and institutional affiliations.

Copyright: © 2022 by the authors. Licensee MDPI, Basel, Switzerland. This article is an open access article distributed under the terms and conditions of the Creative Commons Attribution (CC BY) license (https://creativecommons.org/licenses/by/4.0/).

1. Introduction

Breast cancer is the most commonly diagnosed type of cancer among women that has led to the cause of cancer death in women of all ages [1,2]. This mortality rate can be reduced drastically if those cancers are detected early [1]. Digital mammography (DM) has been a conventional tool for early breast cancer diagnosis [3,4]. Recent research on both randomized controlled trials and observational studies has indicated that regular screening

DM can reduce breast cancer drastically, which has a limitation of inability to image overlap dense breast tissue [5]. Digital breast tomosynthesis (DBT) has been developed to solve the tissue overlap of DM, and DBT acquisition involves an X-ray tube moving in an arc over the compressed breast taking multiple images from different angles. These images are reconstructed or synthesized into three-dimensional (3D) images via a reconstruction algorithm [6]. Several studies have recorded the improved diagnostic accuracy parameter such as sensitivity and specificity of 3D DBT alone or a combination with the DM [7–10]. A promising new technique is the dedicated cone-beam computed tomography (CBBCT) which provides real isotropic spatial resolution 3D images [6]. This modality also provides maximum breast comfortability to patients due to its reduced breast compression, unlike conventional DM and its DBT counterpart. Of particular importance is the CBBCT, which provides high-quality images and real-time 3D visualization of breast imaging and has proven to better visualize overlapping breast tissues than other imaging modalities like DM and ultrasound (US) [11–13]. Few studies have been documented on the review of diagnostic accuracy of DBT [14–17], while few pieces of literature have been recorded on the screening using CBBCT [18]. Contrast-enhanced cone-beam breast CT (CE-CBBCT) may improve the detection of breast cancer with possibly high specificity compared to that of DM, but with the cost of the high radiation exposure due to double scan. Uhlig et al. [19] carried out a meta-analysis study to compare the diagnostic performance of CE-CBBCT and that of non-contrast CBBCT (NC-CBBCT). They found a non-significant difference in sensitivity and specificity of CE-CBBCT, but considerable significance between-study heterogeneity in the NC-CBBCT.

Studies carried out about 10 years ago by Belair et al. [20] and Zuley et al. [21] compared the diagnostic accuracy of CBBCT and DBT, and their results showed that overall confidence in diagnosis was higher for both benign and malignant breast lesions using DBT. The authors suggested that future advances in technology and improvement in the readers' performance might lead to better performance of CBBCT in the future. In the last 7 years, few studies have reported on the diagnostic accuracy of CBBCT, none of these studies has directly compared CBBCT with DBT or used a meta-analysis approach to address this issue by comparing the potential diagnostic ability of these two 3D breast imaging modalities is still a hanging fruit yet to plug. Therefore, this study aims to systematically review and analyze the diagnostic accuracy of existing studies on CBBCT and DBT for breast cancer detection, thereby increasing the statistical power and thus eliminating any disagreement between individual studies.

2. Materials and Methods

This systematic review and meta-analysis was prospectively registered at PROSPERO with the registration number of CRD: 42020180192 [22]. The systematic review was performed by two independent reviewers (TEK and OAO or CZ and GY) using a well-established review protocol adapted from the Cochrane collaborative approach for evaluating diagnostic test accuracy [23] with Preferred Reporting Items for Systematic Reviews and Meta-analyses (PRISMA) guidelines [24], see Supplementary File S1. The two reviewers discussed the discrepancies between the two results, and then a more experienced third reviewer (XY or JZ or ML) was consulted if the interrater consensus was not reached. We searched for women who underwent breast imaging screening using either CBBCT or DBT, which reported the characterization of malignant and benign lesions with well-documented diagnostic accuracy. We searched separately because no available literature reported comparison studies on CBBCT and DBT for diagnostic or screening purposes. This search includes comparative, prospective and retrospective studies, and interrater consensus.

2.1. Data Sources and Search Strategy

PubMed, Inspec, Web of Science and Cochrane Central Register of Controlled Trials (CENTRAL) libraries were searched for relevant literature published from January 2015 up to and including December 2021. We used selected controlled terms extracted from different studies retrieved from each database to build the text words and subject terms as "breast computed tomography", "Sensitivity", "Specificity" for the CBBCT arm, and "Digital breast tomosynthesis", "Sensitivity", "Specificity" for CBBCT arm and DBT arm, respectively, as shown in the complete PRISMA search path (Figure 1). These selected controlled terms gave a wide representation for the review. In PubMed and CENTRAL databases, selected controlled terms were input as MeSH terms while in the Web of Science and Inspec, we used them as text words for detail see Supplementary File S2.

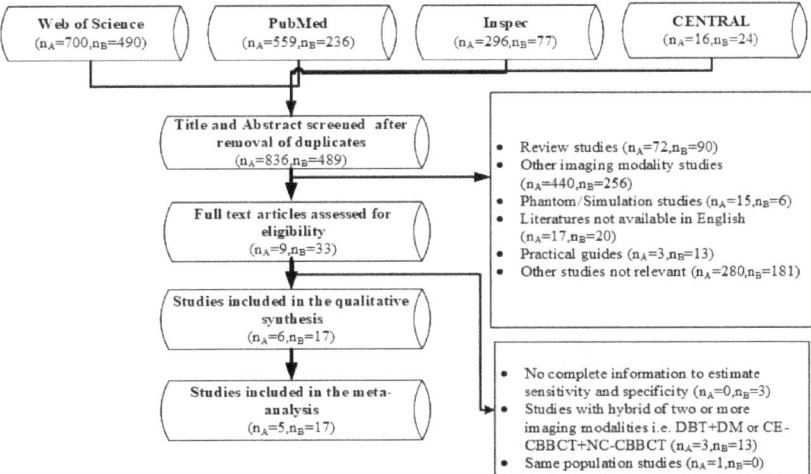

Figure 1. PRISMA flowchart of inclusion and exclusion criteria, n_A = number of literature in the CBBCT arm and n_B = the number of literature in the DBT arm. PRISMA = Preferred Reporting Items for Systematic Reviews and Meta-analyses. DBT = Digital breast tomosynthesis, DM = Digital mammography, CE-CBBCT = Contrast-Enhanced Cone-beam breast computed tomography, and NC-CBBCT = Non-Contrast Cone-beam breast computed tomography.

2.2. Eligibility Criteria

Studies were eligible for inclusion in this meta-analysis if they met eligibility criteria adapted from Cochrane diagnostic test accuracy protocol using PRISMA guidelines [24]. Literature was included in the study if it utilized dedicated CBBCT and DBT to detect breast cancer, with at least the sensitivity and specificity reported. The included studies were retrospective, prospective studies, an observer performance study, clinical trials, and comparative studies in different modalities. The exclusion criteria were studies that involved literature reviews, phantom or simulation studies, other radiation studies apart from CBBCT and DBT like radiotherapy and studies with computer-aided detection (CAD), i.e., machine and deep learning application in diagnostic accuracy.

Additionally, a study that reported two or more hybrid modalities like DBT with DM or contrast-enhanced CBBCT (CE-CBBCT) with non-contrast CBBCT (NC-CBBCT) was excluded. However, if it reports both modalities separately, the data for the modality under consideration will be extracted and vice versa. Likewise, for multiple publications that reported the same study or sub-set, the most detailed study in terms of data availability was used.

2.3. Study Selection

Articles retrieved for both arms were manually sorted, and duplicates were removed using titles/abstracts, then followed by full text according to the predefined search criteria, and final eligible studies were selected.

2.4. Data Collection Process

A standardized extraction sheet was developed, and two independent blinded reviewers (TEK and OAO or CZ and GY) extracted the information needed and resolved the conflict by interrater consensus from eligible studies, which include: study type (prospective or retrospective studies), study clinical settings (diagnostic or screening), number of patients and mean age of the patients, diagnostic equipment model, mean glandular dose, number of radiologists that interpreted the index test and year of experience, sensitivity and specificity. The positive and negative likelihood ratios are computed when they cannot be extracted [25], and other details of formulations of estimated diagnostic test accuracy parameters can be found in [26]. Additionally, the percentage of benign and malignant cases with a brief intervention description is included (Table 1).

2.5. Risk of Bias and Quality Appraisal

The quality of included studies was assessed using Quality Assessment of Diagnostic Accuracy Studies-Comparative (QUADAS-C), a tool for comparative diagnostic accuracy tests with different cohorts [27], a modified version of QUADAS-2 [28] to ensure appropriateness for comparing the two modalities. The domains assessed were patient selection, index tests, reference standard, flow and timing, and applicability. Two reviewers performed an independent quality assessment, and the final result was based on consensus. The overall study quality is shown in Figure 2.

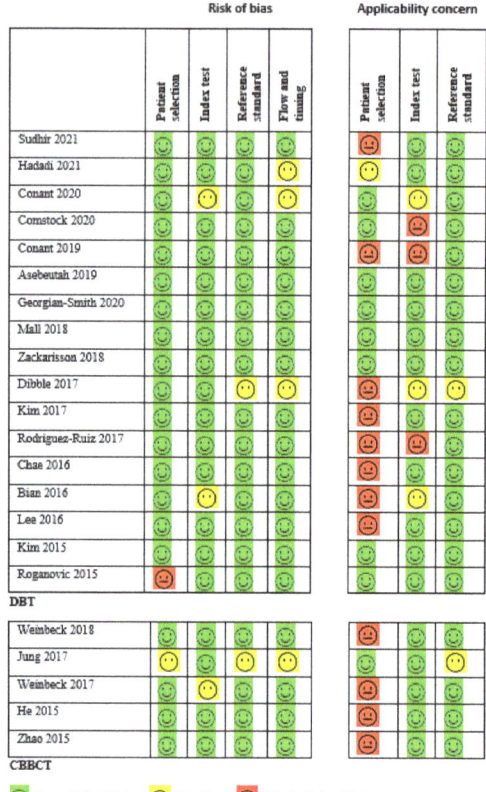

Figure 2. Risk of bias and applicability concerns: reviewers' judgments about each domain for each included study.

2.6. Data Analysis

A univariate meta-analysis was performed separately for sensitivity and specificity in both CBBCT and DBT to estimate the diagnostic accuracy of each modality using the random-effects model (RE) [29]. The primary outcomes were sensitivity, specificity and summary receiver operating characteristic (SROC) curve. We calculated point estimates and 95% confidence intervals (CI) for each study to ensure consistency in sensitivity and specificity. To plot the SROC curve, we used a bivariate meta-analysis of sensitivity and specificity using R version 4.1.2 with RStudio version 2021.09.1 + 372 implementing "mada" and "meta", R-packages to estimate the AUC of SROC [30]. Additionally, secondary outcomes like positive likelihood and negative likelihood ratios were estimated using MetaDiSc 1.4 software [31]. Statistical heterogeneity between studies was evaluated with Cochran's Q test and the I^2 statistic [32]. For the Q statistic, values range 0–40% imply insignificant heterogeneity, 30–60% connote moderate heterogeneity, and 75–100% implies a considerable heterogeneity. Publication bias was evaluated and visualized by constructing a funnel plot [33]. The *p*-values were based on two-sided tests, and the *p*-value < 0.05 was considered statistical significance.

3. Results

3.1. Study Inclusion

For the DBT arm, a total of 489 different studies were found eligible for abstract screening, 33 studies were checked at full-text (Figure 1). Seventeen studies [10,34–49] met our inclusion criteria for synthesis and meta-analysis. Additionally, for the CBBCT, 836 different studies were eligible for the title and abstract screening, nine were assessed for full text, and finally, only five studies met our predefined condition [11–13,48,49]. The meta-analysis was performed separately using univariate analysis for both CBBCT and DBT. Full details about the inclusion and exclusions criteria are given in the Preferred Items for Systematic Reviews and Meta-Analyses (PRISMA) flowchart (Figure 1).

3.2. Overview of Included Studies

For the DBT arm, with 17 studies included, which comprise of retrospective screening studies [34,40,42,44–46,48–51] and prospective studies [35–38], few prospective clinical trials [10,39], above 95% of all included studies are comparative. All the studies reported sensitivity and specificity, in which the (2 × 2) confusion matrix can be derived, other parameters like positive and negative likelihood ratios and AUC of SROC were estimated using MetaDiSc [31] and "mada" package of R, respectively [30]. Most of the studies specified the total number of benign and malignant lesion cases [10,35,37,38,41–47]. Approximately 53 % of the studies data were acquired using the Hologic Selenium Dimension model [10,34,36,40,44–47], 13% goes for Siemens Mammomat Inspiration model [38,39], and 13% also for GE Senographe Essential model [37,42].

The CBBCT arm comprises five studies only, retrospective observers' studies [12,47], prospective study [48], and retrospective diagnostic study [11]. This majorly consists of comparison studies, i.e., CBBCT vs. DM [12,13], CBBCT vs. DM vs. US, or MRI [11,49]. All the studies reported both the sensitivity and specificity of the diagnostic equipment, while the AUC of SROC was estimated separately like that of the DBT arm. All the studies reported the number of benign and malignant cases, 80% of studies acquired data via the Koning Breast CT (KBCT 1000) model [11–13,49].

3.3. Quality Assessment and Publication Bias

In the DBT arm, one study reported a high risk of bias due to inappropriate exclusion and method of patient selection [47]. Two studies (11.8%) reported an unclear risk of bias because the diagnostic threshold was not specified, and no information on whether the readers were blinded to the result of clinical outcomes [34,44]. One study (6.7%) did not give enough information about the pathological findings and, if necessary, follow-up was made, thus providing an unclear risk of bias for a reference standard [40]. Three studies

(17.6%) did not give details information if the patients received the reference standard or if the appropriate time interval between the reference standard and index test, thus providing an unclear risk of bias for flow and timing [34,40,51]. Additionally, eight studies (47.1%) had a high risk of bias for applicability concerns regarding patient selection as the criteria for selecting patients did not match exactly our review questions, three studies (17.6%) provided high risk and unclear risk of bias regarding applicability for index test, only one study (5.9%) gave unclear applicability concerns regarding reference standard. The risk of bias and applicability concern and reviewers' judgment about each domain for all the included study is shown in Figure 2. Likewise, for the CBBCT arm, none of the studies reported a high risk of bias, although the unclear risk of bias exists in patient selection, reference standard, and flow and timing in one study due to scanty information [12,48]. The overview of bias and applicability risk is shown in Figure 3. A visual assessment of funnel plots revealed asymmetrical distribution around inverted funnel for included studies of DBT which signifies publication bias which might be attributed to reporting bias [33], as shown in Figure 4. However, the likelihood of publication bias might also exist in the CBBCT arm due to the small number of studies included in the meta-analysis. More details about the risk of bias and applicability of concerns using QUADASS-2 assessment is shown in Figure 3.

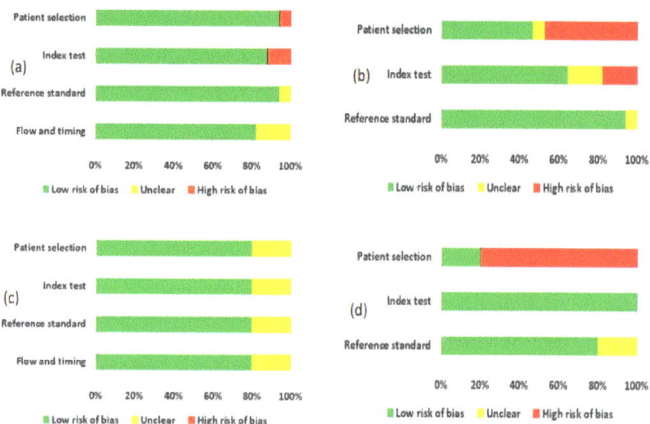

Figure 3. Risk of bias and applicability concerns expressed as percentages across all included studies. (**a**) Risk of bias for DBT; (**b**) Applicability concerns for DBT; (**c**) Risk of bias for CBBCT; (**d**) Applicability concerns for CBBCT.

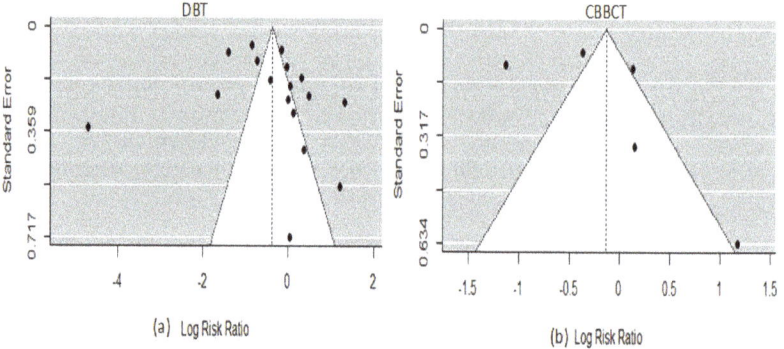

Figure 4. Funnel plots of the likelihood of bias in included studies. (**a**) DBT; (**b**) CBBCT.

3.4. DBT Meta-Analysis

A total of 17 studies with different observations on sensitivity, specificity, and AUC contributed to the meta-analysis of the DBT arm [10,34–49]. The forest plot of sensitivity and specificity with point estimates of 95% confidence intervals across different studies are shown in Figure 5. The pooled sensitivity was 86.7% (95% CI: 80.3–91.2, $I^2 = 89$) and specificity is 87.0% (95% CI: 79.9–91.8, $I^2 = 95$). Since all the within studies had Higgins I^2 for both sensitivity and specificity above 75%, and the p-value of Cochran Q statistic is less than 0.05, which implies there is substantial heterogeneity.

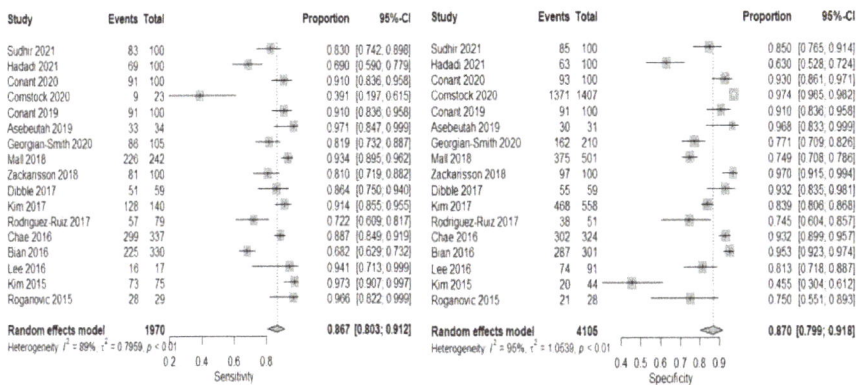

Figure 5. Forest plots using random effect model univariate meta-analysis model for DBT showing pooled sensitivity and pooled specificity.

To show both practical and statistical significance between DBT and CBBCT modalities, the difference in sensitivity and specificity of these modalities were estimated, the result of the difference in effect size for sensitivity is 3% (p-value = 0.7622) and specificity is 16.4% (p-value = 0.0622). The effect size for DBT exceeded CBBCT by 3% and 15.3% for sensitivity and specificity, respectively, which indicate better performance for DBT. Although it is statistically is non-significant since both p-values are greater than 0.05. The pooled positive likelihood ratio (LR^+) is 6.28 (95% CI: 4.40–8.96, $I^2 = 93$), while the pooled negative likelihood ratio (LR^-) is 0.17 (95% CI: 0.12–0.25, $I^2 = 92$), as shown in Figure 6. The pooled AUC of SROC is 0.925, as shown in Figure 7a.

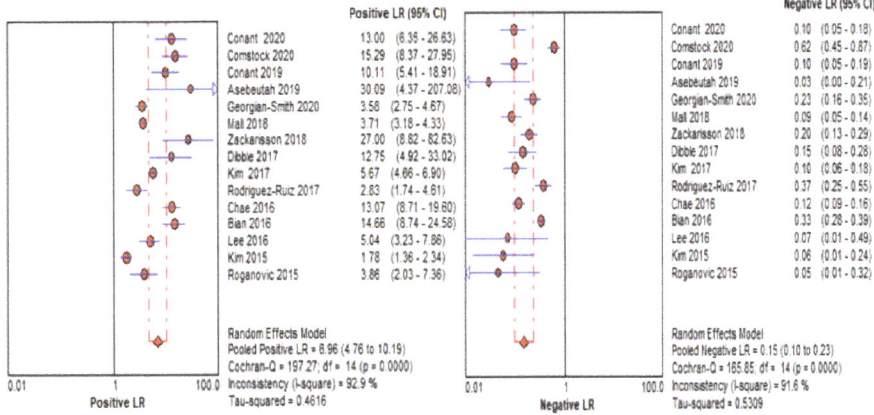

Figure 6. Forest plots of summary of positive (LR^+) and negative (LR^-) likelihood ratios of DBT using random effects bivariate model.

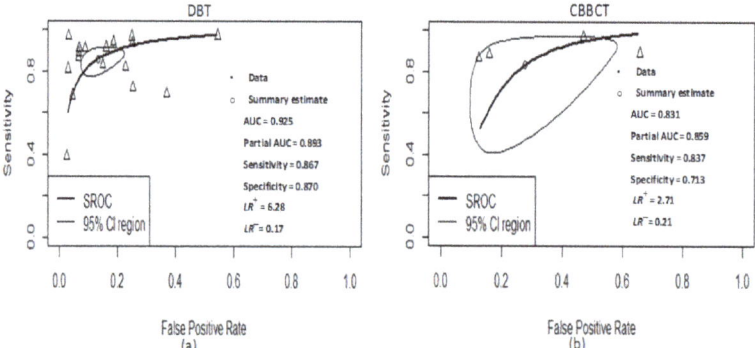

Figure 7. The plot of diagnostic performance using bivariate Summary Receiver Operating Characteristics (SROC) curve. (**a**) SROC of DBT; (**b**) SROC of CBBCT. The prediction region is shown in a dashed dark line, the confidence region shown in a small black ellipse, summary point in black diamond plus ad scaled dataset points for each study in a small triangle. CI: Confidence interval; AUC: area under the curve.

3.5. CBBCT Meta-Analysis

A total of five different observation studies were included in the meta-analysis of the CBBCT arm; the summary of all necessary information is tabulated in Table 1. Pooled sensitivity with 95% confidence intervals across the studies is 83.7% (95% CI: 54.6–95.7, $I^2 = 94$); while the pooled specificity is 71.3% (95% CI: 47.5–87.2, $I^2 = 94$); as shown in Figure 8. There is substantial heterogeneity within studies for both sensitivity and specificity as the value of I^2 is higher than 75% and a p-value less than 0.05. Due to the small number of included studies, further subgroup analyses for evaluating a potential source of heterogeneity were not performed. The pooled positive likelihood ratio (LR^+) is 2.71 (95% CI: 1.39–5.29, $I^2 = 95$), while the pooled negative likelihood ratio (LR^-) is 0.21 (95% CI: 0.07–0.32, $I^2 = 97$), as shown in Figure 9. The pooled AUC of SROC is 0.831, as shown in Figure 7b.

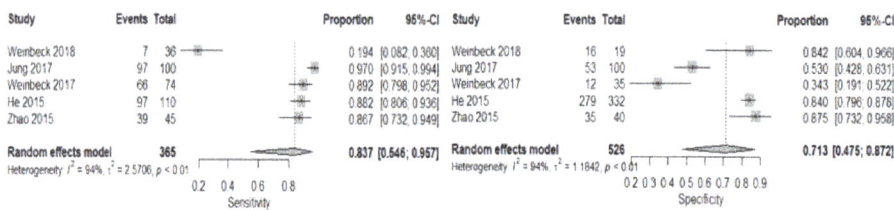

Figure 8. Forest plots using random effects univariate meta-analysis model for CBBCT showing pooled sensitivity and pooled specificity.

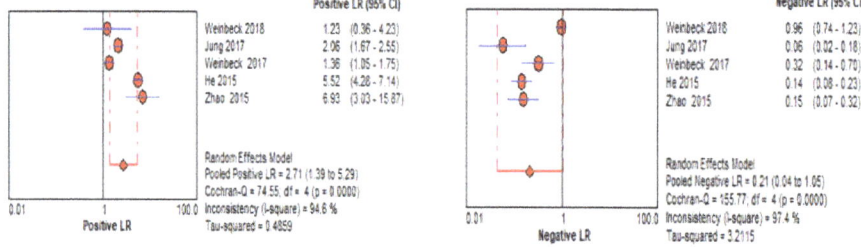

Figure 9. Forest plots of summary of positive and negative likelihood ratios of CBBCT using random effects bivariate model.

Table 1. Characteristics of studies included in digital breast tomosynthesis and cone-beam breast computed tomography.

Study	Country	Equipment	Total No. of Patients	(Mean Age ± SD) Years	No. of Radiol. (Mean Years)	Gland. Dose (mGy)	Sens.	Specf.	Benign Cases (%)	Malig. Cases (%)	Study Intervention
Digital Breast Tomosynthesis											
Sudhir et al. [50]	India	N/A	130	45 ± 12	2 (N/A)	N/A	82.8/100	84.8/100	N/A	N/A	DM vs. DBT vs. US+DBT vs. CEDM [a]
Hadadi et al. [51]	Australia	N/A	35	N/A	7 (2)	N/A	69/100	63/100	N/A	N/A	DBT vs. DM [a]
Conant et al. [34]	USA	Hologic Selenia Dimensions	56839	54 ± NA	N/A	N/A	91.2/100	92.6/100	N/A	N/A	DBT vs. DM [a]
Comstock al. [35]	USA/Germany	N/A	1444	54.9 ± 0.85	2 (N/A)	N/A	9/23	1371/1407	0.6	99.4	One-view DBT vs. DM [b]
Conant et al. [36]	USA	Hologic Selenia Dimensions	50971	54.6 ± 8.9	13 (N/A)	N/A	90.6/100	91.3/100	N/A	N/A	DBT vs. DM [b,e]
Asbeutah et al. [37]	Kuwait	GE Senographe Essential	58	54.3 ± 12.6	1 (>10)	N/A	33/34	30/31	47.7	52.3	DBT vs. DM [b,f]
Georgian-Smith et al. [38]	USA	Siemens Mammomat Inspiration system	330	56.3 ± 9.8	31 (4–38)	N/A	86/105	162/210	63.6	31.8	DBT vs. DM [b,e]
Mall et al. [10]	Australia	Hologic Selenia Dimensions	144	N/A	15 (16)	N/A	226/242	375/501	66.7	33.3	DBT vs. DM [b,d]
Zackrisson et al. [39]	Sweden	Siemens Mammomat Inspiration system	14848	57.0 ± 10.0	7 (2–14)	2.30	81.1/100	97.2/100	N/A	N/A	DBT vs. DM [b,d,f]
Dibble et al. [40]	USA	Hologic Selenia Dimensions	59	58.9 ± N/A	3 (6–16)	N/A	51/59	55/59	N/A	N/A	DBT vs. DM [a]
Kim et al. [41]	Korea	Hologic Selenia Dimensions	698	48.7 ± 11.2	12 (9.3)	1.30	128/140	468/558	79.9	20.1	DBT vs. US [b,f]
Rodriguez-Ruiz et al. [42]	Netherlands	N/A	181	52 ± N/A	6 (23)	2.41	57/79	38/51	39.2	60.8	DBT vs. DM [a,f]
Chae et al. [43]	Korea	GE Senographe Essential	319	49.0 ± N/A	3 (8–18)	N/A	299/337	302/324	11.1	88.9	DBT vs. DM [b,c]
Bian et al. [44]	China	Hologic Selenia Dimensions	631	45.0 ± N/A	3 (3–20)	N/A	225/330	287/301	47.7	52.3	DBT vs. DM [a]
Lee et al. [45]	Korea	Hologic Selenia Dimensions	108	46.3 ± 7.8	3 (N/A)	1.50	17/17	74/91	84.3	15.7	DBT vs. US [a,f]
Kim et al. [46]	Korea	Hologic Selenia Dimensions	113	49.6 ± N/A	3 (>13)	N/A	73/75	20/44	37.0	63.0	DBT vs. US [a,f]
Roganovic et al. [47]	Bosnia and Herzegovina	Hologic Selenia Dimensions	N/A	53.2 ± N/A	1(10)	2.3	29/29	21/28	49.1	50.9	DBT vs. DM vs. MRI [b,f]
Cone-Beam Breast Computed Tomography											
Weinbeck et al. [12]	Germany	Koning (CBCT 1000) Breast CT	41	67.8 ± N/A	2 (>7)	5.85–7.5	7/36	16/19	43.0	51.0	CBCT vs. MRI vs. DM [a,e]
Jung et al. [48]	N/A	N/A	30	30 ± N/A	4 (7)	N/A	97/100	53/100	76.5	23.5	CBCT [a,c]
Weinbeck et al. [11]	Germany	Koning (CBCT 1000) Breast CT	59	N/A	2 (18.5)	5.8–16.6	66/74	12/35	31.3	66.1	CBCT vs. DM [a,c]
He et al. [49]	China	Koning (CBCT 1000) Breast CT	212	48 ± N/A	2 (>10)	8 ± 1.6	97/110	279/332	75.1	24.9	CBCT vs. DM vs. US [b]
Zhao et al. [13]	USA	Koning (CBCT 1000) Breast CT	65	55.6 ± 9.8	2 (>7)	5.8–24.84	39/45	35/40	47.1	52.9	CBCT vs. DM [b,e]

Note: [a] Retrospective study, [b] Prospective studies, [c] Observer performance studies, [d] Clinical trial studies, [e] Diagnostic studies, [f] Screening studies DBT: Digital Breast Tomosynthesis, DM: Digital Mammography, Sens.—Sensitivity, Specf.—Specificity, Gland. Dose—Mean glandular dose, LR^+: Positive likelihood ratio and LR^-: Negative likelihood ratio, CEDM Contrast-enhanced digital mammography.

4. Discussion

The systematic review identified 17 studies for the DBT arm and five studies for the CBBCT arm, comparing the diagnostic accuracy using sensitivity, specificity, mean AUC of SROC, positive and negative likelihood ratios as a figure of merits. Our results showed that the pooled sensitivity of DBT was 86.7% (95% CI: 80.3–91.2) and was higher than that of the pooled sensitivity of CBBCT 83.7% (95% CI: 54.6–95.7), with about 3% with a p-value of 0.7622. Likewise, the pooled specificity of DBT showed an improvement over CBBCT from 87.7% (95% CI: 79.9–91.8) and 71.3% (95% CI: 47.5–87.2) by 16.4%. The pooled LR^+ of DBT is 6.28 (95% CI: 4.40–8.96) and was slightly higher than that of CBBCT with pooled LR^+ of 2.71 (95% CI: 1.39–5.29). The result signifies that DBT is six times more likely to detect patients with breast cancer than patients without breast cancer, as LR^+ is greater than 10 and LR^- is less than 0.1 produces the greatest efficiency [25]. The pooled AUC of SROC of the DBT arm is 0.925 and was significantly higher than that of the CBBCT arm (p-value = 0.016), 0.831. The pooled LR^+ and LR^- of the CBBCT are 2.71 and 0.21, respectively, which cause a small change in the pre-test probability [25]. Although the result presented by Uhlig et al. [19] showed a pooled sensitivity of 78.9%, the specificity of 69.7% and AUC of 0.817, the result of our CBBCT arm showed higher improvement in terms of pooled sensitivity and sensitivity and mean AUC value. The summary of pooled results is shown in Table 2.

Table 2. Summary of all estimated diagnostic test accuracy.

DOR Parameters	Pooled Value at 95% CI (DBT)	Pooled Value at 95% CI (CBBCT)
Sensitivity	86.7% (80.3–91.2, I^2 = 89%)	83.7% (54.6–95.7 I^2 = 94%)
Specificity	87.0% (79.9–91.8, I^2 = 95%)	71.3% (47.5–87.2, I^2 = 94%)
LR^+	6.28 (4.40–8.96, I^2 = 93%)	2.71 (1.39–5.29, I^2 = 95%)
LR^-	0.17 (0.12–0.25, I^2 = 91%)	0.21 (0.04–1.05, I^2 = 97%)
AUC of SROC	0.925	0.831

Note: LR^+ = Positive likelihood ratio, LR^- = Negative likelihood ratio, DBT = Digital breast tomosynthesis, DM = Digital mammography, CBBCT = Cone-beam breast computed tomography, SROC = Summary Receiver Operating Characteristics, CI = Confidence interval; AUC = area under the curve.

We decided to check the effect of the different study protocols (prospective and retrospective studies) on diagnostic performance by conducting a sub-group analysis. The analysis with retrospective studies has a sensitivity of 84.6% (95% CI: 74.6–91.1, I^2 = 84% for 8 studies), while that of prospective studies was 86.7% (95% CI: 80.3–91.3, I^2 = 89% for 9 studies), indicating no significant heterogeneity between the sensitivity as shown in Appendix A (Figure A1). In addition, the specificity is 83.0% (95% CI: 69.2–91.3, I^2 = 93% for 6 studies) for retrospective studies, while the specificity of prospective studies is 87.0% (95% CI: 79.9–91.8, I^2 = 96% for 9 studies) in Appendix A (Figure A1). The result indicates that prospective studies of DBT show a slight non-significantly improvement over retrospective studies in terms of sensitivity and specificity with a p-value of 0.2509.

This increase in mean AUC of DBT might have resulted from the significantly higher value of sensitivity and specificity recorded by most of the included studies [34–36,39,40,42–44]. In contrast, similar lower specificity has been recorded in the CBBCT counterparts [12,48,49], contrarily [11,13] reported higher specificity like that of its DBT counterparts as likely supported by Chappell et al. [30], that an effective diagnostic test should have corresponding high sensitivity and specificity, which significantly contribute to the AUC of the SROC curve. The pooled result of our study has demonstrated the diagnostic potency of DBT over the CBBCT for both sensitivity, specificity, positive and negative likelihood ratio, and AUC. When we compared our pooled sensitivity and specificity with that of Belair et al. [20], which had a sensitivity of 87% (95% CI: 80–92) and 70% (95% CI: 60–79) for DBT and CBBCT and specificity of 81% (95% CI: 72–87) and 67% (95% CI: 57–77), we discovered that our pooled sensitivity for the DBT is within the same range, while the pooled specificity has improved by approximately

7.2%. Comparing Belair et al. [20] with our pooled result for CBBCT showed that sensitivity and specificity have improved by 13.7% and 4.3 %, respectively. According to Zuley et al. [21], for lesion visibility and diagnostic accuracy of CBBCT, DBT, and MRI, the AUC of 0.84 and 0.75 was estimated for DBT and CBBCT pooled AUC result improved by 11.3% and 10.8%. The result shows a statistical significance in the pooled AUC for DBT with p-value = 0.016, as this will provide better diagnostic power compared to univariate sensitivity and specificity. Although the abbreviated 3D breast MRI has been used to screen patients with a high risk of breast cancer due to its high sensitivity between 80–94% and specificity of 80–100% [52,53], however, some small lesions of less than 5 mm in size and ductal carcinoma in situ (DCIS) are not easily visible due to their diffuse pattern of spread [53,54]. Additionally, the cost of an MRI examination and the time cost for each examination has limited its widespread application [55]. Previous studies on the comparison of CBBCT with DM have shown the higher performance of CBBCT on breast masses characterization [12,13], in cancer detection [48] and improved performance and good interrater agreement among readers [47], therefore making CBBCT a potential modality for improved diagnosis of breast cancer.

The studies have several limitations; firstly, the result of both arms was not extracted from the same studies (comparison with a different cohort) according to Yang et al. [27], as no comparison studies between CBBCT and DBT were available within the study's scope and range of year covered, which might have introduced a potential bias between the result. Secondly, the sample size of the CBBCT arm is also one-third of that of the DBT arm, the pooled estimate may not fully represent the statistical power we are looking for; thus, the CBBCT result is underrepresented; therefore, the statistical significance of CBBCT might reduce as more sample size tends to increase the statistical significance of a model. Thirdly, due to the recent introduction of CBBCT as a screening or diagnostic imaging modality, no large multicenter prospective or clinical trial studies are available with no standardized acquisition protocol [19], thus making a direct comparison with the DBT modality a daunting task.

5. Conclusions

Our study demonstrates that DBT shows improved diagnostic performance over CBBCT with pooled sensitivity, specificity AUC, and positive and negative likelihood ratios. This improvement shows a statistical significance for AUC diagnostic parameter, as this parameter would represent higher diagnostic power compared to its derivative sensitivity and specificity. We believe that the diagnostic performance of CBBCT would continue to improve due to more understanding of the underpinned imaging physics of this modality coupled with computer-aided detection application and better experiences of a radiologist. We recommended more prospective studies on the direct comparison of diagnostic accuracy of CBBCT and DBT for breast cancer characterization and detection.

Supplementary Materials: The following are available online at https://www.mdpi.com/article/10.3390/s22093594/s1, File S1: PRISMA checklist table; File S2: Detailed search strategy describing the MeSH and text-word for all the databases.

Author Contributions: T.E.K.: methodology, validation, formal analysis, investigation, data curation and conceptualization, writing—original draft, writing—reviewing, and editing. C.Z.: methodology, validation, formal analysis, investigation, data curation, writing—original draft, writing—reviewing and editing. O.A.O.: methodology, validation, formal analysis, investigation, data curation, writing—original draft, writing—reviewing and editing. G.Y.: methodology, validation, investigation: methodology, validation, formal analysis, investigation, writing—reviewing and editing. Q.D.: methodology, validation, investigation and formal analysis. M.L.: methodology, validation, investigation and formal analysis. J.Z.: methodology, validation, investigation and formal analysis, writing—reviewing and editing. X.Y.: methodology, validation, formal analysis, investigation, data curation and conceptualization, writing—original draft, writing—reviewing and editing, project administration and supervision. All authors have read and agreed to the published version of the manuscript.

Funding: This work was supported in part by the National Key Research and Development Program of China under Grant 2016YFC0104505, in part by the National Natural Science Foundation of China under Grant 61701492, in part by the Jiangsu Science and Technology Department under Grant BK20170392, in part by the Suzhou Municipal Science and Technology Bureau under Grant SYG201825. TEK receives support from the Chinese Government Scholarship for his Ph.D. studies (CSC No. 2017GXZ021382).

Institutional Review Board Statement: Not applicable.

Informed Consent Statement: Not applicable.

Data Availability Statement: All the supporting data are included in the study and Appendix A.

Acknowledgments: The authors acknowledged Kayode Charles Komolafe at Jackson State University, United States of America for proofreading this article and other anonymous reviewers for their constructive criticism.

Conflicts of Interest: The authors declare no competing financial interest or personal relationship that could have appeared to influence the work reported in this paper.

Abbreviations

CI: Confidence interval; CBBCT: Cone-beam breast computed tomography; DBT: Digital breast tomosynthesis; DM: Digital mammography; AUC: Area under the curve; LR^+: Positive likelihood ratio; LR^-: Negative likelihood ratio; PRISMA: Preferred reporting items for systematic reviews and meta-analyses; QUADAS-2: Quality assessment of diagnostic accuracy studies-2; RE: Random effects.

Appendix A

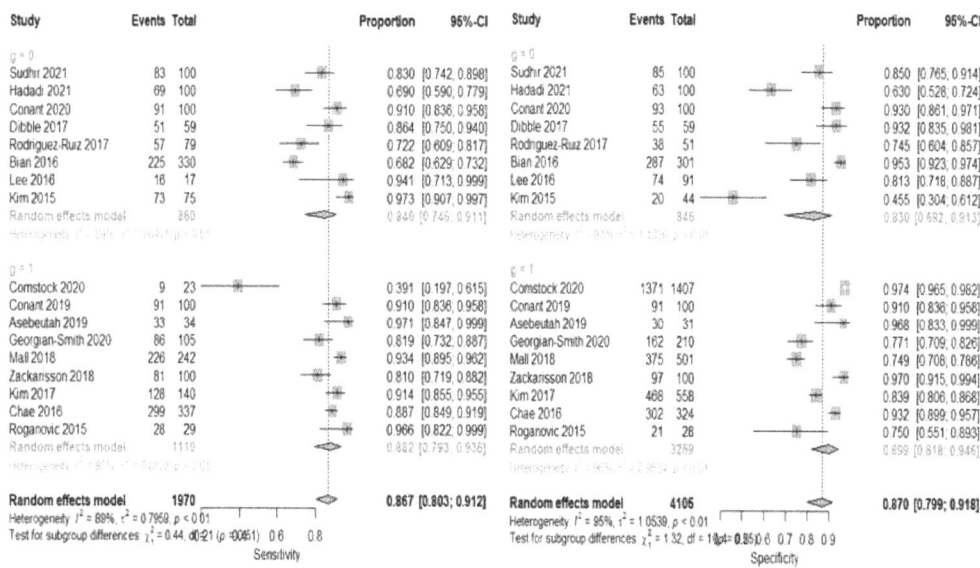

Figure A1. Univariate sub-group analysis of sensitivity and specificity with random model based on the different study protocols. g represents sub-group analysis of data when g = 0 (Retrospective studies) and g = 1 (Prospective studies).

References

1. O'Connell, A.; Conover, D.L.; Zhang, Y.; Seifert, P.; Logan-Young, W.; Lin, C.F.; Sahler, L.; Ning, R. Cone-beam CT for breast imaging: Radiation dose, breast coverage, and image quality. *AJR Am. J. Roentgenol.* **2010**, *195*, 496–509. [CrossRef] [PubMed]
2. DeSantis, C.; Ma, J.; Bryan, L.; Jemal, A. Breast cancer statistics, 2013. *CA Cancer J. Clin.* **2014**, *64*, 52–62. [CrossRef] [PubMed]

3. Bustamante, M.; Rienzo, A.; Osorio, R.; Lefranc, E.; Duarte-Mermoud, M.A.; Herrera-Viedma, E.; Lefranc, G. Algorithm for processing mammography: Detection of microcalcifications. *IEEE Lat. Am. Trans.* **2018**, *16*, 2460–2466. [CrossRef]
4. Mellado, M.; Osa, A.M.; Murillo, A.; Bermejo, R.; Burguete, A.; Pons, M.J.; Erdozain, N. Influencia de la mamografía digital en la detección y manejo de microcalcificaciones [Impact of digital mammography in the detection and management of microcalcifications]. *Radiologia* **2013**, *55*, 142–147. [CrossRef]
5. Mann, R.M.; Hooley, R.; Barr, R.G.; Moy, L. Novel approaches to screening for breast cancer. *Radiology* **2020**, *297*, 266–285. [CrossRef]
6. Zhu, Y.; O'Connell, A.M.; Ma, Y.; Liu, A.; Li, H.; Zhang, Y.; Zhang, X.; Ye, Z. Dedicated breast CT: State of the art—Part II. Clinical application and future outlook. *Eur. Radiol.* **2022**, *32*, 2286–2300. [CrossRef]
7. Conant, E.F.; Beaber, E.F.; Sprague, B.L.; Herschorn, S.D.; Weaver, D.L.; Onega, T.; Tosteson, A.N.A.; McCarthy, A.M.; Poplack, S.P.; Haas, J.; et al. Breast cancer screening using tomosynthesis in combination with digital mammography compared to digital mammography alone: A cohort study within the PROSPR consortium. *Breast Cancer Res. Treat.* **2016**, *156*, 109–116. [CrossRef]
8. Fontaine, M.; Tourasse, C.; Pages, E.; Laurent, N.; Laffargue, G.; Millet, I.; Molinari, N.; Taourel, P. Local Tumor Staging of Breast Cancer: Digital Mammography versus Digital Mammography Plus Tomosynthesis. *Radiology* **2019**, *291*, 594–603. [CrossRef]
9. Iotti, V.; Rossi, P.G.; Nitrosi, A.; Ravaioli, S.; Vacondio, R.; Campari, C.; Marchesi, V.; Ragazzi, M.; Bertolini, M.; Besutti, G.; et al. Comparing two visualization protocols for tomosynthesis in screening: Specificity and sensitivity of slabs versus planes plus slabs. *Eur. Radiol.* **2019**, *29*, 3802–3811. [CrossRef]
10. Mall, S.; Noakes, J.; Kossoff, M.; Lee, W.; McKessar, M.; Goy, A.; Duncombe, J.; Roberts, M.; Giuffre, B.; Miller, A.; et al. Can digital breast tomosynthesis perform better than standard digital mammography work-up in breast cancer assessment clinic? *Eur. Radiol.* **2018**, *28*, 5182–5194. [CrossRef]
11. Wienbeck, S.; Uhlig, J.; Luftner-Nagel, S.; Zapf, A.; Surov, A.; von Fintel, E.; Stahnke, V.; Lotz, J.; Fischer, U. The role of cone-beam breast-CT for breast cancer detection relative to breast density. *Eur. Radiol.* **2017**, *27*, 5185–5195. [CrossRef] [PubMed]
12. Wienbeck, S.; Fischer, U.; Luftner-Nagel, S.; Lotz, J.; Uhlig, J. Contrast-enhanced cone-beam breast-CT (CBBCT): Clinical performance compared to mammography and MRI. *Eur. Radiol.* **2018**, *28*, 3731–3741. [CrossRef] [PubMed]
13. Zhao, B.; Zhang, X.; Cai, W.; Conover, D.; Ning, R. Cone beam breast CT with multiplanar and three dimensional visualization in differentiating breast masses compared with mammography. *Eur. J. Radiol.* **2015**, *84*, 48–53. [CrossRef] [PubMed]
14. Movik, E.; Dalsbø, T.K.; Fagelund, B.C.; Friberg, E.G.; Håheim, L.L.; Skår, Å. *Digital Breast Tomosynthesis with Hologic 3D Mammography Selenia Dimensions System for Use in Breast Cancer Screening: A Single Technology Assessment*; Report from the Norwegian Institute of Public Health No. 2017–08; Knowledge Centre for the Health Services at The Norwegian Institute of Public Health (NIPH): Oslo, Norway, 2017.
15. Thompson, W.; Argaez, C. *Digital Breast Tomosynthesis for the Screening and Diagnosis of Breast Cancer: A Review of the Diagnostic Accuracy, Cost-Effectiveness and Guidelines*; Canadian Agency for Drugs and Technologies in Health: Ottawa, ON, Canada, 2019.
16. Melnikow, J.; Fenton, J.J.; Whitlock, E.P.; Miglioretti, D.L.; Weyrich, M.S.; Thompson, J.H.; Shah, K. Supplemental screening for breast cancer in women with dense breasts: A systematic review for the US Preventive Services Task Force. *Ann. Intern. Med.* **2016**, *164*, 268–278. [CrossRef] [PubMed]
17. Phi, X.A.; Tagliafico, A.; Houssami, N.; Greuter, M.J.W.; de Bock, G.H. Digital breast tomosynthesis for breast cancer screening and diagnosis in women with dense breasts-a systematic review and meta-analysis. *BMC Cancer* **2018**, *18*, 380. [CrossRef]
18. Uhlig, J.; Fischer, U.; Biggemann, L.; Lotz, J.; Wienbeck, S. Pre- and post-contrast versus post-contrast cone-beam breast CT: Can we reduce radiation exposure while maintaining diagnostic accuracy? *Eur. Radiol.* **2019**, *29*, 3141–3148. [CrossRef]
19. Uhlig, J.; Uhlig, A.; Biggemann, L.; Fischer, U.; Lotz, J.; Wienbeck, S. Diagnostic accuracy of cone-beam breast computed tomography: A systematic review and diagnostic meta-analysis. *Eur. Radiol.* **2019**, *29*, 1194–1202. [CrossRef]
20. Belair, J.; Zuley, M.; Ganott, M.; Kelly, A.; Shinde, D.; Shah, R.; Catullo, V.; Mishra, M.D.V.; Gur, D. Non-contrast Cone-Beam CT vs Tomosynthesis: Identification and Classification of Benign and Malignant Breast Lesions. In Proceedings of the Radiological Society of North America 2012 Scientific Assembly and Annual Meeting, Chicago, IL, USA, 25–30 November 2012. Available online: http://archive.rsna.org/2012/12022690.html (accessed on 24 December 2021).
21. Zuley, M.; Guo, B.; Ganott, M.; Bandos, A.; Catullo, V.; Lu, A.; Kelly, A.E.; Anello, M.L.; Abrams, G.S.; Chough, D. Comparison of Visibility and Diagnostic Accuracy of Cone Beam Computed Tomography, Tomosynthesis, MRI and Digital Mammography for Breast Masses. In Proceedings of the Radiological Society of North America 2013 Scientific Assembly and Annual Meeting, Chicago, IL, USA, 1–6 December 2013. Available online: http://archive.rsna.org/2013/13022530.html (accessed on 24 December 2021).
22. Komolafe, T.E.; Olagbaju, O.A.; Li, M.; Zheng, J.; Yang, X. Comparison of Diagnostic Accuracy of Cone-Beam Breast Computed Tomography and Digital Breast Tomosynthesis: A Systematic Review and Meta-Analysis Approach. PROSPERO2020CRD42020180192. Available online: https://www.crd.york.ac.uk/prospero/display_record.php?ID=CRD42020180192 (accessed on 5 January 2022).
23. Deeks, J.J.; Bossuyt, P.M.M. Chapter 3: Evaluating diagnostic tests. In *Cochrane Handbook for Systematic Reviews of Reviews of Diagnostic Test Accuracy Version 2*; Deeks, J.J., Bossuyt, P.M.M., Leeflang, M.M.G., Takwoingi, Y., Eds.; Cochrane: London, UK, 2017.
24. McInnes, M.D.F.; Moher, D.; Thombs, B.D.; McGrath, T.A.; Bossuyt, P.M.; The PRISMA-DTA Group. Preferred Reporting Items for a Systematic Review and Meta-analysis of Diagnostic Test Accuracy Studies: The PRISMA-DTA Statement. *JAMA* **2018**, *319*, 388–396. [CrossRef]

25. Manikandan, R.; Dorairajan, L.N. How to appraise a diagnostic test. *Indian J. Urol.* **2011**, *27*, 513–519.
26. Komolafe, T.E.; Cao, Y.; Nguchu, B.A.; Monkam, P.; Olaniyi, E.O.; Sun, H.; Zheng, J.; Yang, X. Diagnostic test accuracy of deep learning detection of COVID-19: A systematic review and meta-analysis. *Acad. Radiol.* **2021**, *8*, 1507–1523. [CrossRef]
27. Yang, B.; Mallett, S.; Takwoingi, Y.; Davenport, C.F.; Hyde, C.J.; Whiting, P.F.; Deeks, J.J.; Leeflang, D.M.M.; the QUADAS-C Group. QUADAS-C: A Tool for Assessing Risk of Bias in Comparative Diagnostic Accuracy Studies. *Ann. Intern. Med.* **2021**, *174*, 1592–1599. [CrossRef] [PubMed]
28. Whiting, P.F.; Rutjes, A.W.; Westwood, M.E.; Mallett, S.; Deeks, J.J.; Reitsma, J.B.; Leeflang, M.M.; Sterne, J.A.; Bossuyt, P.M.; QUADAS-2 Group. QUADAS-2: A revised tool for the quality assessment of diagnostic accuracy studies. *Ann. Intern. Med.* **2011**, *155*, 529–536. [CrossRef] [PubMed]
29. Shim, S.R.; Kim, S.J.; Lee, J. Diagnostic test accuracy: Application and practice using R software. *Epidemiol. Health* **2019**, *41*, e2019007. [CrossRef] [PubMed]
30. Chappell, F.M.; Raab, G.M.; Wardlaw, J.M. When are summary ROC curves appropriate for diagnostic meta-analyses? *Stat. Med.* **2009**, *28*, 2653–2668. [CrossRef]
31. Zamora, J.; Abraira, V.; Muriel, A.; Khan, K.S.; Coomarasamy, A. Meta-DiSc: A software for meta-analysis of test accuracy data. *BMC Med. Res. Methodol.* **2006**, *6*, 31. [CrossRef]
32. Higgins, J.P.; Thompson, S.G. Quantifying heterogeneity in a meta-analysis. *Stat. Med.* **2002**, *21*, 1539–1558. [CrossRef]
33. Liu, J.L. The role of the funnel plot in detecting publication and related biases in meta-analysis. *Evid.-Based Dent.* **2011**, *12*, 121–122. [CrossRef]
34. Conant, E.F.; Zuckerman, S.P.; McDonald, E.S.; Weinstein, S.P.; Korhonen, K.E.; Birnbaum, J.A.; Tobey, J.D.; Schnall, M.D.; Hubbard, R.A. Five Consecutive Years of Screening with Digital Breast Tomosynthesis: Outcomes by Screening Year and Round. *Radiology* **2020**, *295*, 285–293. [CrossRef]
35. Comstock, C.E.; Gatsonis, C.; Newstead, G.M.; Snyder, B.S.; Gareen, I.F.; Bergin, J.T.; Rahbar, H.; Sung, J.S.; Jacobs, C.; Harvey, J.A.; et al. Comparison of Abbreviated Breast MRI vs Digital Breast Tomosynthesis for Breast Cancer Detection Among Women with Dense Breasts Undergoing Screening. *JAMA* **2020**, *323*, 746–756. [CrossRef]
36. Conant, E.F.; Barlow, W.E.; Herschorn, S.D.; Weaver, D.L.; Beaber, E.F.; Tosteson, A.N.A.; Haas, J.S.; Lowry, K.P.; Stout, N.K.; Trentham-Dietz, A.; et al. Association of Digital Breast Tomosynthesis vs Digital Mammography with Cancer Detection and Recall Rates by Age and Breast Density. *JAMA Oncol.* **2019**, *5*, 635–642. [CrossRef]
37. Asbeutah, A.M.; Karmani, N.; Asbeutah, A.A.; Echreshzadeh, Y.A.; AlMajran, A.A.; Al-Khalifah, K.H. Comparison of Digital Breast Tomosynthesis and Digital Mammography for Detection of Breast Cancer in Kuwaiti Women. *Med. Princ. Pract.* **2019**, *28*, 10–15. [CrossRef] [PubMed]
38. Georgian-Smith, D.; Obuchowski, N.A.; Lo, J.Y.; Brem, R.F.; Baker, J.A.; Fisher, P.R.; Rim, A.; Zhao, W.; Fajardo, L.L.; Mertelmeier, T. Can Digital Breast Tomosynthesis Replace Full-Field Digital Mammography? A Multireader, Multicase Study of Wide-Angle Tomosynthesis. *Am. J. Roentgenol.* **2019**, *212*, 1393–1399. [CrossRef]
39. Zackrisson, S.; Lång, K.; Rosso, A.; Johnson, K.; Dustler, M.; Förnvik, D.; Andersson, I. One-view breast tomosynthesis versus two-view mammography in the Malmö Breast Tomosynthesis Screening Trial (MBTST): A prospective, population-based, diagnostic accuracy study. *Lancet Oncol.* **2018**, *19*, 1493–1503. [CrossRef]
40. Dibble, E.H.; Lourenco, A.P.; Baird, G.L.; Ward, R.C.; Maynard, A.S.; Mainiero, M.B. Comparison of digital mammography and digital breast tomosynthesis in the detection of architectural distortion. *Eur. Radiol.* **2018**, *28*, 3–10. [CrossRef] [PubMed]
41. Kim, W.H.; Chang, J.M.; Lee, J.; Chu, A.J.; Seo, M.; Gweon, H.M.; Förnvik, H.; Sartor, H.; Timberg, P.; Tingberg, A.; et al. Erratum to: Diagnostic performance of tomosynthesis and breast ultrasonography in women with dense breasts: A prospective comparison study. *Breast Cancer Res. Treat.* **2017**, *163*, 197. [CrossRef] [PubMed]
42. Rodriguez-Ruiz, A.; Gubern-Merida, A.; Imhof-Tas, M.; Lardenoije, S.; Wanders, A.J.T.; Andersson, I.; Zackrisson, S.; Lång, K.; Dustler, M.; Karssemeijer, N.; et al. One-view digital breast tomosynthesis as a stand-alone modality for breast cancer detection: Do we need more? *Eur. Radiol.* **2018**, *28*, 1938–1948. [CrossRef]
43. Chae, E.Y.; Kim, H.H.; Cha, J.H.; Shin, H.J.; Choi, W.J. Detection and characterization of breast lesions in a selective diagnostic population: Diagnostic accuracy study for comparison between one-view digital breast tomosynthesis and two-view full-field digital mammography. *Br. J. Radiol.* **2016**, *89*, 20150743. [CrossRef]
44. Bian, T.; Lin, Q.; Cui, C.; Li, L.; Qi, C.; Fei, J.; Su, X. Digital Breast Tomosynthesis: A New Diagnostic Method for Mass-Like Lesions in Dense Breasts. *Breast J.* **2016**, *22*, 535–540. [CrossRef]
45. Lee, W.K.; Chung, J.; Cha, E.S.; Lee, J.E.; Kim, J.H. Digital breast tomosynthesis and breast ultrasound: Additional roles in dense breasts with category 0 at conventional digital mammography. *Eur. J. Radiol.* **2016**, *85*, 291–296. [CrossRef]
46. Kim, S.A.; Chang, J.M.; Cho, N.; Yi, A.; Moon, W.K. Characterization of breast lesions: Comparison of digital breast tomosynthesis and ultrasonography. *Korean J. Radiol.* **2015**, *16*, 229–238. [CrossRef]
47. Roganovic, D.; Djilas, D.; Vujnovic, S.; Pavic, D.; Stojanov, D. Breast MRI, digital mammography and breast tomosynthesis: Comparison of three methods for early detection of breast cancer. *Bosn. J. Basic Med. Sci.* **2015**, *15*, 64–68. [CrossRef] [PubMed]
48. Jung, H.K.; Kuzmiak, C.M.; Kim, K.W.; Choi, N.M.; Kim, H.J.; Langman, E.L.; Yoon, S.; Steen, D.; Zeng, D.; Gao, F. Potential Use of American College of Radiology BI-RADS Mammography Atlas for Reporting and Assessing Lesions Detected on Dedicated Breast CT Imaging: Preliminary Study. *Acad. Radiol.* **2017**, *24*, 1395–1401. [CrossRef] [PubMed]

49. He, N.; Wu, Y.P.; Kong, Y.; Lv, N.; Huang, Z.M.; Li, S.; Wang, Y.; Geng, Z.-J.; Wu, P.-H.; Wei, W.-D. The utility of breast cone-beam computed tomography, ultrasound, and digital mammography for detecting malignant breast tumors: A prospective study with 212 patients. *Eur. J. Radiol.* **2016**, *85*, 392–403. [CrossRef] [PubMed]
50. Sudhir, R.; Sannapareddy, K.; Potlapalli, A.; Krishnamurthy, P.B.; Buddha, S.; Koppula, V. Diagnostic accuracy of contrast-enhanced digital mammography in breast cancer detection in comparison to tomosynthesis, synthetic 2D mammography and tomosynthesis combined with ultrasound in women with dense breast. *Br. J. Radiol.* **2021**, *94*, 20201046. [CrossRef]
51. Hadadi, I.; Rae, W.; Clarke, J.; McEntee, M.; Ekpo, E. Breast cancer detection: Comparison of digital mammography and digital breast tomosynthesis across non-dense and dense breasts. *Radiography* **2021**, *27*, 1027–1032. [CrossRef]
52. Deike-Hofmann, K.; Koenig, F.; Paech, D.; Dreher, C.; Delorme, S.; Schlemmer, H.P.; Bickelhaupt, S. Abbreviated MRI Protocols in Breast Cancer Diagnostics. *J. Magn. Reson* **2019**, *49*, 647–658. [CrossRef]
53. Mann, R.M.; Cho, N.; Moy, L. Breast MRI: State of the Art. *Radiology* **2019**, *292*, 520–536. [CrossRef]
54. Shimauchi, A.; Jansen, S.A.; Abe, H.; Jaskowiak, N.; Schmidt, R.A.; Newstead, G.M. Breast cancers not detected at MRI: Review of false-negative lesions. *Am. J. Roentgenol* **2010**, *194*, 1674–1679. [CrossRef]
55. Mango, V.L.; Morris, E.A.; Dershaw, D.D.; Abramson, A.; Fry, C.; Moskowitz, C.S.; Hughes, M.; Kaplan, J.; Jochelson, M.S. Abbreviated protocol for breast MRI: Are multiple sequences needed for cancer detection? *Eur. J. Radiol.* **2015**, *84*, 65–70. [CrossRef]

MDPI AG
Grosspeteranlage 5
4052 Basel
Switzerland
Tel.: +41 61 683 77 34

Sensors Editorial Office
E-mail: sensors@mdpi.com
www.mdpi.com/journal/sensors

Disclaimer/Publisher's Note: The statements, opinions and data contained in all publications are solely those of the individual author(s) and contributor(s) and not of MDPI and/or the editor(s). MDPI and/or the editor(s) disclaim responsibility for any injury to people or property resulting from any ideas, methods, instructions or products referred to in the content.

www.ingramcontent.com/pod-product-compliance
Lightning Source LLC
LaVergne TN
LVHW070402100526
838202LV00014B/1368